Data Science for Wind Energy

Data Science for Wind Energy

Yu Ding

CRC Press
Taylor & Francis Group
Boca Raton London New York

CRC Press is an imprint of the
Taylor & Francis Group, an **informa** business

CRC Press
Taylor & Francis Group
6000 Broken Sound Parkway NW, Suite 300
Boca Raton, FL 33487-2742

First issued in paperback 2020

© 2020 by Taylor & Francis Group, LLC
CRC Press is an imprint of Taylor & Francis Group, an Informa business

No claim to original U.S. Government works

Printed on acid-free paper

ISBN 13: 9781138590526 (hbk)
ISBN 13: 9780367729097 (pbk)

DOI: 10.1201/9780429490972

Visit the Taylor & Francis Web site at
http://www.taylorandfrancis.com

and the CRC Press Web site at
http://www.crcpress.com

To my parents.

Contents

PART II **Wind Turbine Performance Analysis**

CHAPTER 5 ▪ Power Curve Modeling and Analysis 125

CHAPTER 6 ▪ Production Efficiency Analysis and Power Curve 159

Foreword

Wind power is a rapidly growing source of renewable energy in many parts of the globe. Building wind farms and maintaining turbine assets also provide numerous job opportunities. As a result, the wind energy sector plays an increasingly important role in the new economy. While being scaled up, efficiency and reliability become the key to making wind energy competitive. With the arrival of the data science and machine learning era, a lot of discussions are being made in the related research community and wind industry, contemplating strategies to take full advantage of the potentials and opportunities unleashed by the large amount of data to address the efficiency and reliability challenges.

Data Science for Wind Energy arrives at the right time, becoming one of the first dedicated volumes to bridge the gap, and provides expositions of relevant data science methods and abundant case studies, tailored to address research and practical challenges in wind energy applications.

This book of eleven technical chapters is divided into three parts, unified by a general data science formulation presented in Chapter 1. The overarching formulation entails the modeling and solution of a set of probability density functions, conditional or otherwise, not only to account for the mean estimation or prediction, but also to allow for uncertainty quantification. The first part of the book embodies the modeling of a spatio-temporal random wind field and uses that as a springboard for better forecasting. Chapter 2 recaps the existing methods for modeling data in a univariate time series, and Chapters 3 and 4 bring to the readers many new data science concepts and methods. The asymmetry quantification and asymmetric spatio-temporal modeling introduced in Chapter 3 and the regime-switching methods discussed in Chapter 4 are particularly interesting. The second part of the book concentrates on the system-level, power production-oriented turbine performance assessment. This part starts off with a power curve analysis (Chapter 5), followed by adding physically informed constraints to power curve modeling for devising productive efficiency metrics (Chapter 6). Chapters 7 and 8 further discuss, respectively, the circumstances when a turbine's performance can be enhanced by a purposeful action or diminished due to the wake effect. The third part of the book focuses on reliability management and load analysis for wind turbines, nested within an integrative framework combining models, simulations and data (Chapter 9). The load analysis for reliability assessment involves heavily statistical sampling techniques, as detailed in Chapters 10 and

11, and those methods are useful to general reliability engineering purposes—my own research on electrical power system reliability has been benefited by these data science methodologies. I am pleased to see the anomaly detection and fault diagnosis methods presented in Chapter 12, borrowing experiences and successes from other industries for the benefit of wind energy practice.

One of the reasons I am fond of this book is the author's diligence and generosity in collecting, arranging, and releasing ten important wind farm datasets, more than 150 megabytes in volume, plus another 440 megabytes of simulated data used in reliability verification. On top of that, the author provides computer codes for all eleven technical chapters, most of them in R while some are in MATLAB®, either for reproducing figures and tables in the book or implementing some major algorithm. I am sure that those data and codes will immensely help both academic researchers and practitioners.

To appreciate a book, it is helpful to understand the author. I had the good fortune to get to know Dr. Yu Ding shortly after he joined Texas A&M faculty in 2001. There was a university-wide event celebrating the 125th anniversary of Texas A&M University. Yu and I happened to sit next to each other at the same table, and at that moment, I had been with the university for 23 years, while Yu had been for about 25 days. In the ensuing years, Yu's path and mine have crossed often. We served on the committees of each other's students, co-authored papers and co-directed research projects, and because of these, I am reasonably familiar with most of the materials presented in this book. I have witnessed Yu's quick ascent to a leading and authoritative researcher on the intersection of data science and wind energy. Yu's unique multidisciplinary training and penetrating insights allow him and his research team to produce many influential works, contributing to methodology development and benefiting practices. Yu's work on turbine performance assessment in particular leads to large-scale fleet-wide implementations, rendering multi-million-dollar extra revenues. Not surprisingly, Yu was recognized with a Research Impact Award by Texas A&M College of Engineering in May 2018 *"for innovations in data and quality science impacting the wind energy industry."*

It is thus a great pleasure for me to introduce this unique and timely book and a dear colleague to the academia and practitioners who want to know more about data science for wind energy.

Chanan Singh
Regents Professor and Irma Runyon Chair Professor
Electrical & Computer Engineering Department
Texas A&M University, College Station, Texas

June 2019

Preface

All models are wrong but some are useful.

— George E. P. Box

My introduction to the field of wind energy started from a phone call taking place sometime in 2004. Dr. Jiong Tang of the University of Connecticut called and asked if I would be interested in tackling some wind turbine reliability problems.

I got to know Jiong when we were both mechanical engineering graduate students at the Pennsylvania State University. I later left Penn State for my doctoral study at the University of Michigan. My doctoral research was oriented towards a specialty area of data science—the quality science, which employs and develops statistical models and methods for quality improvement purpose. Prior to that phone call, my quality science applications were exclusively in manufacturing. I reminded Jiong that I knew almost nothing about wind turbines and wondered how I could be of any help. Jiong believed that data available from turbine operations had not been taken full advantage of and thought my data science expertise could be valuable. I was intrigued by the research challenges and decided to jump at the opportunity.

The first several years of my wind energy research, however, involved little data. Although the industry had gathered a large amount of operational data through the supervisory control and data acquisition systems of turbines, we had a hard time persuading any turbine manufacturer or owner/operator to share their data. Our luck turned around a few years later, after we aligned ourselves with national labs and several wind companies. Through the academia-government-industry partnership, my research group was able to collect over 100 gigabytes wind turbine testing data and wind farm operational data. Working with the vast amount of real-world data enabled me to build a rewarding career that developed data science methods to address wind energy challenges and it is still going strong.

While working in the wind energy area, I benefited from having a mechanical engineering background. The majority of wind energy research is carried out, for understandable reasons, by domain experts in aerospace, mechanical, civil, or electrical engineering. My engineering training allows me to communicate with domain experts with ease. Maybe this is why Jiong thought of involving me in his wind turbine project in the first place.

As I got involved more and more in the field of wind energy, I observed a disconnection between this typical engineering field and the emerging field of data science. Wind engineers or wind engineering researchers routinely handle data, but most of the domain experts are not exposed to systematic data science training while in schools because the engineering curricula, until very recently, offered only basic engineering statistics. This did not keep pace with the fast development of new ideas and methods introduced by data science in the past twenty years. On the other hand, wind engineering, like most other substantial engineering fields, finds a relatively small number of trained data scientists from computer science or statistics disciplines working in the area, probably because the entry barrier associated with mastering domain knowledge appears intimidating. This may explain that while there are plenty of generic data science and machine learning books, books that can bridge the two distinctive fields and offer specific and sophisticated data science solutions to wind energy problems are, in fact, scarce.

I had been thinking of writing a book filling precisely this void. I came to realize in early 2017 that I may have enough materials when I was leading a research team and preparing a National Science Foundation proposal to its BIG DATA program. In fact, the structure of this book closely mirrors the structure of that proposal, as it embodies three main parts discussing, respectively, wind field analysis, wind turbine performance analysis, and wind turbine load and reliability management. The 2017 NSF proposal was funded at the end of the summer, and, I decided to submit the book proposal to Chapman & Hall/CRC Press later in 2017.

I am grateful for the opportunities and privilege to work with many talented individuals on a problem of national importance. A few of those individuals played pivotal roles in my wind energy research career. The first is obviously Dr. Jiong Tang—without him, I wouldn't be writing this preface. Then, there is Dr. Eunshin Byon, a former Ph.D. student of mine and now a faculty member at the University of Michigan. Eunshin was the first student who worked with me on wind energy research. She came to my group during that aforementioned "data-light" period. Understandably, it was a difficult time for those of us who work with data. Eunshin was instrumental in sustaining our research at that time, finding data through public sources and testing innovative ideas that lay the foundation for the subsequent collaborations with several industry members. I am delighted to see that Eunshin becomes a recognized expert herself in the intersecting area of data science and wind energy.

I appreciate immensely Mr. Brian Hayes, Executive Vice President of EDP Renewables, North America, for his vision in starting the Texas A&M-EDP Renewables partnership and his generous support in funding our research and sharing their wind farm operational data. I am deeply grateful to Dr. Shuangwen (Shawn) Sheng at the National Renewable Energy Laboratory for engaging my research team at the national or international level and for countless hours of stimulating discussions that drive my research to new levels. Of course, I am indebted to my Ph.D. advisor, Dr. Jianjun Shi, then at the University of Michigan and now with the Georgia Institute of Technology, for bringing me to the data science world and for teaching me how to be an independent researcher.

Last but not least, I would like to thank my wife, Ying Li, and our daughter, Alexandra, for their love and support.

Yu Ding
Texas A&M University
College Station, Texas

June 2019

Acknowledgments

Special thanks goes to the following former and current students who help arrange the datasets and provide computer codes for producing many tables and figures or implementing certain algorithms:

- Hoon Hwangbo helped arrange most of the datasets used in this book, except the `Wind Spatial-Temporal Dataset1`, `Wind Spatial-Temporal Dataset2`, and `Simulated Bending Moment Dataset`, which were prepared by others. Hoon also provided the code for generating Tables 6.1–6.3, Table 7.4, Tables 8.2–8.5, and Tables 10.2–10.6, and for generating Figures 6.3–6.6, Figures 6.12–6.14, Figure 7.5, Figure 8.4, Figure 8.7, Figure 8.9, Figures 10.2–10.4, and Figures 10.6–10.10. Furthermore, Hoon created the illustrations in Figures 6.7, 6.9, and 6.11.

- Abhinav Prakash provided the code for generating Tables 2.1–2.8 and Tables 5.2–5.8, and for generating Figures 2.1–2.5, Figure 3.1 and Figure 3.2. Additionally, Abhinav created the illustrations in Figures 12.2, 12.4, and 12.6.

- Arash Pourhabib arranged the `Wind Spatial-Temporal Dataset1` and provided the code for generating Tables 3.1–3.3 and Figure 3.3.

- Ahmed Aziz Ezzat arranged the `Wind Spatial-Temporal Dataset2` and provided the code for generating Tables 3.4–3.8, Tables 4.4–4.6, and for generating Figures 3.4–3.7, Figure 4.4, and Figure 4.9. Aziz also created the illustrations in Figure 4.3 and Figures 4.5–4.7.

- Giwhyun Lee developed the original code for implementing the methods in Section 5.2, Section 7.3, and Chapter 10. Giwhyun also created the illustration in Figure 5.7.

- Eunshin Byon provided the `Simulated Bending Moment Dataset`, the code for Algorithm 9.1 (generating Figure 9.6), and the code for establishing the generalized additive model in Section 11.4.1 as the conditional probability of exceedance function. Eunshin also created the illustrations in Figures 9.2, 9.3, 9.7, 9.8, 10.5, and 11.6.

- Imtiaz Ahmed provided the code for Algorithm 12.3.

I would like to acknowledge the contribution of many people to the research work that forms the backbone of this book. Marc Genton, Jianhua Huang, Andrew Johnson, Mikyoung Jun, Bani K. Mallick, Lewis Ntaimo, Chanan Singh, Le Xie, and Li Zeng are faculty members who collaborated with me on wind energy research topics. Mithun P. Acharya, Daniel Cabezon-Martinez, Andrew Cordes, Aldo Dagnino, Oliver Eisele, Travis Galoppo, Ron Grife, Jaimeet Gulati, Ulrich Lang, Georgios Pechlivanoglou, and Guido Weinzierl were our industrial collaborators. Many of the former and current students, in addition to those mentioned above, contributed to various aspect of my wind energy research: Jason Lawley, Briana Niu, David Pérez, Eduardo Pérez, and Yei-Eun Shin. Randi Cohen and the team at Chapman & Hall/CRC Press have done a fantastic job in managing the book project and assisted me in numerous occasions. I also gratefully acknowledge NSF for its support of this work and Mike and Sugar Barnes for their generous endowment.

The author's following publications are reused, in part or in whole, in the respective chapters.

- **Chapter 3**
  Pourhabib, Huang, and Ding. "Short-term wind speed forecast using measurements from multiple turbines in a wind farm." *Technometrics*, 58:138–147, 2016.

  Ezzat, Jun, and Ding. "Spatio-temporal asymmetry of local wind fields and its impact on short-term wind forecasting." *IEEE Transactions on Sustainable Energy*, 9:1437–1447, 2018.

- **Chapter 4**
  Ezzat, Jun, and Ding. "Spatio-temporal short-term wind forecast: A calibrated regime-switching method." *The Annals of Applied Statistics*, in press, 2019.

- **Chapter 5**
  Lee, Ding, Genton, and Xie. "Power curve estimation with multivariate environmental factors for inland and offshore wind farms." *Journal of the American Statistical Association*, 110:56–67, 2015.

- **Chapter 6**
  Hwangbo, Johnson, and Ding. "A production economics analysis for quantifying the efficiency of wind turbines." *Wind Energy*, 20:1501–1513, 2017.

  Niu, Hwangbo, Zeng, and Ding. "Evaluation of alternative efficiency metrics for offshore wind turbines and farms." *Renewable Energy*, 128:81–90, 2018.

Hwangbo, Johnson, and Ding. "Power curve estimation: Functional estimation imposing the regular ultra passum law." *Working Paper*, 2018. Available at SSRN:http://ssrn.com/abstract=2621033.

- **Chapter 7**
Lee, Ding, Xie, and Genton. "Kernel Plus method for quantifying wind turbine upgrades." *Wind Energy*, 18:1207–1219, 2015.

Hwangbo, Ding, Eisele, Weinzierl, Lang, and Pechlivanoglou. "Quantifying the effect of vortex generator installation on wind power production: An academia-industry case study." *Renewable Energy*, 113:1589–1597, 2017.

Shin, Ding, and Huang. "Covariate matching methods for testing and quantifying wind turbine upgrades." *The Annals of Applied Statistics*, 12:1271–1292, 2018.

- **Chapter 8**
Hwangbo, Johnson, and Ding. "Spline model for wake effect analysis: Characteristics of single wake and its impacts on wind turbine power generation." *IISE Transactions*, 50:112–125, 2018.

- **Chapter 9**
Byon and Ding. "Season-dependent condition-based maintenance for a wind turbine using a partially observed Markov decision process." *IEEE Transactions on Power Systems*, 25:1823–1834, 2010.

Byon, Ntaimo, and Ding. "Optimal maintenance strategies for wind turbine systems under stochastic weather conditions." *IEEE Transactions on Reliability*, 59:393–404, 2010.

Byon, Pérez, Ding, and Ntaimo. "Simulation of wind farm operations and maintenance using DEVS." *Simulation–Transactions of the Society for Modeling and Simulation International*, 87:1093–1117, 2011.

- **Chapter 10**
Lee, Byon, Ntaimo, and Ding. "Bayesian spline method for assessing extreme loads on wind turbines." *The Annals of Applied Statistics*, 7:2034–2061, 2013.

- **Chapter 12**
Ahmed, Dagnino, and Ding. "Unsupervised anomaly detection based on minimum spanning tree approximated distance measures and its application to hydropower turbines." *IEEE Transactions on Automation Science and Engineering*, 16: 654–667, 2019.

Introduction

DOI: 10.1201/9780429490972-1

W ind energy has been used as far back as Roman Egypt [51] (or even earlier [194]). The well-preserved windmills that dotted the Dutch coastline or along the Rhine River have become symbols of usage before the modern age. Although outdated, those windmills are top tourist attractions nowadays. As widespread as those windmills were, wind energy played a rather minor role in commercial electricity generation until the end of the last century. In 2000, the wind power generation in the United States was 5.59 billion kilowatt-hours (kWh), accounting for about 0.15% of the total electricity generated by the US in that year [219]. In the past decade, however, wind energy witnessed a rapid development and deployment. By the end of 2016, the annual wind power production increased 40-fold relative to the amount of wind power in 2000, to nearly 227 billion kWh, and accounted for 5.6% of the total electricity generation in that year [220]. The US Department of Energy even contemplates scenarios in which wind may generate 10% of the nation's electricity by 2020, 20% by 2030, and 35% by 2050 [217].

Remarkable progress has been made in wind turbine technology, which enables the design and installation of larger turbines and allows wind farms to be built at locations where wind is more intermittent and maintenance equipment is less accessible. All these brought new challenges to operational reliability. In an effort to maintain high reliability, with the help of advancement in micro-electronics, modern wind farms are equipped with a large number and variety of sensors, including, at the turbine level, anemometers, tachometers, accelerometers, thermometers, strain sensors, and power meters, and at the farm level, anemometers, vanes, sonars, thermometers, humidity meters, pressure meters, among others. These sensors churn out a lot of data at a fast pace, presenting unprecedented opportunities for data science to play a crucial role in addressing technical challenges in wind energy.

Like solar energy, wind energy faces an intermittent nature of its source. People commonly refer to wind and solar energy as *variable* renewable energy sources. The intermittency makes wind and solar power different from most other types of energy, even hydropower, as reservoirs built for hydropower plants smooth out the impact of irregularity and variability in precipitation on hydropower production.

The intermittency in wind presents a number of challenges to wind energy operations. The non-steady mechanical load yields excessive wear in a turbine's drive train, especially the gearbox and bearings, and makes the wind turbines prone to fatigue failures—wind turbines operate just like a car being driven in a busy city with plenty of traffic lights and rarely any freeway. Meanwhile, the randomness in wind power output makes it difficult to accommodate a substantial level of wind power in the power grid. All these lead to an increased cost and a decreased market competitiveness for wind energy. No wonder that as of 2016, the federal production tax credit (PTC) for wind was still valued at 23 cents per kWh, roughly 30% of the levelized cost of energy for onshore wind. Undoubtedly, this tax credit considerably boosts the marketability of wind energy, but without it, the competitiveness of wind energy will be called into question.

As data continues to be accumulated, data science innovations, providing profound understanding of wind stochasticity and enabling the design of countermeasures, have the potential of generating ground-breaking advancements in the wind industry. The commercial competitiveness of wind energy can benefit a great deal from a good understanding of its production reliability, which is affected by the unpredictability of wind and the productivity of wind turbines. The latter, furthermore, depends on a turbine's ability to convert wind into power during its operation and the availability or reliability of wind turbines. Data science solutions are needed in all of these aspects.

1.1   WIND ENERGY BACKGROUND

The focus of this book is data analytics at the wind turbine and wind farm level. A thorough coverage of such a scope entails a wide variety of data and a broad array of research issues. While data analytics at the power grid level is also an important part of wind energy research, the author's research has yet to be extended to that area. Hence, the scope of this book does not include data analytics at the power grid level. Nevertheless, a great deal of the turbine-level and farm-level data analytics is related to grid-level data analytics. For example, power predictions have a significant impact on grid integration.

The wind turbines considered here are the utility-scale, horizontal axis turbines. As illustrated in Fig. 1.1, a turbine, comprising thousands of parts, has three main, visible components: the blades, the nacelle, and the tower. The drive train and control system, including the gearbox and the generator, are inside the nacelle. While the vast majority of horizontal axis wind turbines use a gearbox to speed up the rotor speed inside the generator, there are also direct drive wind turbines in which the gearbox is absent and the rotor directly drives the generator. An anemometer or a pair of them can be found sitting on top of the nacelle, towards its rear end, to measure wind speed, whereas a vane is for the measurement of wind direction. Responding to changes in wind direction, yaw control is to rotate and point the nacelle to where the wind comes from. Responding to changes in wind speed, pitch control turns the blades in relation to the direction of the incoming air flow, adjusting the capability of the turbine to absorb the kinetic energy in the wind or the turbine's efficiency in doing so.

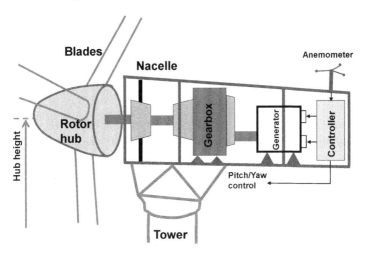

FIGURE 1.1 Schematic of major parts in a wind turbine.

A commercial wind farm can house several hundred wind turbines. For instance, the Roscoe Wind Farm, the largest wind farm in Texas as of this writing, houses 627 wind turbines. Other than turbines, meteorological masts are installed on a wind farm, known as the met towers or met masts. A number of instruments and sensors are installed on the met towers, measuring additional environmental conditions, such as temperature, air pressure, humidity, precipitation, among others. Anemometers and vanes are usually installed at multiple heights of a met tower. The multi-height measurements allow the calculation of vertical wind shear, which characterizes the change in wind speed with height, as well as the calculation of vertical wind veer, which characterizes the change in wind direction with height. The wind speed and direction measured at the nacelle during a commercial operation are typically only at the hub height.

Throughout the book, denote by x the input vector whose elements are the environmental variables, which obviously include wind speed, V, in the unit of meters per second (m/s), and wind direction, D, in degrees ($°$). The zero degree corresponds to due north. Sometimes analysts combine the speed and direction information of wind and express them in two wind velocities along the longitudinal and latitudinal directions, respectively. Other environmental variables include air density, ρ, humidity, H, turbulence intensity, I, and wind shear, S. Not all of these environmental variables are directly measured. Some of them are computed, such as,

- Turbulence intensity, I: first compute the standard deviation of the wind speeds in a short duration and denote it as $\hat{\sigma}$. Then, $I = \hat{\sigma}/\bar{V}$, where $\bar{V}$ is the average wind speed of the same duration. It is worth noting that the concept of turbulence intensity in air dynamics is similar to the coefficient of variation concept in statistics [58].

- Wind shear, S: wind speeds, V_1 and V_2, are measured at heights h_1 and h_2, respectively. Then, the vertical wind shear between the two heights is $S = \ln(V_2/V_1)/\ln(h_2/h_1)$ [175]. When anemometers are installed at locations both above and below the rotor hub, then two wind shears, the above-hub wind shear, S_a, and the below-hub wind shear, S_b, can be calculated.

- Air density, ρ, in the unit of kilograms per cubic meter (kg/m^3): given air temperature, T, expressed in Kelvin and air pressure, P, expressed in Newtons per square meter (N/m^2), $\rho = P/(\varrho \cdot T)$, where $\varrho = 287$ Joule/(kg·Kelvin) is the gas constant [216].

Using the above notation, the input vector to a turbine can be expressed as $\boldsymbol{x} = (V, D, \rho, H, I, S_a, S_b)^T$. But the input vector is not limited to the aforementioned variables. The hours in a day when a measurement is recorded, the power output of a nearby turbine, wind directional variation and wind veer if either or both are available, could also be included in the input vector, $\boldsymbol{x}$. On the other hand, while the wind speed, wind direction, and temperature measurements are commonly available on commercial wind farms, the availability of other measurements may not be.

Two types of output of a wind turbine are used in this book: one is the active power measured at a turbine, denoted by y and in the unit of kilowatts (kW) or megawatts (MW), and the other one is the bending moment, a type of mechanical load, measured at critical structural spots, denoted by z and in the unit of kiloNewtons-meter (kN-m) or million Newtons-meter (MN-m). The power output measures a turbine's power production capability, while the bending moment measurements are pertinent to a turbine's reliability and failure management. The power measurement is available for each and every turbine. Analysts may also aggregate the power outputs of all turbines in an entire wind farm when the whole farm is treated as a single power production unit. The bending moment measurements are currently not available on commercially operated turbines. They are more commonly collected on testing turbines and used for design purposes.

The input and output data can be paired into a data record. For the power response, it is the pair of $(\boldsymbol{x}, y)$, whereas for the mechanical load response, it is $(\boldsymbol{x}, z)$.

Turbine manufacturers provide a wind speed versus power functional curve, referred to as the *power curve*. Fig. 1.2 presents such a power curve. As shown in the power curve, a turbine starts to produce power after the wind reaches the cut-in speed, V_{ci}. A nonlinear relation between y and V then ensues, until the wind reaches the rated wind speed, V_r. When the wind speed is beyond V_r, the turbine's power output will be capped at the rated power output, y_r, also known as the nominal power capacity of the turbine, using control mechanisms such as pitch control and rotor speed regulation. The turbine will be halted when the wind reaches the cut-out speed, V_{co}, because high wind is deemed harmful to the safety of a turbine. The power curve shown here is an ideal power curve, also known as the nominal power curve. When the actual measurements of wind speed and power output are used, the V-versus-y plot will not appear as slim and smooth as the nominal power curve; rather, it will be a data scattering plot, showing considerable amount of noise and variability.

In order to protect the confidentiality of the data providers, the wind power data used in this book are normalized by the rated power, y_r, and expressed as a standardized value between 0 and 1.

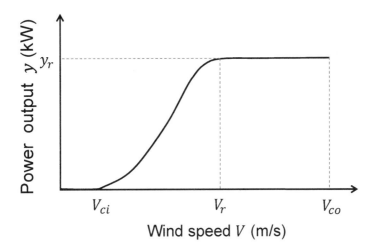

FIGURE 1.2 Nominal power curve of a wind turbine. (Reprinted with permission from Lee et al. [132].)

The raw data on wind turbines are recorded in a relatively fast frequency, in the range of a couple of data points per second to a data point per a couple of seconds. The raw data are stored in a database, referred to as the data historian. When the data are used in the turbine's supervisory control and data acquisition (SCADA) system, the current convention in the wind industry is to average the measurements over 10-minute time blocks because wind speed is assumed stationary over this 10-min duration and other environmental variables are assumed nearly constant. These assumptions are, of course, not always true. In this book, however, we choose to follow this industrial standard practice. With 10-min blocks, a year's worth of data has about

52,560 data pairs if there is no missing data at all. In reality, even with auto-mated measurement devices, missing data is common, almost always making the actual data amount fewer than 50,000 for a year.

Even though the wind speed used is mostly a 10-min average, we decide to drop the overline while representing this average, for the sake of notational simplicity. That is to say, we use V, instead of $\bar{V}$, to denote the average wind speed in a 10-min block. When $\bar{V}$ is used, it refers to the average of 10-min averaged wind speeds.

Fig. 1.3 shows the arrangement of the multi-turbine, multi-year data for a wind farm. In the top panel, the whole dataset is shown as a cube, in which each cross section represents the spatial layout of turbines on a farm and the horizontal axis represents the time. The longitudinal data are the time-series of a turbine's power output, y, and environmental measurements, $\boldsymbol{x}$. The cross-sectional data, or the snapshot data, are of multiple turbines but are for a particular point in time. A cross section could be a short time period, for instance, a couple of days or weeks, during which the turbine's innate condition can be assumed unchanged. The power curve of a turbine is visualized as the light-colored (yellow) curve in the bottom panel (see also Color eBook), with the actual measurements in the background. As mentioned earlier, the actual measurements are noisy, and the nominal power curve averages out the noise.

1.2 ORGANIZATION OF THIS BOOK

We organize this book based on a fundamental data science formulation for wind power production:

$$f_t(y) = \int_{\boldsymbol{x}} f_t(y|\boldsymbol{x}) f_t(\boldsymbol{x}) d\boldsymbol{x}, \tag{1.1}$$

where $f(\cdot)$ denotes a probability density function and the subscript t, the time indicator, signifies the dynamic, time-varying aspect of the function.

This formulation implies that in order to understand $f_t(y)$, namely the stochasticity of power output y, it is necessary to understand the distribution of wind and other environmental variables, $f_t(\boldsymbol{x})$, as well as the turbine's power production conditioned on a given wind and environmental condition $\boldsymbol{x}$. We use a conditional density function, $f_t(y|\boldsymbol{x})$, to characterize the conditional distribution.

When the power output, y, is replaced by the mechanical load response (namely the bending moment), z, the above formulation is still meaningful, with $f(z|\boldsymbol{x})$ representing the conditional load response for a given environ-mental condition.

The use of conditional density functions is a natural result of wind inter-mittency. When the driving force to a turbine changes constantly, the turbine's response, regardless of being the power or the load, ought to be characterized under a given wind and environmental condition.

This book aims to address three aspects related to the aforementioned

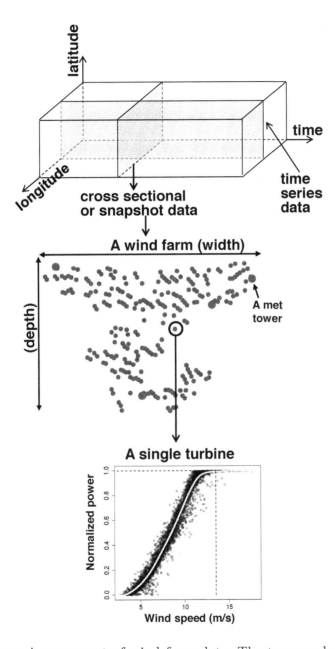

FIGURE 1.3 Arrangement of wind farm data. The top panel shows the spatio-temporal arrangement of wind farm data; the middle panel shows the layout of a wind farm, where the small dots are wind turbines and the big dots are the met towers; and the bottom panel presents the data from a single turbine, where the light-colored (yellow) curve (see Color eBook) is the nominal power curve and the circles in the background are the actual power measurements.

general formulation of wind power production. Thus, we divide the rest of this book into three parts:

1. The first part consists of three chapters. It is about the modeling of $f_t(\boldsymbol{x})$, which begets an analysis of the wind field. Based on the modeling and analysis of the wind field, a wind forecast can be made. If a whole wind farm is simplified as a single location, or the forecast at a single turbine is of concern, the need for a temporal analysis arises. If multiple turbines at different sites are to be studied, or multiple wind farms at different geographic locations are involved, the modeling of $f_t(\boldsymbol{x})$ becomes a spatio-temporal analysis. Both temporal and spatio-temporal methods will be described but the focus is on the spatio-temporal analysis.

2. The second part consists of four chapters. It discusses power response modeling and shows how the power response model can be used for performance evaluation of wind turbines. The general expression, $f(y|\boldsymbol{x})$, depicts a multivariate, probabilistic power response surface. The power curve is in fact the conditional expectation, $\mathbb{E}(y|\boldsymbol{x})$, when $\boldsymbol{x}$ is reduced to a univariate input, the wind speed, V. The modeling of $f(y|\boldsymbol{x})$ or $\mathbb{E}(y|\boldsymbol{x})$ falls into the area of density regression or nonparametric regression analysis.

3. The third part consists of four chapters. It provides a reliability and load analysis of wind turbines. Using Eq. 1.1 to assess power production assumes, implicitly, an up-running wind turbine, namely a non-zero $f_t(y|\boldsymbol{x})$. But wind turbines, under non-steady wind forces, are prone to failures and downtime. To factor in a turbine's reliability impact, it is important to assess a turbine's load response under various wind conditions. The statistical learning underlying the analysis in this part is related to sampling techniques, including importance sampling and Markov chain Monte Carlo sampling.

1.2.1 Who Should Use This Book

The book is intended to be a research monograph, but it can be used for teaching purposes as well. We expect our readers to have basic statistics and probability knowledge, and preferably a bachelor's degree in STEM (Science, Technology, Engineering, and Math). This book provides an in-depth discussion of how data science methods can improve decision making in several aspects of wind energy applications, from near-ground wind field analysis and wind forecast, turbine power curve fitting and performance analysis, turbine reliability assessment, to maintenance optimization for wind turbines and wind farms. A broad set of data science methods are covered, including time series models, spatio-temporal analysis, kernel regression, decision trees, splines, Bayesian inference, and random sampling. The data science methods are described in the context of wind energy applications with examples and case studies. Real data and case studies from wind energy research and industrial practices are used in this book. Readers who may benefit from reading this book include practitioners in the wind industry who look for data science solutions and faculty members and students who may be interested in the research of data science for wind energy in departments such as industrial and systems engineering, statistics, and power engineering.

There are a few books on renewable energy forecasting [117], which overlap, to a certain degree, with the content of Part I. A topic related to wind energy but left out in the book is about grid integration, for which interested readers can refer to the book by Morales et al. [148].

1.2.2 Note for Instructors

This book can be used as the textbook for a stand-alone course, with the course title the same as or similar to the title of this book. It can also be used to as a reference book that provides supplementary materials for certain segments of either a data science course (supplementing wind energy application examples) or a power engineering course (supplementing data science methods). These courses can come from the offerings of a broad set of departments, including Industrial Engineering, Electrical Engineering, Statistics, Aerospace Engineering, or Computer Science.

We recommend that the first chapter be read before later chapters are covered. The three parts after the first chapter are more or less independent of each other. It does not matter in which sequence the three parts are read or taught. Within each part, however, we recommend following the order of the chapters. It will take two semesters to teach the whole book. One can, nevertheless, sample one or two chapters from each part to form the basis for a one-semester course.

Most of the examples are solved using the R programming language, while some are solved using the MATLAB® programming language. At the end of a chapter, acronyms and abbreviations used in that chapter are summarized and explained in the Glossary section.

1.2.3 Datasets Used in the Book

In this book, the following datasets are used:

1. **Wind Time Series Dataset**. This dataset comes from a single turbine on an inland wind farm. The dataset covers the duration of one year, but data at some of the time instances are missing. Two time resolutions are included in the dataset: the 10-min data and the hourly data; the latter is the further average of the former. For each temporal resolution, the data is arranged in three columns. The first column is the time stamp, the second column is the wind speed, and the third column is the wind power.

2. **Wind Spatial Dataset**. This dataset comes from ten turbines in an offshore wind farm. Only the hourly wind speed data are included. The duration of the data covers two months. The longitudinal and latitudinal coordinates of each turbine are given, but those coordinates are shifted by an arbitrary constant, so that the actual locations of these turbines are protected. The relative positions of the turbines, however, remain truthful to the physical layout. The data is arranged in the following fashion. Under the header row, the next two rows are the coordinates of each turbine. The third row under the header is purposely left blank. From the fourth row onwards are the wind speed data. The first column is the time stamp. Columns 2-11 are the wind speed values measured in meters per second.

3. **Wind Spatio-Temporal Dataset1**. This dataset comprises the average and standard deviation of wind speed, collected from 120 turbines in an inland wind farm, for the years of 2009 and 2010. Missing data in the original dataset are imputed by using the iterative singular value decomposition [139]. Two data files are associated with each year—one contains the hourly average wind speed, used in Eq. 3.18, and the other contains the hourly standard deviation of wind speed, used in Eq. 3.25. The naming convention makes it clear which year a file is associated with and whether it is for the average speed (**Ave**) or for the standard deviation (**Stdev**). The data arrangement in these four files is as follows—the columns are the 120 turbines and the rows are times, starting from 12 a.m. on January 1 of a respective year as the first data row, followed by the subsequent hours in that year. The fifth file in this dataset contains the coordinates of the 120 turbines. To protect the wind farm's identity, the coordinates have been transformed by an undisclosed mapping, so that their absolute values are no longer meaningful but the turbine-to-turbine relative distances are maintained.

4. **Wind Spatio-Temporal Dataset2**. The data used in this study consists of one year of spatio-temporal measurements at 200 randomly selected turbines on a flat terrain inland wind farm, between 2010 and 2011. The data consists of turbine-specific hourly wind speeds measured by the anenometers mounted on each turbine. In addition, one year of hourly wind speed and direction measurements are available at three met masts on the same wind farm. Columns B through OK are the wind speed and wind power associated with each turbine, followed by Columns OL through OQ, which are for wind speed and wind direction associated with each mast. The coordinates of the turbines and masts are listed in the top rows, preceding the wind speed, direction, and power data. The coordinates are shifted by a constant, so that while the relative positions of the turbines and the met masts remain faithful to the actual layout, their true geographic information is kept confidential. This anemometer

network provides a coverage of a spatial resolution of one mile and a temporal resolution of one hour.

5. **Inland Wind Farm Dataset1** and **Offshore Wind Farm Dataset1**. Data included in these two datasets are generated from six wind turbines and three met masts and are arranged in six files, each of which is associated with a turbine. The six turbines are named WT1 through WT6, respectively. The layout of the turbines and the met masts is shown in Fig. 5.6. On the offshore wind farm, all seven environmental variables as mentioned above are available, namely $x = (V, D, \rho, H, I, S_a, S_b)$, whereas on the inland wind farm, the humidity measurements are not available, nor is the above-hub wind shear, meaning that $x = (V, D, \rho, I, S_b)$. Variables in x were measured by sensors on the met mast, whereas y was measured at the wind turbines. Each met mast has two wind turbines associated with it, meaning that the x's measured at a met mast are paired with the y's of two associated turbines. For WT1 and WT2, the data were collected from July 30, 2010 through July 31, 2011 and for WT3 and WT4, the data were collected from April 29, 2010 through April 30, 2011. For WT5 and WT6, the data were collected from January 1, 2009 through December 31, 2009.

6. **Inland Wind Farm Dataset2** and **Offshore Wind Farm Dataset2**. The wind turbine data in these two datasets include observations during the first four years of the turbines' operations. The inland turbine data are from 2008 to 2011, whereas the offshore data are from 2007 to 2010. The measurements for the inland wind farm include the same x's as in the **Inland Wind Farm Dataset1** and those for the offshore wind farm include the same x's as in the **Offshore Wind Farm Dataset1**. Most of the environmental measurements x are taken from the met mast closest to the turbine, with the exception of wind speed and turbulence intensity which are measured on the wind turbine. The mast measurements are used either because some variables are only measured at the mast (such as air pressure and ambient temperature, which are used to calculate air density) or because the mast measurements are considered more reliable (such as wind direction).

7. **Turbine Upgrade Dataset**. This dataset includes two sets, corresponding, respectively, to an actual vortex generator installation and an artificial pitch angle adjustment. Two pairs of wind turbines from the same inland wind farm, as used in Chapter 5, are chosen to provide the data, each pair consisting of two wind turbines, together with a nearby met mast. The turbine that undergoes an upgrade in a pair is referred to as the *experimental turbine*, the *reference turbine*, or the *test turbine*, whereas the one that does not have the upgrade is referred to as the *control turbine*. In both pairs, the test turbine and the control turbine

are practically identical and were put into service at the same time. This wind farm is on a reasonably flat terrain.

The power output, y, is measured on individual turbines, whereas the environmental variables in x (i.e., the weather covariates) are measured by sensors at the nearby mast. For this dataset, there are five variables in x and they are the same as those in the Inland Wind Farm Dataset1. For the vortex generator installation pair, there are 14 months' worth of data in the period before the upgrade and around eight weeks of data after the upgrade. For the pitch angle adjustment pair, there are about eight months of data before the upgrade and eight and a half weeks after the upgrade.

Note that the pitch angle adjustment is not physically carried out, but rather simulated on the respective test turbine. The following data modification is done to the test turbine data. The actual test turbine data, including both power production data and environmental measurements, are taken from the actual turbine pair operation. Then, the power production from the designated test turbine on the range of wind speed over 9 m/s is increased by 5%, namely multiplied by a factor of 1.05, while all other variables are kept the same. No data modification of any kind is done to the data affiliated with the control turbine in the pitch angle adjustment pair.

The third column of a respective dataset is the upgrade status variable, of which a zero means the test turbine is not modified yet, while a one means that the test turbine is modified. The upgrade status has no impact on the control turbine, as the control turbine remains unmodified throughout. The vortex generator installation takes effect on June 20, 2011, and the pitch angle adjustment takes effect on April 25, 2011.

8. **Wake Effect Dataset**. This dataset includes data from six pairs of wind turbines (or, 12 wind turbines in total) and three met masts. The turbine pairs are chosen such that no other turbines except the pair are located within 10 times the turbine's rotor diameter. Such arrangement is to find a pair of turbines that are free of other turbines' wake, so that the wake analysis result can be reasonably attributed to the wake of its pair turbine. The operational data for the six pairs of turbines are taken during roughly a yearlong period between 2010 and 2011. The datasets include wind power output, wind speed, wind direction, air pressure, and temperature, of which air pressure and temperature data are used to calculate air density. The wind power outputs and wind speeds are measured on the turbine, and all other variables are measured at the met masts. The data from Mast 1 are associated with the data for Turbine Pairs 1 and 2, Mast 2 with Pairs 3 and 4, and Mast 3 with Pairs 5 and 6. Fig. 8.6 shows the relative locations of the six pairs of turbines and three met masts.

9. **Turbine Bending Moment Dataset.** This dataset includes two parts. The first part is three sets of physically measured blade-root flapwise bending moments on three respective turbines, courtesy of Risø-DTU (Technical University of Denmark) [180]. The basic characteristics of the three turbines can be found in Table 10.1. These datasets include three columns. The first column is the 10-min average wind speed, the second column is the standard deviation of wind speed within a 10-min block, and the third column is the maximum bending moment, in the unit of MN-m, recorded in a 10-min block. The second part of the dataset is the simulated load data used in Section 10.6.5. This part has two sets. The first set is the training data that has 1,000 observations and is used to fit an extreme load model. The second set is the test data that consists of 100 subsets, each of which has 100,000 observations. In other words, the second dataset for testing has a total of 10,000,000 observations, which are used to verify the extreme load extrapolation made by a respective model. Both simulated datasets have two columns: the first is the 10-min average wind speed and the second is the maximum bending moment in the corresponding 10-min block. While all other datasets are saved in CSV file format, this simulated test dataset is saved in a text file format, due to its large size. The data simulation procedure is explained in Section 10.6.5.

10. **Simulated Bending Moment Dataset.** This dataset includes two sets. One set has 600 data records, corresponding to the training set referred to in Section 11.4.1, whereas the other set has 10,000 data records, which are used to produce Fig. 11.1. Each set has three columns of data (other than the serial number). The first column is the wind speed, simulated using a Rayleigh distribution, and the second and third columns are, respectively, the simulated flapwise and edgewise bending moments, in the unit of kN-m. The flapwise and edgewise bending moments are simulated from TurbSim [112] and FAST [113], following the procedure discussed in [149]. TurbSim and FAST are simulators developed at the National Renewable Energy Laboratory (NREL) of the United States.

GLOSSARY

CSV: Comma-separated values Excel file format

DTU: Technical University of Denmark

NREL: National Renewable Energy Laboratory

PTC: Production tax credit

SCADA: Supervisory control and data acquisition

STEM: Science, technology, engineering, and mathematics

US: United States of America

FURTHER READING

C. F. J. Wu. Statistics = Data Science? *Presentation at the H. C. Carver Professorship Lecture*, 1997. https://www2.isye.gatech.edu/~jeffwu/presentations/datascience.pdf

D. Donoho. 50 Years of Data Science. *Journal of Computational and Graphical Statistics*, 26: 745–766, 2017.

I

Wind Field Analysis

CHAPTER 2

A Single Time Series Model

DOI: 10.1201/9780429490972-2

Part I of this book is to model $f_t(x)$. The focus is on wind speed, V, because wind speed is much more volatile and difficult to predict than other environmental variables such as air density or humidity. In light of this thought, $f_t(x)$ is simplified to $f_t(V)$.

A principal purpose of modeling $f_t(V)$ is to forecast wind speed or wind power. Because it is impossible to control wind, forecasting becomes an essential tool in turbine control and wind power production planning. Modeling the time-varying probability density function $f_t(V)$ directly, however, is difficult. In practice, what is typically done is to make a point forecast first and then assess the forecasting uncertainty, which is to attach a confidence interval to the point forecast. The point forecast is a single value used to represent the likely wind speed or power at a future time, corresponding, ideally but not necessarily, to the mean, median, or mode of the probability distribution of wind speed or power at that future time.

The forecasting can be performed either on wind speed or on wind power. Wind power forecasting can be done by forecasting wind speed first and then converting a speed forecast to a power forecast through the use of a simple power curve, as explained in Chapter 1, or the use of a more advanced power curve model, to be explained in Chapter 5. Wind power forecasting can also be done based purely on the historical observations of power output, without necessarily accounting for wind speed information. In the latter approach, the methods developed to forecast wind speed can be used, almost without any changes, to forecast wind power, so long as the wind speed data are replaced with the wind power data. For this reason, while our discussion in this chapter mainly refers to wind speed, please bear in mind its direct applicability to wind power forecast.

In Chapter 2, we consider models that ignore the spatial information and are purely based on the time series data. In Chapters 3 and 4, we discuss various types of spatial or spatio-temporal models.

2.1 TIME SCALE IN SHORT-TERM FORECASTING

One essential question in forecasting is concerning the time-scale requirements of forecast horizons. Turbine control typically requires an instantaneous response in seconds or sub-seconds. Production planning for grid integration and market response is in a longer time scale. Two energy markets, the real-time market and the day-ahead market, demand different response times. The real-time market updates every five minutes, requiring a response in the level of minutes, whereas the day-ahead market is for trading on the next day, requiring forecasting up to 24 hours ahead. Between these two time scales, there are other planning actions that may request a forecast from a few minutes to a few hours ahead. For instance, when the wind power supply is insufficient to meet the demand, the system operators would bring up reserve powers. The spinning reserve, which has been synchronized to the grid system, can be ready for dispatch within 10 minutes, whereas the delivery of contingency reserves may encounter a delay, up to an hour or more, thereby needing a longer lead time for notification. For various planning and scheduling purposes, a common practice for wind owners/operators is to create forecasts, for every hour looking ahead up to 24 hours, and then update that hourly ahead forecast at the next hour for the subsequent 24 hours, using the new set of data collected in between.

When it comes to wind forecasting, there are two major schools of thought. One is the physical model-based approach, collectively known as the Numerical Weather Prediction (NWP) [138], which is the same scientific method used behind our daily weather forecast, and the second is the data-driven, statistical modeling-based approach. By calling the second approach "data-driven," we do not want to leave readers with the impression that NWP is data free; both approaches use weather measurement data. The difference between the two approaches is that NWP involves physical atmospheric models, while the pure data-driven models do not.

Because NWP is based on physical models, it has, on the one hand, the capability to forecast into a relatively longer time horizon, from a few hours ahead to several days ahead. On the other hand, the intensive computation required to solve the complicated weather models limits the temporal and spatial resolutions for NWP, making analysts tend to believe that for a short-term forecast on a local wind field, the data-driven models are advantageous. There is, however, no precise definition of how short is a "short term." Giebel et al. [71] deem six hours as the partition, shorter than which, the data-driven models perform better, while longer than that, NWP ought to be used. Analysts do sometimes push the boundary of data-driven models and make forecasting over a longer horizon, but still, the horizon is generally shorter than 12 hours.

In this book, our interest is to make short-term forecasting on local wind fields. We follow the same limits for short-term as established in the literature, which is usually a few hours ahead and no more than 12 hours ahead.

2.2 SIMPLE FORECASTING MODELS

We first consider the situation that the historical wind data is arranged in a single time series, from time 1 to time t, denoted by $V_i, i = 1, \ldots, t$. The single time series is appropriate to describe the following application scenarios:

- The wind speed or power data measured on a single turbine is used to forecast future wind speed or power on the same turbine.

- The wind speed on a single met tower is used to forecast wind speed, and used as the representation of wind speed for a wind farm.

- The aggregated wind power of a wind farm, namely the summation of wind power output of all individual turbines on the farm, is used to forecast the future aggregated power output of the wind farm.

- Although wind speed is measured at multiple locations, the average wind speed over the locations is used to forecast the future average wind speed.

2.2.1 Forecasting Based on Persistence Model

The simplest point forecasting is based on the *persistence* (PER) model, which says the wind speed or power at any future time, $t + h, h > 0$, is simply the same as what is observed at the current time, t, namely,

$$\hat{V}_{t+h} = V_t, \quad h > 0, \tag{2.1}$$

where the hat notation (^) is used to indicate a forecast (or an estimate). The persistence forecast should, and can easily, be updated when a new observation of V arrives at the next time point.

When the persistence model is used, there is no uncertainty quantification procedure directly associated with it. In order to associate a confidence interval, one needs to establish a probability distribution for wind speed.

2.2.2 Weibull Distribution

Wind speeds are nonnegative and their distribution is right skewed. They do not strictly follow a normal distribution. Understandably, probability densities that are right skewed with nonnegative domain, such as Weibull, truncated normal, or Rayleigh distributions, are common choices for modeling wind speed; for a comprehensive list of distributions, please refer to a survey paper on this topic [32].

There is no consensus on which distribution best describes the data of wind speed, although Weibull distribution is arguably the most popular one. Analysts can try a few of the widely used distributions and test which one fits the data the best. This practice entails addressing two statistical problems— one is to estimate the parameters in the chosen distribution and the other is to assess the goodness-of-fit of the chosen distribution and see if the chosen distribution provides a satisfactory fit to the data.

Consider the Weibull distribution as an example. Its probability density function (pdf) is expressed as

$$f(x) = \begin{cases} \left(\frac{\beta}{\eta}\right)\left(\frac{x}{\eta}\right)^{\beta-1}\exp\left\{-\left(\frac{x}{\eta}\right)^{\beta}\right\} & x \geq 0, \\ 0 & x < 0, \end{cases} \tag{2.2}$$

where $\beta > 0$ is the shape parameter, affecting the skewness of the distribution, and $\eta > 0$ is the scale parameter, affecting the concentration of the distribution. When $\beta \leq 1$, the Weibull density is a decaying function, monotonically going downwards from the origin. When $\beta > 1$, the Weibull density first rises up, passes a peak and then goes down. For commercial wind farms, it makes no practical sense to expect its wind speed to follow a Weibull distribution of $\beta \leq 1$, as what it suggests is that most frequent winds are all low-speed winds. If a wind farm planner does a reasonable job in selecting the farm's location, it is expected to see $\beta > 1$.

The probability density function in Eq. 2.2 is known as the two-parameter Weibull distribution, whose density curve starts at the origin on the x-axis. A more general version, the three-parameter Weibull distribution, is to replace x by $x - \nu$ in Eq. 2.2, where ν is the location parameter, deciding the starting point of the density function on the x-axis. When $\nu = 0$, the three-parameter Weibull density simplifies to the two-parameter Weibull density. The two-parameter Weibull is the default choice, unless one finds that there is an empty gap in the low wind speed measurements close to the origin.

2.2.3 Estimation of Parameters in Weibull Distribution

To estimate the parameters in the Weibull distribution, a popular method is the maximum likelihood estimation (MLE). Given a set of n wind speed measurements, $V_i, i = 1, \ldots, n$, the log-likelihood function, $\mathcal{L}(\beta, \eta|V)$, can be expressed as:

$$\mathcal{L}(\beta, \eta|V) = n\ln\beta - \beta n\ln\eta + (\beta - 1)\sum_{i=1}^{n}\ln V_i - \sum_{i=1}^{n}\left(\frac{V_i}{\eta}\right)^{\beta}. \tag{2.3}$$

Maximizing the log-likelihood function can be done by using an optimization solver in a commercial software, such as nlm in R. Because nlm is for minimization, one should multiply a (-1) to the returned values of the above log-likelihood function while using nlm or a similar minimization routine in other software packages. With the availability of the MASS package in R, fitting

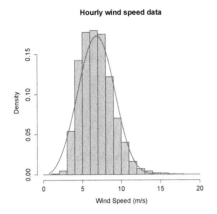
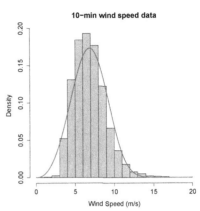

FIGURE 2.1 Fit a Weibell distribution to the wind speed data in the `Wind Time Series Dataset`. The left panel is the fit to the hourly data. The estimated parameters are: $\hat{\eta} = 7.60$, $\hat{\beta} = 3.40$, mean $= 6.84$, median $= 6.69$, mode $= 6.5$, and the standard deviation $= 2.09$. The right panel is the fit to the 10-min data. The estimated parameters are: $\hat{\eta} = 7.61$, $\hat{\beta} = 3.41$, mean $= 6.86$, median $= 6.67$, mode $= 6.5$, and the standard deviation $= 2.06$. The values of mean, median, mode and standard deviation are estimated directly from the data, rather than calculated using $\hat{\eta}$ and $\hat{\beta}$.

a Weibull distribution can be done more directly by using the `fitdistr` function. Suppose that the wind speed data is stored in the vector named `wsdata`. The following R command can be used for fitting a Weibull distribution,

```
fitdistr(wsdata, "weibull").
```

Fig. 2.1 presents an example of using a Weibull distribution to fit the wind speed data in the `Wind Time Series Dataset`. The Weibull distribution parameters are estimated by using the MLE. Fig. 2.1 presents the Weibull fit to the wind speed data of two time resolutions: the 10-min data and the hourly data. The estimates of the shape and scale parameters are rather similar despite the difference in time resolution.

2.2.4 Goodness of Fit

Once a Weibull distribution is fit to a set of data, how can we tell whether or not it is a good fit? This question is answered through a goodness-of-fit test, such as the χ^2 test. The idea of the χ^2 test is simple. It first bins the observed data, like in a histogram. For the j-th bin, one can count the number of actual observations falling into that bin; denote this as O_j. Should the data follow a specific type of distribution, the expected amount of data points in the same bin can be computed from the cumulative distribution function (cdf) of that distribution; denote this quantity as E_j. Suppose that we have a total of B bins. Then, the test statistic, defined below, follows a χ^2 distribution with a degree of freedom of $B - p - 1$, i.e.,

$$\chi^2 := \sum_{j=1}^{B} \frac{(O_j - E_j)^2}{E_j} \sim \chi^2_{B-p-1}, \tag{2.4}$$

where p is the number of parameters associated with the distribution.

The Weibull distribution has a closed form cdf. The fitted Weibull distribution function, by plugging in the estimated parameters, $\hat{\beta}$ and $\hat{\eta}$, is

$$F_{\hat{\beta},\hat{\eta}}(x) = 1 - \exp\left\{-\left(\frac{x}{\hat{\eta}}\right)^{\hat{\beta}}\right\}. \tag{2.5}$$

Of the j-th wind speed bin, the left boundary wind speed value is V_{j-1} and the right boundary value is V_j, so E_j can be calculated by

$$E_j = n[F_{\hat{\beta},\hat{\eta}}(V_j) - F_{\hat{\beta},\hat{\eta}}(V_{j-1})]. \tag{2.6}$$

Once the χ^2 test statistic is calculated, one can compute the p-value of the test by using, for example, the R command, $1-\text{pchisq}(\chi^2, B - p - 1)$. The null hypothesis says that the distribution under test provides a good fit. When the p-value is small enough, say, smaller than 0.05, analysts say that the null hypothesis is rejected at the significance level of 95%, implying that the theoretical distribution is less likely a good fit to the data. When the p-value is not small enough and the null hypothesis cannot be rejected, then the test implies a good fit.

We can apply the χ^2 test to one month of data of the Wind Time Series Dataset and the respective fitted Weibull distributions. The number of parameters in the two-parameter Weibull distribution is $p = 2$. While binning the wind speed data, one needs to be careful about some of the tail bins in which the expected data amount could be too few. The general guideline is that E_j should be no fewer than five; otherwise, several bins should be grouped into a single bin.

The test statistic and the corresponding p-values are shown in Table 2.1. As shown in the table, it looks like using the Weibull distribution to fit the wind speed data does not pass the goodness-of-fit test. This is particularly true when the data amount increases, as in the case of using the 10-min data. Nonetheless, the Weibull distribution still stays as one of the most popular distributions for modeling wind speed data. The visual inspection of Fig. 2.1 leaves analysts with the feeling of a reasonable fit. Passing the formal statistical test in the presence of abundant data appears tough. Analysts interested in a distribution alternative can refer to [32] for more choices.

TABLE 2.1 Goodness-of-fit test statistics and p-values.

	Hourly data	10-min data
Month selected	February	November
Data amount, n	455	3,192
Bin size	0.2 m/s	0.1 m/s
Number of bins, B	66	100
Test statistic	62.7	329.8
p-value	0.012	almost 0

2.2.5 Forecasting Based on Weibull Distribution

Assuming that the distribution of wind speed stays the same for the next time period, i.e., the underlying process is assumed stationary, analysts can use the mean as a point forecast, and then use the distribution to assess the uncertainty of the point forecast. We want to note that such approach is, in spirit, also a persistence forecasting, but it is conducted in the sense of an unchanging probability distribution.

The mean and the standard deviation of a Weibull distribution, if using the estimated distribution parameters, are

$$
\begin{cases}
\hat{\mu} &= \hat{\eta}\Gamma(1 + \frac{1}{\hat{\beta}}), \\
\hat{\sigma} &= \hat{\eta}\sqrt{\Gamma(1 + \frac{2}{\hat{\beta}}) - (\Gamma(1 + \frac{1}{\hat{\beta}}))^2},
\end{cases}
\tag{2.7}
$$

where $\Gamma(\cdot)$ is the gamma function, defined such as $\Gamma(x) = \int_0^\infty t^{x-1}e^{-t}dt$.

While the mean $\hat{\mu}$ is used as the point forecast, one can employ a normal approximation to obtain the $100(1 - \alpha)\%$ confidence interval of the point forecast, as

$$
[\hat{\mu} - z_{\alpha/2} \cdot \hat{\sigma}, \quad \hat{\mu} + z_{\alpha/2} \cdot \hat{\sigma}],
\tag{2.8}
$$

where z_α is the α-quantile point of a standard normal distribution. When $\alpha = 0.05$, $z_{0.05/2} = 1.96$.

Sometimes analysts think that using the mean may not make a good forecast, due to the skewness in the Weibull distribution. Alternatively, median and mode can be used. Their formulas, still using the estimated parameters, are

$$
\begin{cases}
\text{median} &= \hat{\eta}(\ln 2)^{1/\hat{\beta}}, \\
\text{mode} &= \hat{\eta}\left(1 - \frac{1}{\hat{\beta}}\right)^{1/\hat{\beta}} \quad \text{for } \hat{\beta} > 1.
\end{cases}
\tag{2.9}
$$

The mode of a Weibull distribution when $\beta \leq 1$ is zero. As mentioned earlier, the circumstances under which $\beta \leq 1$ are of little practical relevance in wind speed modeling at commercial wind farms.

Analysts may worry that using the normal approximation to obtain the confidence interval may not be accurate enough. If one has a sufficiently large

TABLE 2.2 Estimate of mean and 95% confidence interval of wind speed data. The total data amount is 7,265 for the hourly data and 39,195 for the 10-min data.

	Based on Eq. 2.8		Directly from sample statistics	
	Mean	C.I.	Mean	C.I.
Hourly data	6.83	[2.48, 11.47]	6.84	[3.54, 11.33]
10-min data	6.84	[2.50, 11.18]	6.86	[3.62, 11.49]

amount of wind speed data, say more than 1,000 data points, a simple way is to estimate the mean and its confidence interval directly from the data, following the two steps below.

1. Compute the sample average wind speed, $\bar{V}$,

$$\bar{V} = \frac{1}{n} \sum_{i=1}^{n} V_i,$$

and use it as the point forecast.

2. Order the wind speed data from the smallest to the largest. Denote the ordered sequence as $V_{(1)}, V_{(2)}, \ldots, V_{(n)}$. Then, the $100(1-\alpha)\%$ confidence interval is estimated to be $[V_{[n\alpha/2]}, V_{[n(1-\alpha/2)]}]$, where $[\cdot]$ returns the nearest integer number.

Table 2.2 presents the estimates of mean and confidence interval, either based on the Weibull distribution or directly from the data. One observes that the point forecasts are rather close, but the lower confidence intervals are noticeably different.

2.3 DATA TRANSFORMATION AND STANDARDIZATION

Before the wind speed data is fed into time series models, many of which assume Gaussianity, data preprocessing may be needed. Two common preprocessing tasks are: (1) normalizing the wind data, so that the transformed data behaves closer to a normal distribution, and (2) removing the diurnal nonstationarity or other seasonalities from the data.

A general power transformation is used [23] for the purpose of normalization, such as

$$V_t' = V_t^m, \forall i, \tag{2.10}$$

where V_t' is the transformed wind speed, and m is the power coefficient, with the convention that $m = 0$ refers to the logarithm transformation. Apparently, $m = 1$ means no transformation.

Suppose that the wind data indeed follow a Weibull distribution. A nice property of Weibull distribution is that a Weibull random variable remains Weibull when it is raised to a power $m \neq 0$, with its parameters becoming β/m and η^m, respectively. Dubey [52] points out that when the shape parameter is close to 3.6, a Weibull distribution is closer in shape to a normal distribution. The general advice is to estimate the shape parameter from the original wind data and then solve for m in the power transformation in Eq. 2.10 as

$$m = \frac{\hat{\beta}}{3.6}. \tag{2.11}$$

Alternatively, Hinkley [93] suggests checking the following measure of symmetry, based on sample statistics,

$$sym = \frac{\text{sample mean} - \text{sample median}}{\text{sample scale}}, \tag{2.12}$$

where the sample scale can be the sample standard deviation or the sample inter-quartile range; Hinkley himself prefers the latter. Given this symmetry measure, one could first choose a candidate set of m values (including $m = 0$) and apply the respective transformation on the wind data. Then, calculate the corresponding symmetry measure. To approximate the symmetric normal distribution, the symmetry value is desired to be zero. Whichever m produces a zero sym value is thus chosen as the power in the transformation. If no m in the candidate set produces a sym close to zero, then one can interpolate the computed (m, sym) points and find the m leading to a zero sym. One convenience allowed by Eq. 2.12 is that the logarithm transformation can be tested, together with other power transformations, whereas in using Eq. 2.11, $m = 0$ is not allowed.

Torres et al. [214] show that using Eq. 2.11 on wind data from multiple sites for every month in a whole year, the resulting m values are in the range of $[0.39, 0.70]$, but many of them are close to 0.5. Brown et al. [23] apply both aforementioned approaches on one set of wind data—Eq. 2.11 produces an $m = 0.45$, while the sym measure in Eq. 2.12 selects $m = 1/2$, implying a square-root transformation. It seems that the resulting m values are often not too far from $1/2$. But this may not always be the case. When applying Eq. 2.11 to the data of each month in the Wind Time Series Dataset (see Table 2.3 for the corresponding m values), we find that most m's are around one. This is not surprising. The shape of the density curves in Fig. 2.1 looks rather normal-like, and the corresponding $\hat{\beta}$'s are close to 3.6. For the sake of convenience, analysts still use $m = 1/2$ as the default setting. This square-root transformation is in fact one of the popular normalizing transformations and applying it reduces the right skewness to make the resulting data closer to a normal distribution. When applied to wind data, the square-root transformation can take any wind speed values, since wind speed is supposedly non-negative. In contrast, if one applies the logarithm transformation, the zero wind speed values need to be removed first.

TABLE 2.3 Monthly values of m using Eq. 2.11.

	Jan	Feb	Mar	Apr	May	Jun
Hourly data	0.74	1.00	1.00	1.04	1.10	1.02
10-min data	0.74	0.94	1.02	1.05	1.10	0.98

	Jul	Aug	Sep	Oct	Nov	Dec
Hourly data	1.01	1.14	1.06	0.98	1.02	0.99
10-min data	1.01	1.13	1.06	0.99	1.03	1.00

Wind exhibits diurnal and seasonal nonstationarity. The seasonality is typically handled by carefully choosing the training period, making sure that the seasonal pattern of the training period is consistent with that in the forecasting period. This can be done by using the wind data in a short period of time immediately prior to the forecasting period, say, a few days or a couple of weeks, but usually no more than one month. To remove the diurnal nonstationarity, a simple treatment is to standardize the wind data by using its hourly average and standard deviation.

We show how this is done using the transformed wind data, V_t', but obviously the same procedure can be applied to the original wind data. We first arrange the data such that the time index t is in an hourly increment. If the raw data is in the 10-min format, then, one can get the hourly data by averaging the six 10-min wind data points within the same hourly block. For notational convenience, let us deem that $t = 0$ coincides with 12 a.m. (midnight) of the first day, $t = 1$ with one a.m., and so on. The time repeats itself as the same time on a different day in an increment of 24. We compute 24 hourly averages and standard deviations by pooling the data from the same time on different days in the training period. Suppose that there are a total of d days. Then, we can compute them as

$$\begin{cases} \bar{V}_\ell' = \frac{1}{d} \sum_{j=0}^{d-1} V_{24j+\ell}', \\ s_\ell = \sqrt{\frac{1}{d-1} \sum_{j=0}^{d-1} (V_{24j+\ell}' - \bar{V}_\ell')^2}. \end{cases} \quad \ell = 0, \ldots, 23. \quad (2.13)$$

The standardization of wind speed data is then carried out by

$$V_t'' = \frac{V_t' - \bar{V}_{(t \bmod 24)}'}{s_{(t \bmod 24)}}, \quad (2.14)$$

where mod means a modulo operation, so that $(t \bmod 24)$ returns the remainder when t is divided by 24.

Fig. 2.2 presents the original hourly wind speed data and the standardized hourly data. Although the standardization is conducted for the whole year hourly data in the Wind Time Series Dataset, Fig. 2.2 plots only three

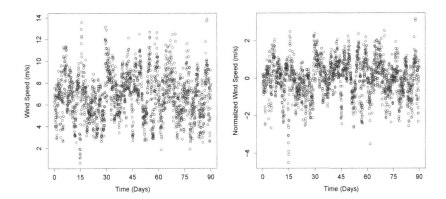

FIGURE 2.2 Left panel: original hourly wind speed data. Right panel: standardized hourly wind speed data. The data amount of the three months is $n = 1,811$. Eq. 2.10, with $m = 1/2$, and Eq. 2.14 are used for standardization.

months (October to December) for a good visualization effect. The difference between the two subplots is not very striking, because the original data, as we explained above, is already close to a normal distribution.

Gneiting et al. [75] introduce a trigonometric function to model the diurnal pattern, as in the following,

$$\Delta_t = c_0 + c_1 \sin\left(\frac{2\pi t}{24}\right) + c_2 \cos\left(\frac{2\pi t}{24}\right) + c_3 \sin\left(\frac{4\pi t}{24}\right) + c_4 \cos\left(\frac{4\pi t}{24}\right), \quad (2.15)$$

where $c_0, c_1, \ldots, c_4$ are the coefficients to be estimated from the data. The estimation is to assume $V_t' = \Delta_t + \varepsilon_t$, and then, use a least squares estimation to estimate the coefficients from the wind data. Subtracting the diurnal pattern from the original wind data produces the standardized wind speed,

$$V_t'' = V_t' - \Delta_t. \quad (2.16)$$

2.4 AUTOREGRESSIVE MOVING AVERAGE MODELS

In this section, we apply a time series model like the autoregressive moving average (ARMA) model to the normalized and standardized wind data. For notational simplicity, we return to the original notation of wind speed, V_t, without the primes.

An autoregressive (AR) model of order p is to regress the wind variable on its own past values, up to p steps in the history, such as

$$V_t = a_0 + a_1 V_{t-1} + \ldots + a_p V_{t-p} + \varepsilon_t, \quad (2.17)$$

where $a_i, i = 1, \ldots, p$, are the AR coefficients and ε_t is the residual error, assumed to be a zero mean, identically, independently distributed (i.i.d) noise. Specifically, $\varepsilon_t \sim \mathcal{N}(0, \sigma_\varepsilon^2)$.

The autoregressive mechanism makes intuitive sense, as the inertia in air movement suggests that the wind speed at the present time is related to its immediate past. The actual relationship, however, may not necessarily be linear. The linear structure assumed in the AR model is for the sake of simplicity, making the model readily solvable.

A general ARMA model is to add a moving average (MA) part to the AR model, which is to model the residual as a linear combination of the i.i.d noises, going back in history for up to q steps. Including the MA part, the ARMA model reads

$$
\begin{aligned}
V_t &= a_0 + a_1 V_{t-1} + \ldots + a_p V_{t-p} + \varepsilon_t + b_1 \varepsilon_{t-1} + \ldots + b_q \varepsilon_{t-q} \\
&= a_0 + \sum_{i=1}^{p} a_i V_{t-i} + \varepsilon_t + \sum_{j=1}^{q} b_j \varepsilon_{t-j},
\end{aligned}
\tag{2.18}
$$

where $b_j, j = 1, \ldots, q$, are the MA coefficients. The overall model in Eq. 2.18 is referred to as $\mathrm{ARMA}(p, q)$, where p is the AR order and q is the MA order.

2.4.1 Parameter Estimation

For the model in Eq. 2.17, the AR parameters can be estimated through a least squares estimation, expressed in a closed form. Suppose that we have the historical data going back n steps. For each step in the past, one can write down an AR model. The following are the n equations,

$$
\begin{aligned}
V_t &= a_0 + a_1 V_{t-1} + \ldots + a_p V_{t-p} + \varepsilon_t, \\
V_{t-1} &= a_0 + a_1 V_{t-2} + \ldots + a_p V_{t-1-p} + \varepsilon_{t-1}, \\
&\quad \cdots \quad \cdots \quad \cdots \quad \cdots \\
V_{t-n} &= a_0 + a_1 V_{t-n-1} + \ldots + a_p V_{t-n-p} + \varepsilon_{t-n}.
\end{aligned}
\tag{2.19}
$$

Express $\mathbf{V} = (V_t, V_{t-1}, \ldots, V_{t-n})_{n \times 1}^T$, $\mathbf{a} = (a_0, a_1, \ldots, a_p)_{(p+1) \times 1}^T$, $\boldsymbol{\varepsilon} = (\varepsilon_t, \ldots, \varepsilon_{t-n})_{n \times 1}^T$, and

$$
\mathbf{W} = \begin{pmatrix}
1 & V_{t-1} & \cdots & V_{t-p} \\
1 & V_{t-2} & \cdots & V_{t-1-p} \\
\vdots & \vdots & \ddots & \vdots \\
1 & V_{t-n} & \cdots & V_{t-n-p}
\end{pmatrix}_{n \times (p+1)}.
$$

Then, Eq. 2.19 can be written in a matrix form, such as

$$
\mathbf{V} = \mathbf{W} \cdot \mathbf{a} + \boldsymbol{\varepsilon}.
\tag{2.20}
$$

As such, the least squares estimate of the parameter vector, $\mathbf{a}$, is

$$
\hat{\mathbf{a}} = (\mathbf{W}^T \mathbf{W})^{-1} \mathbf{W}^T \mathbf{V}.
$$

The fitted wind speed value, $\hat{\mathbf{V}}$, is therefore $\hat{\mathbf{V}} = \mathbf{W}\hat{\mathbf{a}}$. The variance of the residual error term can be estimated by

$$\hat{\sigma}_\varepsilon^2 = \frac{(\mathbf{V} - \hat{\mathbf{V}})^T (\mathbf{V} - \hat{\mathbf{V}})}{n - p - 1} = \frac{(\mathbf{V} - \mathbf{W}\hat{\mathbf{a}})^T (\mathbf{V} - \mathbf{W}\hat{\mathbf{a}})}{n - p - 1}. \tag{2.21}$$

With the MA part included in a general ARMA(p, q) model, the least squares estimation of both AR and MA coefficients does not have a closed form expression anymore. The estimation problem needs to be solved iteratively through a numerical procedure. Analysts use the maximum likelihood estimation method to estimate the parameters. Denote by $\mathbf{b} = (b_1, b_2, \ldots, b_q)_{q \times 1}^T$. The log-likelihood function of an ARMA model, denoted as $\mathcal{L}(\mathbf{a}, \mathbf{b}, \sigma_\varepsilon^2 | \mathbf{V})$, is a bit involved. We choose not to write down its expression here. In practice, it is advised to use the `arima` function in R's `stats` package to carry out the estimation task. The `arima` function is named after the autoregressive integrated moving average model, considered as a generalization of the ARMA model and expressed as ARIMA(p, k, q), which has one extra parameter than an ARMA model has. To handle an ARMA(p, q) model using the three-parameter `arima` function, one can simply set $k = 0$. By default, the `arima` uses the maximum likelihood method for parameter estimation.

To use the `arima` function, one needs to specify p and q. For instance, the command,

```
fit<-arima(wsdata, order = c(3,0,1)),
```

fits an ARMA(3,1) model. Typing `fit` in the R program displays the values of $\hat{a}_0, \hat{a}_1, \hat{a}_2, \hat{a}_3, \hat{b}_1$, the standard deviations of the respective estimates, as well as $\hat{\sigma}_\varepsilon^2$. It also displays a few other things, such as the log-likelihood value and AIC, which we explain next.

2.4.2 Decide Model Order

When using the `arima` function, the model orders p and q need to be specified. In the `forecast` package, there is an `auto.arima` function, which can decide the model order on its own. If one is curious about how `auto.arima` selects its model order or wants to have more control on model selection by oneself, this section explains the thought process.

Popular model selection criteria used for time series models include the Akaike Information Criterion (AIC) [7] and Bayesian Information Criterion (BIC) [197]. Both criteria follow the same philosophy, which is to trade off between a model's training error and its complexity, in order to select a simple enough model that in the meanwhile fits well enough to the training data. The difference between AIC and BIC is in the specific weighting used to trade off the two objectives, which is going to be clear below.

The AIC is defined as

$$\text{AIC} = 2 \times \text{number of parameters} - 2\hat{\mathcal{L}}, \tag{2.22}$$

where $\hat{\mathcal{L}}$ is the log-likelihood value of the ARMA model, evaluated at the estimated parameters. The log-likelihood value is one of the outputs from the `arima` function. The number of parameters in an ARMA(p, q) model is $p + q + 1$. Hence, AIC $= 2(p + q + 1) - 2\hat{\mathcal{L}}$ for an ARMA(p, q) model.

The BIC is defined as

$$\text{BIC} = \ln(n) \times \text{number of parameters} - 2\hat{\mathcal{L}} \\ = \ln(n) \cdot (p + q + 1) - 2\hat{\mathcal{L}}. \quad (2.23)$$

Using AIC or BIC, one would select the model that minimizes either of the criteria.

The log-likelihood value indicates how well an ARMA model fits the training data—the greater, the better. Because the data are noisy, a model that fits too well to the training data could have read too much into the noise part, a problem known as *overfitting* [86]. An overfit model loses its predictive ability and has actually a worse forecasting accuracy. Analysts come to realize that an effective way to avoid overfitting is to select a simpler model. The number of parameters in an ARMA model measures its model complexity—the fewer the parameters, the simpler a model is.

AIC deems that one unit increase in the model complexity, namely one more parameter included in the model, is equivalent to one unit decrease in the log-likelihood. In using AIC, this trade-off is independent of the data amount, n. BIC, instead, considers the weighting coefficient to be dependent on the data amount. Specifically, it uses $\ln(n)$ to quantify the model complexity. When $n = 7.4$, meaning the training data points are seven or eight, $\ln(n) = 2$, making AIC and BIC equivalent. When $n \geq 8$, BIC tends to choose a simpler model than AIC. In practical situations, n is much greater than eight, suggesting that BIC yields a simpler ARMA model that tends to forecast more accurately on future data.

Aware of the shortcoming of the original AIC, analysts propose a corrected AIC [34], referred to as AICc and defined in the context of ARMA(p, q) as

$$\text{AICc} = \text{AIC} + 2 \times \frac{(p + q + 1)^2 + (p + q + 1)}{n - p - q}. \quad (2.24)$$

AICc is virtually AIC with an extra penalty term for model complexity. When n is far greater than the square of the number of parameters in a model, AIC and AICc behave almost the same.

The `arima` function returns the values of AIC. One can use the `BIC` function to compute the BIC value, and use the formula in Eq. 2.24 to calculate AICc. When using `auto.arima`, one can set its argument `ic` to be either `aicc`, `aic`, or `bic`, so that the respective information criterion is used in selecting p and q in the model. For instance,

```
fit<-auto.arima(wsdata, ic=c('bic'))
```

uses the BIC for model selection. The default setting in `auto.arima` is AICc.

TABLE 2.4   The log-likelihood, BIC, AIC, and AICc values of 18 candidate models, up to ARMA(6, 3), based on the hourly data of April in the `Wind Time Series Dataset`, where $n = 433$. Boldface values are either the largest log-likelihood or the smallest values of a respective information criterion.

Model	Log-likelihood	BIC	AIC	AICc
ARMA (1 , 1)	−293.7	**605.5**	593.3	593.4
ARMA (1 , 2)	−293.0	610.4	594.1	594.2
ARMA (1 , 3)	−292.9	616.1	595.8	595.9
ARMA (2 , 1)	−293.3	610.9	594.7	594.7
ARMA (2 , 2)	−292.7	615.8	595.4	595.6
ARMA (2 , 3)	−292.8	622.0	597.6	597.8
ARMA (3 , 1)	−292.7	615.8	595.4	595.6
ARMA (3 , 2)	−290.5	617.3	**593.0**	**593.2**
ARMA (3 , 3)	−289.7	622.0	593.5	593.8
ARMA (4 , 1)	−293.0	622.3	597.9	598.1
ARMA (4 , 2)	−289.8	622.0	593.5	593.8
ARMA (4 , 3)	−289.7	628.0	595.5	595.8
ARMA (5 , 1)	−293.0	628.4	599.9	600.2
ARMA (5 , 2)	−289.7	627.9	595.4	595.7
ARMA (5 , 3)	−289.1	632.9	596.2	596.7
ARMA (6 , 1)	−290.7	630.1	597.5	597.8
ARMA (6 , 2)	−289.0	632.7	596.1	596.5
ARMA (6 , 3)	**−288.6**	638.0	597.3	597.8

We want to note that certain software packages, like these in R, count the variance estimate, $\hat{\sigma}_\varepsilon^2$, as a parameter estimated. Hence, the number of parameters in an ARMA(p, q) model becomes $p + q + 2$. Using this parameter number does change the AIC and BIC values but they do not change the model selection outcome, as all AIC's or BIC's are basically offset by a constant. When this new number of parameters is used with AICc, however, it could end up choosing a different model.

When applying to one month (April) of hourly data in the `Wind Time Series Dataset`, the BIC produces the simplest ARMA model, which is ARMA(1,1), namely $p = q = 1$. Had AIC or AICc been used on the same set of data, ARMA(3,2) would have been chosen, which is more complicated than ARMA(1,1). For the detailed information, please refer to Table 2.4. The estimated parameters for this ARMA(1,1) model are: $\hat{a}_0 = 0.0727$, $\hat{a}_1 = 0.8496$, $\hat{b}_1 = 0.0871$, and $\hat{\sigma}_\varepsilon^2 = 0.2265$.

2.4.3   Model Diagnostics

In addition to using the information criteria, described above, to choose an appropriate time series model, analysts are encouraged to use graphical plots to check the model's fitting quality—this is referred to as model diagnostics or diagnostic checking. For ARMA models, the two most commonly used plots are the autocorrelation function (ACF) plot and the partial autocorrelation function (PACF) plot.

The model diagnostics is performed on the residuals after a model is fitted. The purpose is to check whether the model assumptions regarding the error term hold. The plots are supposed to show that the residuals, after the model part is removed from the data, appear random and contain no systematic patterns; otherwise, it suggests the model fitting is not properly done. Some diagnostics also tests if the residual follows a normal distribution.

Based on Eq. 2.18, we can compute the residuals recursively, using the estimated parameters, such as

$$
\hat{\varepsilon}_t = V_t - \hat{a}_0 - \sum_{i=1}^{p} \hat{a}_i V_{t-i} - \sum_{j=1}^{q} \hat{b}_j \hat{\varepsilon}_{t-j}, \quad t = 1, \ldots, n,
$$

$$
V_\ell = 0, \ \hat{\varepsilon}_\ell = 0, \quad \forall \ell \leq 0.
$$

(2.25)

The autocorrelation function of ε_t is just the correlation function of the random variable with its own past. Denote by $Cov(X, Y)$ the covariance of two random variables, X and Y. Then, the autocovariance function between two time points, t and $t - h$, in the stochastic process of ε_t, is denoted as $Cov(\varepsilon_t, \varepsilon_{t-h})$. When $h = 0$, $Cov(\varepsilon_t, \varepsilon_t) = \sigma_\varepsilon^2$ is the variance of the underlying process. Define by $\rho(X, Y)$ the correlation between two random variables, X and Y. Then, the autocorrelation function of ε_t is

$$
\rho(\varepsilon_t, \varepsilon_{t-h}) = \frac{Cov(\varepsilon_t, \varepsilon_{t-h})}{Cov(\varepsilon_t, \varepsilon_t)} = \frac{Cov(\varepsilon_t, \varepsilon_{t-h})}{\sigma_\varepsilon^2}.
$$

Considering that the residuals should be stationary (after all these modeling steps), then the autocorrelation function does not depend on the starting point in time but only on the time lag h. As such, its notation can be simplified as ρ_h. With the residuals computed in Eq. 2.25, the sample autocorrelation can be computed through

$$
\hat{\rho}_h = \frac{\sum_{t=h+1}^{n} (\hat{\varepsilon}_t - \bar{\hat{\varepsilon}})(\hat{\varepsilon}_{t-h} - \bar{\hat{\varepsilon}})}{\sum_{t=1}^{n} (\hat{\varepsilon}_t - \bar{\hat{\varepsilon}})^2} \approx \frac{\sum_{t=h+1}^{n} \hat{\varepsilon}_t \hat{\varepsilon}_{t-h}}{\sum_{t=1}^{n} \hat{\varepsilon}_t^2},
$$

(2.26)

where $\bar{\hat{\varepsilon}}$ is the sample mean of the residuals, which is supposed to be zero (or near zero), so that they can be omitted from the equation. Applying Bartlett's formula [20, Eq. 6.2.2], the standard error (se) for testing the significance of $\hat{\rho}_h$ is approximated by

$$
se_\rho = \sqrt{\frac{1 + 2\sum_{i=1}^{h-1} \hat{\rho}_i^2}{n}}.
$$

The 95% confidence interval for $\hat{\rho}_h$ is approximated by $\pm 1.96 \cdot se_\rho$. Under the null hypothesis that the residuals are uncorrelated, meaning $\hat{\rho}_h = 0, \forall h > 0$, the standard error is then simplified to $se_\rho = \sqrt{1/n}$, and correspondingly, the 95% confidence interval becomes simply $\pm 1.96/\sqrt{n}$.

One could plot $\hat{\rho}_h$ against a series of time lags, h, and observe how much, if any at all, the residuals are still correlated with their own past. This can be done by using the R function `acf` in the `forecast` package. The default setting in `acf` draws an autocorrelation plot, on which there are two dashed lines (blue in color print). These lines correspond to the 95% confidence interval under the null hypothesis, which are at the values of $\pm 1.96/\sqrt{n}$, as explained above. With an autocorrelation plot, analysts can quickly inspect if there is any $\hat{\rho}_h$ exceeding the line of $\pm 1.96/\sqrt{n}$, and if yes, that suggests still strong enough autocorrelation.

The autocorrelation between ε_t and ε_{t-2} presumably comes from two sources—one is a lag-1 propagation via the correlation between ε_t and ε_{t-1} and then the correlation between ε_{t-1} and ε_{t-2}, while the other is the correlation directly between ε_t and ε_{t-2}. The autocorrelation, ρ_2, as defined and computed above, is the summation of the two sources. When one sees a large ρ_2, one may wonder if its large value is caused by a large lag-1 autocorrelation and its propagation or if it is caused by the direct correlation. The concept of partial autocorrelation is therefore introduced to quantify this direct correlation, which is the amount of correlation between a variable and a lag of itself that is not explained by correlations at all lower-order lags.

Consider the AR model of order p in Eq. 2.17. Applying the correlation operation with V_{t-1} on each term in both sides gives us the following equation, where we replace the coefficient, a_i, in Eq. 2.17 by ϕ_{pi}, such as

$$\rho_1 = \phi_{p1} + \phi_{p2}\rho_1 + \ldots + \phi_{pp}\rho_{p-1}. \tag{2.27}$$

In the above equation, we replace ρ_0 by its value, which is one. Here we use a double index subscript on ϕ to signify that this set of coefficients is obtained when we use an AR model of order p. Do the correlation operation with V_{t-j}, for $j = 1, \ldots, p$. We end up with the set of Yule-Walker equations [20] as,

$$\underbrace{\begin{pmatrix} \rho_1 \\ \rho_2 \\ \vdots \\ \rho_p \end{pmatrix}}_{\rho} = \underbrace{\begin{pmatrix} 1 & \rho_1 & \cdots & \rho_{p-1} \\ \rho_1 & 1 & \cdots & \rho_{p-2} \\ \vdots & \vdots & \ddots & \vdots \\ \rho_{p-1} & \rho_{p-2} & \cdots & 1 \end{pmatrix}}_{\mathbf{R}} \underbrace{\begin{pmatrix} \phi_{p1} \\ \phi_{p2} \\ \vdots \\ \phi_{pp} \end{pmatrix}}_{\phi}, \tag{2.28}$$

or in the matrix format,

$$\rho = \mathbf{R}\phi.$$

Because $\mathbf{R}$ is a full-rank and symmetric matrix, we can solve for ϕ as

$$\hat{\phi} = \mathbf{R}^{-1}\rho.$$

The partial autocorrelation function is estimated by the sequence of

$\hat{\phi}_{11}, \hat{\phi}_{22}, \ldots$, which can be obtained by solving the Yule-Walker equations for $p = 1, 2, \ldots$. Here it becomes apparent why we replace the single index coefficient, a_i, in the AR model with the double index coefficient, ϕ_{pi}, in the Yule-Walker equations; it is otherwise difficult to express the partial autocorrelation function.

The Yule-Walker equations can be solved recursively using the Levinson-Durbin formula [54],

$$\hat{\phi}_{pp} = \frac{\hat{\rho}_p - \sum_{j=1}^{p-1} \hat{\phi}_{(p-1)j} \hat{\rho}_{p-j}}{1 - \sum_{j=1}^{p-1} \hat{\phi}_{(p-1)j} \hat{\rho}_j}, \quad p = 2, 3, \ldots$$

$$\hat{\phi}_{pj} = \hat{\phi}_{(p-1)j} - \hat{\phi}_{pp} \hat{\phi}_{(p-1)(p-j)},$$

$$\hat{\phi}_{11} = \hat{\rho}_1. \tag{2.29}$$

Using the above equations, we figure out that the partial autocorrelation of lag 2, $\hat{\phi}_{22}$, is

$$\hat{\phi}_{22} = \frac{\hat{\rho}_2 - \hat{\rho}_1^2}{1 - \hat{\rho}_1^2}. \tag{2.30}$$

Recall the example mentioned earlier about the autocorrelation between ε_t and ε_{t-2}. The two-step sequential propagation of the lag-1 autocorrelation is $\hat{\rho}_1^2$, whereas $\hat{\rho}_2$ is the full lag-2 autocorrelation. If $\hat{\rho}_2 = \hat{\rho}_1^2$, the correlation between ε_t and ε_{t-2} that is not explained by correlations at the lower lag-1 order is zero. As such, it is reflected in the partial autocorrelation function as $\hat{\phi}_{22} = 0$, according to Eq. 2.30. If $\hat{\rho}_2 \neq \hat{\rho}_1^2$, their difference, scaled by $1 - \hat{\rho}_1^2$, is the partial autocorrelation of lag 2, or the direct correlation between ε_t and ε_{t-2}.

The partial autocorrelation function is useful in identifying the model order of an autoregressive process. If the original process is autoregressive of order k, then for $p > k$, we should have $\phi_{pp} = 0$. This can again be done in a partial autocorrelation function plot by inspecting, up to which order, PACF becomes zero or near zero. By setting `type=c('partial')` in one of its arguments, the `acf` function computes PACF values and draws a PACF plot. Alternatively, the `pacf` function in the `tseries` package can do the same. The dashed line on a PACF plot bears the same value as the same line on an ACF plot.

Fig. 2.3 presents the ACF and PACF plots using the April data in the `Wind Time Series Dataset`. The ARMA(1,1) model is fit to the set of data, and the model residuals are thus computed. The ACF and PACF plots of the residuals are presented in Fig. 2.3 as well.

2.4.4 Forecasting Based on ARMA Model

Suppose that our final model selected is an ARMA(p, q) and their parameters are estimated using the training data. Then, for the h-step ahead point

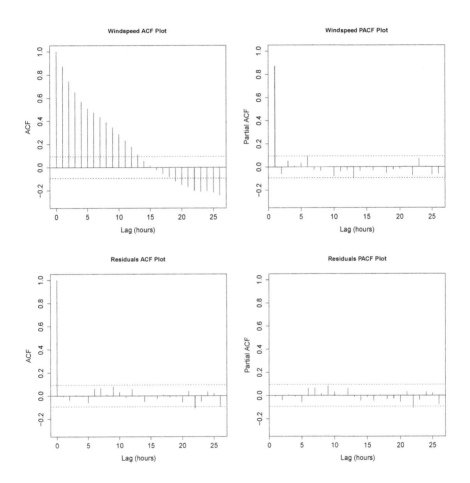

FIGURE 2.3 Top panel: ACF and PACF plots of the original hourly wind data; bottom panel: ACF and PACF plots of the residuals after an ARMA(1,1) model is fit.

forecasting, which is to obtain $\hat{V}_{t+h}$, we use the following formula,

$$
\begin{aligned}
\hat{V}_{t+h} &:= \mathbb{E}(V_{t+h}|V_1, V_2, \dots, V_n) \\
&= \hat{a}_0 + \sum_{i=1}^{p} \hat{a}_i \hat{V}_{t-i+h} + \sum_{j=1}^{q} \hat{b}_j \hat{\varepsilon}_{t-j+h}.
\end{aligned}
\tag{2.31}
$$

In the above equation, when the time index on a $\hat{V}$ is prior to t, meaning that the wind data has been observed, then $\hat{V}$ is replaced by its observed value at that time and $\hat{\varepsilon}$ is estimated in Eq. 2.25, whereas when the time index on a $\hat{V}$ is posterior to t, then $\hat{V}$ is the forecasted value at that time and $\mathbb{E}(\hat{\varepsilon}) = 0$.

To assess the uncertainty of the forecast, we need to calculate the variance of the forecasting error. For that, we use the Wold decomposition [8]. The Wold decomposition says that the ARMA model in Eq. 2.18 can be expressed as an infinite summation of all the error terms, such as

$$
V_{t+h} = a_0 + \varepsilon_{t+h} + \psi_1 \varepsilon_{t+h-1} + \dots \psi_{h-1} \varepsilon_{t+1} + \psi_h \varepsilon_t + \psi_{h+1} \varepsilon_{t-1} + \dots, \tag{2.32}
$$

where ψ_i's can be decided from a_i's and b_j's in Eq. 2.18. We here omit the detailed expression for ψ_i's.

With the expression in Eq. 2.32, the h-step ahead forecast is

$$
\hat{V}_{t+h} := \mathbb{E}(V_{t+h}|V_1, V_2, \dots, V_n) = \hat{a}_0 + \psi_h \hat{\varepsilon}_t + \psi_{h+1} \hat{\varepsilon}_{t-1} + \dots. \tag{2.33}
$$

Therefore, the h-step ahead forecast error at time t, denoted by $e_t(h)$, can be expressed as

$$
e_t(h) = V_{t+h} - \hat{V}_{t+h} = \varepsilon_{t+h} + \psi_1 \varepsilon_{t+h-1} + \dots + \psi_{h-1} \varepsilon_{t+1}. \tag{2.34}
$$

The expectation of $e_t(h)$ is zero, namely $\mathbb{E}(e_t(h)) = 0$, and its variance is expressed as

$$
Var(e_t(h)) = Var\left(\sum_{\ell=0}^{h-1} \psi_\ell \varepsilon_{t+h-\ell} \right) = \sigma_\varepsilon^2 \sum_{\ell=0}^{h-1} \psi_\ell^2, \tag{2.35}
$$

where we define $\psi_0 = 1$. Combining the point forecast and the variance, the $100(1 - \alpha)\%$ prediction interval for the h-step ahead forecasting is

$$
\hat{V}_{t+h} \pm z_{\alpha/2} \cdot \sqrt{Var(e_t(h))} = \hat{V}_{t+h} \pm z_{\alpha/2} \cdot \sigma_\varepsilon \cdot \sqrt{\sum_{\ell=0}^{h-1} \psi_\ell^2}. \tag{2.36}
$$

From the above formula, it is apparent that farther in the future the forecast is, the greater the forecasting variance becomes.

In R, one can use the function `forecast` in the `forecast` package to make forecasting. The basic syntax is `forecast(wsdata, `h`, model = fit)`, which

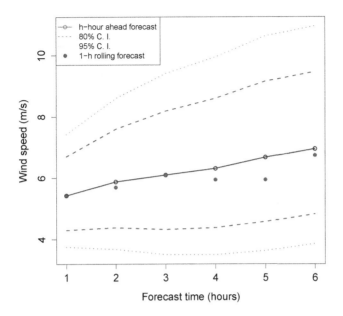

FIGURE 2.4 Wind speed forecasting based on the ARMA(1,1) model. $h = 1, 2, \ldots, 6$.

makes an h-step ahead forecasting using the fitted ARMA model whose parameters are stored in `fit`. The `forecast` function plots both the point forecast and the confidence intervals. By default, the `forecast` function draws two confidence intervals, which are the 80% and 95% confidence intervals. The confidence levels can be adjusted by setting the input argument `level` to other values. For example, `level = c(95, 99)` sets the two confidence intervals at 95% and 99%, respectively.

Fig. 2.4 presents the forecasting outcome based on the ARMA(1,1) model estimated in the previous subsections and using the hourly data of April. The solid line is the h-hour ahead forecast, assuming that the data is available only up to time t. The solid dots represent a one-hour ahead rolling forward forecasting by using the new wind speed observation, at $t + 1, t + 2, \ldots, t + 5$, respectively. For the rolling forward forecasting, the ARMA(1,1) model is refit every time. It is understandable that the two forecasts are the same at $t + 1$ but they differ starting from $t + 2$ when the one-hour ahead rolling forward forecasting uses the actual wind speed observations V_{t+h} at $h > 0$, while the h-hour ahead forecasting uses the forecasted wind speed $\hat{V}_{t+h}$ at $h > 0$.

2.5 OTHER METHODS

Several other methods, some of machine learning flavor, have been developed for short-term forecasting. In this section, we discuss the use of the Kalman filter (KF), support vector machine (SVM) and artificial neural network (ANN). We defer discussion on regime switching techniques [6, 75] to Chapter 4.

2.5.1 Kalman Filter

The Kalman filter [116] was initially developed for linear dynamic systems described by a state space model. Using the notations introduced in this chapter, the state space model for wind speed forecasting can be expressed as,

$$
\begin{aligned}
\text{state equation} \quad & \mathbf{a}_t = \boldsymbol{\Phi}\mathbf{a}_{t-1} + \boldsymbol{\omega}_{t-1}, \\
\text{observation equation} \quad & V_t = \mathbf{h}_t^T \mathbf{a}_t + \varepsilon_t,
\end{aligned}
\tag{2.37}
$$

where $\boldsymbol{\Phi}$ is known as the state matrix, $\mathbf{a}_t = (a_{1,t}, \ldots, a_{p,t})^T$ is the state vector, $\mathbf{h}_t = (V_{t-1}, \ldots, V_{t-p})^T$ is the observation vector, and, ε_t and $\boldsymbol{\omega}_t$ are random noises. The first equation is referred to as the state equation, whereas the second equation is referred to as the observation equation. The observation equation is essentially an AR model, which is to predict the future wind speed (or power) as a linear combination of its past observations. Unlike the AR model, the Kalman filter model treats the coefficients, $\mathbf{a}_t$, as variables rather than constants, and updates them as new observations arrive, so as to catch up with the dynamics in the wind data. The two noise terms are often assumed to be normal variables, namely $\varepsilon_t \sim \mathcal{N}(0, (\sigma_\varepsilon^2)_t)$ and $\boldsymbol{\omega}_t \sim \mathcal{N}(\mathbf{0}, \mathbf{Q}_t)$, where $(\sigma_\varepsilon^2)_t$ is the time-varying variance of ε_t and $\mathbf{Q}_t$ is the time-varying covariance matrix of $\boldsymbol{\omega}_t$. The state vector, $\mathbf{a}_t$, is a random vector. It also has a covariance matrix, which we define by $\mathbf{P}_t$.

In the wind application, the observation vector, as expressed in Eq. 2.37, is the past n observations of wind speed, immediately before the current time t. But some analysts use the output from an NWP [44, 136] as their observation vector, and in this way, the Kalman filter serves to enhance the predictive resolution and accuracy of the heavy-computing, slow-running NWP.

The state matrix, $\boldsymbol{\Phi}$, is often assumed an identity matrix, namely $\boldsymbol{\Phi} = \mathbf{I}$, unless the underlying process dictates a different evolution dynamics of the state vector, $\mathbf{a}_t$. A further simplification is to assume that $\mathbf{Q}_t$ is a diagonal matrix—random variables in $\boldsymbol{\omega}_t$ are uncorrelated—and has an equal variance. As such, we can express $\mathbf{Q}_t = (\sigma_\omega^2)_t \cdot \mathbf{I}$, where $(\sigma_\omega^2)_t$ is known as the variance of the *system noise*, whereas the $(\sigma_\varepsilon^2)_t$ is known as the variance of the *observation noise*.

Before introducing the Kalman filter prediction and updating mechanism, we need to articulate the meaning of time instance t here. When we say "at time t," we mean that we have observed the wind data at that time. The Kalman filter has an update step between two time instances, $t - 1$ and t, or more specifically, after the wind data at $t - 1$ has been observed but before the observation at t. To denote this update step, analysts use the notation, $t|t-1$. For example, $\mathbf{a}_{t|t-1}$ is the predicted value of the state vector after the observations up to $t-1$ but before the observation at t.

The Kalman filter runs through two major steps in iteration—prediction and update. Suppose that we stand between $t-1$ and t, and have the historical observations in $\mathbf{h}_t$ as well as previous estimations, $\hat{\mathbf{a}}_{t-1}$ and $\mathbf{P}_{t-1}$. At this moment, before we observe V_t, we can predict

$$\hat{\mathbf{a}}_{t|t-1} = \mathbf{\Phi}\hat{\mathbf{a}}_{t-1}, \tag{2.38}$$

$$\mathbf{P}_{t|t-1} = \mathbf{\Phi}\mathbf{P}_{t-1}\mathbf{\Phi}^T + (\sigma_\omega^2)_{t-1} \cdot \mathbf{I}, \tag{2.39}$$

$$\hat{V}_{t|t-1} = \mathbf{h}_t^T \hat{\mathbf{a}}_{t|t-1}, \tag{2.40}$$

$$(\hat{\sigma}_V^2)_{t|t-1} = \mathbf{h}_t^T \mathbf{P}_{t|t-1} \mathbf{h}_t + (\sigma_\varepsilon^2)_t. \tag{2.41}$$

The last two equations are used to make a one-step ahead forecasting. The $100(1-\alpha)\%$ predictive confidence interval for V_t, before V_t is observed, is

$$[\hat{V}_{t|t-1} - z_{\alpha/2} \cdot (\hat{\sigma}_V^2)_{t|t-1}, \quad \hat{V}_{t|t-1} + z_{\alpha/2} \cdot (\hat{\sigma}_V^2)_{t|t-1}].$$

If the desire is to make multiple-hour ahead forecasting, then the state space model should be built on a coarse temporal granularity. The default temporal resolution is an hour, meaning that one hour passes from $t-1$ to t. If we increase the temporal granularity to two hours, meaning that two hours pass from $t-1$ to t, then, the above one-step ahead forecasting makes a 2-hour ahead forecast. The downside is that the historical data is thinned and the data point between the two chosen time instances for the Kalman filter are ignored—this apparently is a drawback.

At time t, after V_t is observed, $\hat{\mathbf{a}}_t$ and $\mathbf{P}_t$ get an update through the following steps,

$$\mathbf{K}_t = \frac{1}{(\hat{\sigma}_V^2)_{t|t-1}} \mathbf{P}_{t|t-1}\mathbf{h}_t, \tag{2.42}$$

$$\hat{\mathbf{a}}_t = \hat{\mathbf{a}}_{t|t-1} + \mathbf{K}_t(V_t - \hat{V}_{t|t-1}), \tag{2.43}$$

$$\mathbf{P}_t = (\mathbf{I} - \mathbf{K}_t\mathbf{h}_t^T)\mathbf{P}_{t|t-1}, \tag{2.44}$$

where $\mathbf{K}_t$ is known as the *Kalman gain*. To start the process, analysts can set the initial values for $\hat{\mathbf{a}}_t$ and $\mathbf{P}_t$ as

$$\mathbf{a}_0 = (1, 0, \dots, 0)^T \quad \text{and} \quad \mathbf{P}_0 = \begin{pmatrix} 1 & 0 \\ 0 & 1 \end{pmatrix}.$$

The above $\mathbf{a}_0$ means that at the beginning, the prediction uses only the immediate past observation. Another parameter to be decided in the Kalman filter is p, the size of the state vector. This p can be decided by fitting an AR model and choosing the best p based on BIC.

To run the above Kalman filter, the variances of the observation noise and system noise are also needed. Crochet [44] suggests using the Smith algorithm and Jazwinski algorithm to dynamically estimate $(\sigma_\varepsilon^2)_t$ and $(\sigma_\omega^2)_t$, respectively. The basic idea for estimating the observation noise, $(\sigma_\varepsilon^2)_t$, is to treat it as the product of a nominal value, $(\sigma_\varepsilon^2)_0$, and a coefficient, ζ_t, where ζ_t is further assumed to follow an inverse gamma distribution with a shape parameter κ_t. Then, the Smith algorithm [199] updates the observation noise variance by

$$(\sigma_\varepsilon^2)_t = \zeta_{t-1} \cdot (\sigma_\varepsilon^2)_0,$$

$$\zeta_t = \frac{\zeta_{t-1}}{\kappa_{t-1}+1}\left(\kappa_{t-1} + \frac{(V_t - \hat{V}_{t|t-1})^2}{(\sigma_V^2)_{t|t-1}}\right), \tag{2.45}$$

$$\kappa_t = \kappa_{t-1} + 1.$$

The variance of the system noise, $(\sigma_\omega^2)_t$, can be estimated through the Jazwinski algorithm [107] as

$$(\sigma_\omega^2)_t = \left(\frac{(V_t - \hat{V}_{t|t-1})^2 - \mathbf{h}_t^T \boldsymbol{\Phi} \mathbf{P}_{t-1} \boldsymbol{\Phi}^T \mathbf{h}_t - (\sigma_\varepsilon^2)_t}{\mathbf{h}_t^T \mathbf{h}_t}\right)_+, \tag{2.46}$$

where $(\cdot)_+$ returns the value in the parenthesis if it is positive, or zero otherwise. The initial values used in Eq. 2.45 are set as $(\sigma_\varepsilon^2)_0 = 1$, $\zeta_0 = 1$, and $\kappa_0 = 0$. The initial value, $(\sigma_\omega^2)_0$, is also set to zero.

Fig. 2.5 presents an illustrative example, which compares the Kalman filter forecast with AR(1) model forecast, when both are applied to the hourly data of April. The order of the AR model is chosen based on BIC. The best order, corresponding to the smallest BIC, is $p = 1$. Because the Kalman filter updates its one-hour ahead forecast with the new observation, to make a fair comparison, we use the AR(1) model to conduct a one-hour ahead forecast on a rolling forward basis from $t+1$ to $t+6$, the same as what is done for the solid dots in Fig. 2.4. The difference is that the model used then is ARMA(1,1), whereas the model used here is AR(1). The actual difference is, however, negligible, because $\hat{b}_1 = 0.0871$ in the ARMA(1,1) model, and as such, ARMA(1,1) behaves nearly identically to AR(1) with the same autoregressive coefficient. The point forecast of both methods are similar here, but the confidence interval of the Kalman filter is narrowing as more data are accumulated, while the confidence interval of the AR(1) one-hour ahead forecast stays much flatter.

2.5.2 Support Vector Machine

Support vector machine is one of the machine learning methods that are employed in wind speed forecasting. Support vector machine was initially developed for the purpose of classification, following and extending the work of optimal separating hyperplane. Its development is largely credited to Vladimir Vapnik [221].

Two important ideas are employed in a support vector machine. The first

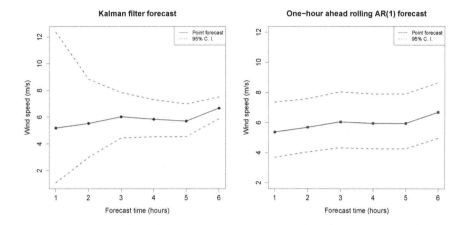

FIGURE 2.5 One-hour ahead forecasting plots from $t+1$ to $t+6$: Kalman filter (left panel) and AR(1) model (right panel).

is to use a small subset of the training data, rather than the whole set, in the task of learning. This subset of data points was called the *support vector* by its original developers, namely Vapnik and his co-authors. This is where the name Support Vector Machine comes from. In the case of a two-class classification, the data points constituting the support vector are those close to the boundary separating the two competing classes. The data points that are more interior to a data class and farther away from the separating boundary do not affect the classification outcome.

The second idea is to transform the data from its original data space to a potentially high-dimensional space for a better modeling ability. This type of transformation is nonlinear, so that a complicated response surface or a complex feature in the original space may become simpler and easier to model in the transformed space. The theoretical foundation for such transformation lies in the theory of reproducing kernel Hilbert space (RKHS) [86].

The use of the first idea helps the use of the second idea. One key reason for SVM to do well in a higher dimensional space without imposing too much computational burden is because the actual number of data points involved in its learning task, which is the size of the support vector, is relatively small.

The application of SVM to wind speed data is to solve a regression problem, in which the response is a real value, albeit nonnegative, instead of a categorical value. SVM is applicable to regression problems but a different loss function ought to be used. We will discuss those next.

Support vector machine falls into the class of supervised machine learning methods, in which a set of data pairs, $\{\boldsymbol{x}_i, y_i\}_{i=1}^n$, is collected and used to train a model (model training is the same as to decide the model order and estimate the model parameters). In the data pairs, $\boldsymbol{x}_i$ is the input and y_i

is the corresponding output. In the context of wind speed forecasting, what analysts use to forecast a future value is the historical observations. At time t, the input vector comprises the wind speed data p-step back in the history, and the response y is the h-step ahead to be forecasted. In other words, $\boldsymbol{x}_t$ and y_t can be expressed as

$$\boldsymbol{x}_t = (V_t, \ldots, V_{t-p+1})^T \quad \text{and} \quad y_t = V_{t+h}.$$

This $\boldsymbol{x}_t$ is essentially the same as the observation coefficient vector, $\mathbf{h}_{t+1}$, in the Kalman filter. We group the data in the collection of historical observations running from time 1 to time $n + h$ and label them as $\boldsymbol{x}$'s and y's accordingly. Wind speed V_ℓ for $\ell \leq 0$ is set to zero. Like in the Kalman filter, p can be chosen by fitting an AR model to the wind data.

SVM finds the relationship between $\boldsymbol{x}$ and y, so that a forecast can be made for h-step ahead whenever a new set of wind speed observations are available. Unlike in AR models and the Kalman filter, the y-to-$\boldsymbol{x}$ relationship found by SVM is not necessarily linear. In fact, it is generally nonlinear. Analysts believe that a nonlinear functional relationship is more flexible and capable, and could hence lead to an enhanced forecasting capability. When using SVM, for a different h, a different SVM predictive model needs to be built, or needs to be trained. This aspect appears different from the recursive updating nature of the Kalman filter or the ARMA model.

The general learning problem of SVM can be formulated as

$$\hat{\boldsymbol{\alpha}} = \arg\min \left\{ L(\mathbf{y}, \mathbf{K}\boldsymbol{\alpha}) + \frac{\gamma}{2}\boldsymbol{\alpha}^T \mathbf{K}\boldsymbol{\alpha} \right\}, \tag{2.47}$$

where $L(\cdot, \cdot)$ is a loss function that can take different forms, depending on whether this is a regression problem or a classification problem, $\mathbf{y} = (y_1, \ldots, y_n)^T$ is the output vector, $\mathbf{K}$ is the Gram matrix (or the kernel matrix), to be explained below, $\boldsymbol{\alpha}$ is the model parameters to be learned in the training period, using the training dataset, $\{\boldsymbol{x}_i, y_i\}_{i=1}^n$, and γ is the penalty constant to regulate the complexity of the learned functional relationship. A large γ forces a simpler, smooth function, while a small γ allows a complicated, more wiggly function. Recall the overfitting issue discussed in Section 2.4.2. An overly complicated function leads to overfitting, which in turn harms a model's predictive capability. The inclusion of γ is to help select a simple enough model that has good predictive performances.

The above formulation appears to be different from many of the SVM formulations presented in the literature. This is because the above SVM formulation is expressed under the reproducing kernel Hilbert space framework. The RKHS theory is too involved to be included here—after all, the main purpose of this book is not machine learning fundamentals. The benefit to invoke this RKHS framework is that doing so allows the SVM formulation to be presented in a clean and unified way and also be connected easily with other learning methods, such as Gaussian process regression [173] or smoothing splines [86].

In the kernel space formulation, one key element is the Gram matrix $\mathbf{K}$, which is created by a kernel function $K(\cdot, \cdot)$, such that the (i, j)-th element of $\mathbf{K}$ is $(\mathbf{K})_{i,j} = K(\boldsymbol{x}_i, \boldsymbol{x}_j)$. This is how the input vector information $\boldsymbol{x}$'s get incorporated in the learning equation of Eq. 2.47; otherwise, it may appear strange that SVM learns $\boldsymbol{\alpha}$ by using $\mathbf{y}$ only, as on the surface, $\boldsymbol{x}$ does not appear in Eq. 2.47.

There are several commonly used kernel functions in SVM. A popular one is the radial basis function kernel, defined as

$$K(\boldsymbol{x}_i, \boldsymbol{x}_j) = \exp\left\{-\phi \|\boldsymbol{x}_i - \boldsymbol{x}_j\|_2^2\right\}, \tag{2.48}$$

where $\|\cdot\|_2$ defines a 2-norm; for more discussions on norm, please refer to Section 12.3.1. The radial basis kernel is also known as the Gaussian kernel, as its function form resembles the density function of a Gaussian (normal) distribution. Using the radial basis kernel, it introduces one extra parameter, ϕ, which will be decided in a similar fashion as how γ in Eq. 2.47 is decided. This is to be discussed later.

Once the parameters in $\boldsymbol{\alpha}$ are learned, analysts can use the resulting SVM to make forecasting. For instance, we train an SVM using data from 1 to n. Then, with a new observation, V_{n+1}, we would like to make a forecast of V_{n+h+1}. We first form a new input vector, denoted by $\boldsymbol{x}_{\text{new}} = (V_{n+1}, V_n, \ldots, V_{n-p+2})^T$. Then, the forecasting model is

$$\hat{V}_{n+h+1}(\boldsymbol{x}_{\text{new}}) = \sum_{i=1}^{n} \hat{\alpha}_i K(\boldsymbol{x}_{\text{new}}, \boldsymbol{x}_i). \tag{2.49}$$

For a general h-step ahead forecasting where $h > 1$, it is important to make sure that $\boldsymbol{x}_{\text{new}}$ properly includes the new observations that matter to the forecasting. Then, the same formula can be used to obtain a general h-step ahead forecast $\hat{V}_{t+h}$.

SVM for classification and SVM for regression use different loss functions. First, let us define a general prediction function for SVM as $g(\boldsymbol{x})$. Similar to the prediction expressed in Eq. 2.49, the general prediction function takes the form of

$$g(\boldsymbol{x}) = \sum_{i=1}^{n} \alpha_i K(\boldsymbol{x}, \boldsymbol{x}_i). \tag{2.50}$$

The loss function can be denoted by $L(y, g(\boldsymbol{x}))$. For classification, a hinge loss function,

$$L(y, g(\boldsymbol{x})) = \sum_{i=1}^{n} (1 - y_i g(x_i))_+, \tag{2.51}$$

is used. As illustrated in Fig. 2.6, left panel, this loss function, expressed in yg, looks like a hinge comprising two straight lines. For regression, an ϵ-sensitive error loss function,

$$L(y, g(\boldsymbol{x})) = \sum_{i=1}^{n} (|y_i - g(x_i)| - \epsilon)_+, \tag{2.52}$$

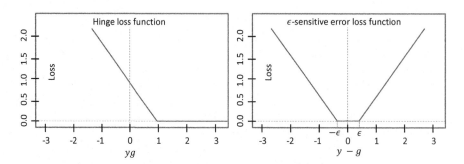

FIGURE 2.6 The loss functions used by support vector machine in classification (left panel) and in regression (right panel).

is used, which is illustrated in Fig. 2.6, right panel.

SVM regression can be made equivalent to Gaussian process regression, if (a) the loss function uses a squared error loss function, (b) $\gamma/2$ is set to σ_ε^2, which is the variance of the i.i.d noise term, (c) when the kernel function, $K(\cdot, \cdot)$, is set to be a covariance function. This connection becomes clearer after we discuss the Gaussian process regression in Section 3.1.3 (see also Exercise 3.2).

To run an SVM regression, the needed input is the training dataset $\{\boldsymbol{x}_i, y_i\}_{i=1}^n$ and three exogenous parameters, γ in Eq. 2.47, ϕ in Eq. 2.48, and ϵ in Eq. 2.52 (not to confuse this ϵ with the i.i.d noise term ε). These exogenous parameters can be decided by using a cross-validation strategy [86]. A five-fold cross validation is carried out through the steps in Algorithm 2.1.

In R's e1071 package, a number of functions can help execute the SVM regression and make forecast. The svm function performs both classification and regression. It performs a regression, if it detects real values in y. By default, the svm uses a radial basis function and sets $\gamma = 1$, $\phi = 1/p$, and $\epsilon = 0.1$. Please note that γ in our formulation is the reciprocal of the cost argument used in the standard SVM package in R, and the radial kernel coefficient, ϕ, is called gamma in the SVM package.

The following command can be used to perform a model training,

```
svm.model <- svm(Y~ X, data = trainset).
```

To apply the SVM to the test dataset,

```
svm.pred <- predict(svm.model, testset).
```

To select the exogenous parameters, analysts can use the tune function to run a grid search. Suppose that we have fixed $\phi = 1$ but want to see which combination of γ and ϵ produces a better model, we may use

```
outcome<-tune(svm, Y~ X, data = trainset, ranges =
    list(epsilon = seq(0,1,0.1), cost = 10^(-4:4))).
```

Algorithm 2.1 A five-fold cross-validation procedure.

---

1. Choose a value for γ, ϕ, and ϵ, respectively.

2. Split the whole training dataset into five subsets of nearly equal data amount.

3. Use four subsets of the data to train an SVM regression model.

4. Use the remaining unused data subset to evaluate the performance of the model, using one of the performance metrics that are to be discussed in Section 2.6.

5. Repeat Steps 3 and 4 five times. Each time, always use four subsets to train a model and use the unused fifth subset to evaluate the model's forecasting performance.

6. Use the average of the performance metric values as the final model performance.

7. Repeat from Step 1 by trying other combinations of γ, ϕ, and ϵ. Select whichever combination produces the best forecasting model.

Because the tune function runs an exhaustive search, it could take a long time, especially if all three parameters are to be optimized. To speed up, analysts can optimize one factor at a time or employ a meta-heuristic optimization routine such as the genetic algorithm.

Using the same April wind data as used in the previous subsections, we explore which parameter combination produces the best SVM. Here, a radial basis kernel is used, $p = 1$ as in the Kalman filter example, and $\phi = 1$. To ease the computation, we use a greedy search strategy, which is to fix the value of $\epsilon = 0.1$, vary cost in a broad range. It turns out that cost $= 1$ is preferred. Then, fix cost $= 1$ and vary ϵ from 0 to 1. This process chooses $\epsilon = 0.2$.

2.5.3 Artificial Neural Network

Artificial neural network is another machine learning method that is widely employed in wind speed forecasting. ANN can be used for both classification and regression, too. Like in the case of SVM, the application of ANN to wind speed forecasting is a regression problem. The problem setting is similar to that described in the SVM section:

- A set of training data points, $\{x_i, y_i\}_{i=1}^n$, is collected, where x_i and y_i are defined likewise as in SVM.

- ANN aims to find the relationship between x and y, and the resulting relationship is nonlinear, as in the case of SVM and unlike the linear relationship assumed in AR models and the Kalman filter.

- To make a forecast at $t + h$, one chooses V_{t+h} as the corresponding y_i. ANN can train a model with multiple outputs, meaning that the outputs of an ANN can make forecasts, all at once, at a number of h-step ahead times with different h's. This is, at least conceptually, a convenience provided by ANN. On the flip side, training a multi-output model takes more care than training a single-output model.

Neural networks consist of an input layer and an output layer, which are connected through one or many hidden layers in between. Fig. 2.7, left panel, presents a multiple-input and single-output neural network, which has only one hidden layer. Each layer comprises a number of nodes. The nodes on the input layer are basically the input variables, whereas the nodes on the output layer are the response variables, namely the forecast to be made in the wind applications. By letting $y = V_{t+h}$, the neural net in Fig. 2.7 is to make an h-step ahead forecast for the given h. As mentioned above, it is straightforward for an ANN to have multiple outputs, so as to make simultaneous forecasts at multiple future time instances.

The information flow in an ANN goes as follows. The input layer takes in the input data. The connection between the input nodes and a node on the hidden layer feeds a linear combination of the inputs to the hidden node and outputs a value after a nonlinear transformation. The final output of the network is a linear combination of the values of the hidden nodes. Denote by Z the node on the hidden layer and assume that there are M hidden nodes, i.e., $Z_1, \ldots, Z_M$. As such, a neural net is described mathematically as,

$$Z_m(\boldsymbol{x}) = \sigma(\alpha_{0m} + \boldsymbol{\alpha}_m^T \boldsymbol{x}), m = 1, \ldots, M, \qquad (2.53)$$

$$\hat{y} = g(\boldsymbol{x}) = \beta_0 + \sum_{m=1}^{M} \beta_m Z_m(\boldsymbol{x}), \qquad (2.54)$$

where α_{0m}, $\boldsymbol{\alpha}_m$, and β_i, $i = 0, 1, \ldots, M$ are the model parameters to be learned from the training data, and $\sigma(\cdot)$ is the sigmoid function, taking the form $\sigma(x) = 1/(1 + e^{-x})$. For an illustration, please take a look at Fig. 2.6, right panel. This sigmoid function is the nonlinear transformation, referred to a short while ago, that takes place at the hidden nodes. Because of this nonlinear transformation, the resulting ANN model is inherently nonlinear. This sigmoid function is called an *activation* function, as what it does is to tame an input if its value is negative, but let the input pass if its value is positive. This function is adopted to mimic the activation of a biological neuron responding to a stimulus—this analogy earns the method its name. In Eq. 2.53, analysts sometimes use the radial basis function as the $\sigma(\cdot)$ function, instead of a sigmoid function. If so, the resulting ANN is referred to as a radial basis function neural net.

If we choose an identity function as $\sigma(\cdot)$, namely $\sigma(x) = x$, then the ANN

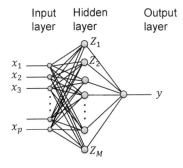
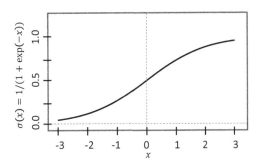

FIGURE 2.7 Left panel: a single hidden layer, a single-output neural network. Right panel: a sigmoid function.

model simplifies to a linear model. In this way, an ANN can be thought of as a two-stage, nonlinear generalization of the linear model. A general ANN also has multiple layers. It has been long believed that having multiple layers increases the data modeling capability of the resulting neural net, but the difficulty surrounding the optimization for parameter estimation made a multiple-layer neural net initially less practical. This optimization problem was addressed about a decade ago, and consequently, the many-layered neural nets become popular nowadays. The many-layered neural nets are referred to as deep neural nets, or commonly, *deep learning* models. The single layer one, by contrast, is called a shallow neural net.

An ANN is parameterized by α_{0m}, $\boldsymbol{\alpha}_m$, and β_i, $i = 0, 1, \ldots, M$, known as the *weights* in the language of neural nets, as they can be viewed as the weights associated with the links between an input node and a hidden node, or between a hidden node and an output node. For regression, the loss function used in an ANN training is the squared error loss, i.e., $\sum_{i=1}^{n}(y_i - g(\boldsymbol{x}_i))^2$.

For a single-layer, single-output ANN, the number of inputs and that of the hidden nodes need to be decided before the training stage. Concerning the number of inputs, we recommend using the same number of inputs as in the Kalman filter or the support vector machine, for which the choice of p can be hinted by fitting an AR model. Please be aware that the inputs to an ANN can be easily expanded. Analysts have included wind power, time in a day, temperature, among other things, as inputs. Due to the flexibility of an ANN, the training is supposed to take care of the y-to-$\boldsymbol{x}$ relationship, depending much less on the nature of the inputs. Concerning the number of hidden nodes, Hastie et al. [86] recommend the range of 5 to 100 and using more hidden nodes if there are more input nodes, and offer the following rule of thumb—"*Generally speaking it is better to have too many hidden units [nodes] than too few.*"

When training a neural net, the starting values of the parameters are typically chosen to be random values near zero [86]. When inputs of different physical units are used, it is advised to standardize the inputs to have a zero mean and a standard deviation of one.

The R package `neuralnet` can facilitate the process of building a neural net. Suppose that we choose to have 10 hidden nodes on a single hidden layer. The following R command can be used,

```
nn <- neuralnet(Y~ X,data=trainset, hidden=10, linear.output=T),
```

where `linear.output=T` means that this is a regression problem. By default, the `neuralnet` function uses the resilient back-propagation with weight back-tracking algorithm [178] to solve the optimization problem and estimate the parameters. If one chooses a multi-layer neural net, then the `hidden` argument needs to be set accordingly. For instance, setting `hidden = c(5, 4, 3)` means that the resulting ANN has three hidden layers, having 5, 4, and 3 nodes, respectively. To visualize the resulting neural net, one can use `plot(nn)`. To test the resulting ANN on a set of test data, one can use

```
test.nn <- compute(nn, testset).
```

Using the April wind data and $p = 1$, we test a single hidden layer ANN with four different choices for the number of the hidden nodes, which are 5, 10, 15, and 30. A ten-fold cross validation settles at five hidden nodes.

2.6 PERFORMANCE METRICS

In order to assess the forecasting quality, a number of performance metrics are used. Consider the case that we have a set of n_{test} test data points, V_i, $i = 1, \ldots, n_{\text{test}}$, the corresponding forecast of each of which is $\hat{V}_i$. The most popular two metrics are the root mean squared error (RMSE) and the mean absolute error (MAE), defined, respectively, as

$$\text{RMSE} = \sqrt{\frac{1}{n_{\text{test}}} \sum_{i=1}^{n_{\text{test}}} (\hat{V}_i - V_i)^2}, \quad \text{and} \quad (2.55)$$

$$\text{MAE} = \frac{1}{n_{\text{test}}} \sum_{i=1}^{n_{\text{test}}} |\hat{V}_i - V_i|. \quad (2.56)$$

Both metrics evaluate the performance of a point forecast. RMSE is based on the squared error loss function, and thus sensitive to the existence of outliers, whereas MAE is based on the absolute error loss, and thus less sensitive to outliers.

Both RMSE and MAE count the absolute amount of forecasting error, regardless of the base value to be predicted. Some may argue that an error of 1 m/s, when predicting at the base wind speed of 3 m/s versus predicting at 15 m/s, has different impacts. To measure the relative error, the mean absolute percentage error (MAPE) is used. MAPE is defined as

$$\text{MAPE} = \frac{100}{n_{\text{test}}} \sum_{i=1}^{n_{\text{test}}} \left| \frac{\hat{V}_i - V_i}{V_i} \right|. \tag{2.57}$$

Note that MAPE is given as a percentage quantity but its value can exceed 100%.

We want to point out that in some literature, for instance, in [75], MAE is called the mean absolute prediction error, the acronym of which is also MAPE. This confusion can be cleared in the context by looking at the spelled-out version of the acronym or the definition.

Hering and Genton [91] favor measuring the impact on the final power response affected by wind speed forecast. This is because the impact of a forecast error in wind speed on wind power is not uniform. Recall the power curve in Fig. 1.2. For a wind speed smaller than the cut-in wind speed or larger than the rated wind speed, an error in wind speed forecast has a smaller impact on wind power than the same amount of forecasting error has when the wind speed is between the cut-in speed and the rated speed, where the power curve has a steeper slope. To factor in the impact on a turbine's power response, Hering and Genton [91] propose the following power curve error (PCE), defined as

$$\text{PCE}_i = \begin{cases} \xi \left(g(V_i) - g(\hat{V}_i) \right) & \text{if} \quad \hat{V}_i \leq V, \\ (1 - \xi) \left(g(\hat{V}_i) - g(V_i) \right) & \text{if} \quad \hat{V}_i > V, \end{cases} \tag{2.58}$$

$$\text{PCE} = \frac{1}{n_{\text{test}}} \sum_{i=1}^{n_{\text{test}}} \text{PCE}_i, \tag{2.59}$$

where $g(\cdot)$ is the power curve function and $\xi \in (0, 1)$ is introduced to penalize underestimation and overestimation differently. In practice, underestimating incurs more cost than overestimating. Therefore, for practical purposes $\xi > 0.5$. Hering and Genton [91] recommend setting $\xi = 0.73$. Generally speaking, using the PCE ensures that the optimal forecast is a ξ-quantile [73]. If $\xi = 0.5$, PCE is the same as MAE.

The above metrics all measure the quality of a point forecast. If the forecasting is a probability density, to measure the quality of a density estimation or prediction, we use the mean continuous ranked probability score (CRPS) [76]. CRPS compares the estimated cumulative distribution function with the observations, and it is computed as

$$\text{CRPS} = \frac{1}{n_{\text{test}}} \sum_{i=1}^{n_{\text{test}}} \int \left(\hat{F}(V) - \mathbb{1}(V > V_i) \right)^2 dV, \tag{2.60}$$

where $\hat{F}(V)$ is the estimated cdf and $\mathbb{1}(\cdot)$ is an indicator function, such that $\mathbb{1}(\texttt{logic}) = 1$ if $\texttt{logic}$ is true and zero otherwise. When the cdf, $F(\cdot)$, is replaced by a point forecast, CRPS reduces to MAE [75].

TABLE 2.5 Model parameters of SVM and ANN selected by cross validation. The ϕ parameter in SVM is set to be the reciprocal of p.

		SVM			ANN	
h	p	cost	ϵ		p	# of hidden nodes
1	1	100	0.3		1	10
2	1	10	0.4		1	5
3	4	1	0.5		1	5
4	4	1	0.6		1	10
5	3	1	0.3		1	5
6	3	1	0.5		1	5

2.7 COMPARING WIND FORECASTING METHODS

In this section, we conduct a comparison study using the yearlong hourly data in the Wind Time Series Dataset and see how individual forecasting models work.

For each month, we split the wind speed data into two portions as follows. We reserve the last six hours of data points as one of the test sets and take the remaining data in that month as one of the training datasets. We then group all 12 monthly training sets into an aggregated training set for the whole year.

Five different forecasting methods are considered—the persistence model, forecasting based on Weibull distribution (WEB), ARMA model, SVM, and ANN. For ARMA, BIC is used to decide the best model order. When training SVM and ANN, the cross-validation strategy is used to decide the exogenous parameters. For the first four models, the training data in the yearlong dataset are used to find the best model order, if applicable, and estimate the respective model parameters. For ANN, the convergence of the R package while using the yearlong dataset is very slow. We instead use only one month of data (April) in a cross validation to decide the number of hidden nodes for a single-layer neural net. Once that is decided, the remaining parameters in the ANN model are estimated still based on the yearlong training data.

For WEB, the mean of the estimated distribution is used as the forecast for all six h-hour ahead forecasts. For ARMA, an ARMA(2,2) model is chosen for making h-hour ahead forecasts at $h = 1, 2, \ldots, 6$. For SVM and ANN, six different models of each kind are trained to cover all six h values. For instance, when $h = 1$, we train an SVM and an ANN for one-hour ahead forecasting; when $h = 2$, we will train another SVM or ANN model for two-hour ahead forecasting; and so forth. Recall that this is a feature of the machine learning methods mentioned on page 42. The parameters of the selected SVM and ANN models are presented in Table 2.5.

The trained models are used to make forecasts at each month's test data. For each h, there are 12 test data points, i.e., $n_{\text{test}} = 12$, one per month. The

TABLE 2.6   RMSE (m/s) of five different forecasting methods.

Method	$h = 1$	$h = 2$	$h = 3$	$h = 4$	$h = 5$	$h = 6$
PER	0.826	1.597	2.055	2.336	2.659	3.005
WEB	3.237	3.439	3.177	3.474	2.703	2.322
ARMA(2,2)	0.984	1.541	1.777	2.394	2.348	2.488
SVM	1.065	1.504	2.661	2.487	2.154	2.905
ANN	1.074	1.727	1.857	2.666	2.595	2.429

TABLE 2.7   MAE (m/s) of five different forecasting methods.

Method	$h = 1$	$h = 2$	$h = 3$	$h = 4$	$h = 5$	$h = 6$
PER	0.631	1.194	1.626	1.744	2.227	2.442
WEB	2.405	2.452	2.457	2.839	2.099	1.813
ARMA(2,2)	0.769	1.163	1.303	1.962	2.055	2.024
SVM	0.864	1.258	2.007	1.959	1.780	2.235
ANN	0.856	1.452	1.441	2.125	2.148	2.010

12 test data points are used to compute three performance metrics—RMSE, MAE, and MAPE—for each forecasting method.

Tables 2.6–2.8 present the three metrics for the five methods. We observe the following:

1. For very short terms, like $h = 1$ or $h = 2$, the persistence model and ARMA model are clear winners. The method based on Weibull distribution is the worst by a noticeable margin.

2. Despite the bad performance for very near-term forecasting, WEB holds steady its performance as the forecasting horizon projects into the future, while the performances of all other methods deteriorate quickly. Eventually, WEB becomes the best forecasting at $h = 6$. PER suffers the greatest performance degradation when h increases from one hour to six hours.

3. The two machine learning methods, SVM and a single-layer ANN in this comparison, perform rather similarly. It is difficult to conclude which method is better. A many-layered ANN, or a deep neural net might, however, win over SVM. That remains to be studied.

4. If this study is used as a guide, then analysts are advised to use PER for one-hour or two-hour ahead forecasting, WEB for six-hour ahead or longer forecasting (before switching to NWP), and use ARMA models or machine learning methods for forecasting in between.

TABLE 2.8   MAPE (percentage) of five different forecasting methods.

Method	$h=1$	$h=2$	$h=3$	$h=4$	$h=5$	$h=6$
PER	8.2	16.0	21.4	19.4	27.9	30.3
WEB	26.6	27.0	29.8	29.4	24.6	21.7
ARMA(2,2)	9.3	16.2	18.1	20.9	25.6	27.0
SVM	9.8	16.5	23.7	20.7	23.0	26.8
ANN	9.6	18.1	19.1	21.9	26.0	26.6

GLOSSARY

ACF: Autocorrelation function

AIC: Akaike information criterion

AICc: Akaike information criterion corrected

ANN: Artificial neural network

AR: Autoregressive

ARMA: Autoregressive moving average

BIC: Bayesian information criterion

cdf: Cumulative distribution function

CRPS: Continuous ranked probability score

i.i.d: Identically, independently distributed

KF: Kalman filter

MA: Moving average

MAE: Mean absolute error

MAPE: Mean absolute percentage error

MLE: Maximum likelihood estimation

NWP: Numeric weather prediction

PACF: Partial autocorrelation function

PCE: Power curve error

pdf: Probability density function

PER: Persistence model or forecasting

RKHS: Reproducing kernel Hilbert space

RMSE: Root mean squared error

SVM: Support vector machine

WEB: Weibull distribution-based forecasting

EXERCISES

2.1 Find the probability density function for a three-parameter Weibull distribution.

   a. Derive the corresponding log-likelihood function.

   b. Use the three-parameter Weibull distribution to fit the hourly data in the Wind Time Series Dataset and report the estimated parameters.

   c. Suppose the turbine cut-in speed is 4 m/s. Remove the wind speed data below the cut-in speed and fit both the two-parameter Weibull distribution and the three-parameter distribution. Please discuss the differences in your estimation outcomes.

2.2 Evaluate what impact different bin widths may have on the χ^2 goodness-of-fit test.

   a. Use one month of the hourly data in the Wind Time Series Dataset and try the following bin widths: 0.2, 0.5, 1, 2 m/s.

   b. Switch to one week of the 10-min data and try the same set of bin widths.

2.3 Use Hinkley's method to select the power transformation coefficient, m.

   a. Try this on the hourly data in the Wind Time Series Dataset and try the following m values: 0, 0.5, 1, 2. Which m produces a $sym = 0$? Interpolation may be needed.

   b. Switch to the 10-min data and try the same set of m values.

2.4 Remove the diurnal trend in the hourly data in the Wind Time Series Dataset by using Gneiting's trigonometric function in Eq. 2.15. Plot the original time series and the standardized time series. Compare them with the standardization using Eq. 2.14 and note any difference that you may have observed.

2.5 For the linear model in Eq. 2.20, the objective function leading to a least-squares estimation is

$$\min \left\{ (\mathbf{V} - \hat{\mathbf{V}})^T (\mathbf{V} - \hat{\mathbf{V}}) = (\mathbf{V} - \mathbf{W}\hat{\mathbf{a}})^T (\mathbf{V} - \mathbf{W}\hat{\mathbf{a}}) \right\}.$$

The least-squares estimation can be attained by taking the first derivative of this objective function, with respect to $\hat{a}$, and setting it to zero. Please derive the least-squares estimation formula.

2.6 Use the hourly data in the `Wind Time Series Dataset` and conduct an ARMA modeling exercise. First, select the data from one of its months, and use this specific month data and do the following.

    a. Fit a series of AR models, with $p = 1, 2, \ldots, 6$, respectively. When applying the three information criteria, do they select the same model order? Which criterion selects the simplest model?

    b. Use the simplest AR model order selected in (a) and denote it as p_0. Compare the model $AR(p_0)$ with $ARMA(p_0, q)$ for $q = 1, 2, 3$, and select the model order q in a similar fashion as in (a) that selects p_0. Denote the resulting MA model order as q_0.

    c. Conduct some model diagnostics of this $ARMA(p_0, q_0)$ model by plotting its ACF and PACF. Do the ACF and PACF plots confirm a good model fit?

2.7 Derive Eq. 2.27 and Eq. 2.28 from Eq. 2.17.

2.8 When the loss function is a squared error loss function in Eq. 2.47, find the closed-form expression for the optimal $\hat{\alpha}$.

2.9 Take the January hourly data from the `Wind Time Series Dataset` and use that as the historical training data. In the presence of missing data, please simply skip time stamps where data are missing and continue with the next available data.

    a. Fit a series of AR models, with $p = 1, 2, \ldots, 6$, respectively. Use BIC to select the best model order p.

    b. Use the resulting AR model to do an h-hour ahead forecast, for $h = 1, 2, \ldots, 100$. One hundred hours is a little bit over four days. Call this forecast 1.

    c. Use the resulting AR model in (a) to do a one-hour ahead forecast. Shift the data sequence in (a) by one hour, namely that adding one new observation and dropping the oldest observation. Repeat, for the next 100 time instances, both the model fitting (including the determination of p) and the forecasting. Call this forecast 2.

    d. Use a Kalman filter to do the one-hour forecasting but continue running the Kalman filter for the next 100 time instances. Set the p in the Kalman filter as that found in (a). Call this forecast 3.

e. For each forecasting, record both the forecasting result and the corresponding wind speed observation at every time instance. Compute RMSE and MAE for each forecast. Compare the performance metrics for all three forecasts and discuss pros and cons of each approach.

2.10 For the hourly data in the `Wind Time Series Dataset`, take wind power data, instead of wind speed data, and repeat the comparison study conducted in Section 2.7. Compute the three performance metrics for five different methods for each $h = 1, 2, \ldots, 6$.

CHAPTER 3

Spatio-temporal Models

DOI: 10.1201/9780429490972-3

When building predictive models for short-term wind forecast, spatial information is less frequently used than temporal information. Chapter 2 uses data obtained from a single turbine on a wind farm, which can also be applied to a single time-series data aggregating wind power outputs from the whole farm. Analysts have noticed that valuable information may be elicited by considering spatial measurements in a local region, as wind characteristics at a site may resemble those at neighboring sites. This gives rise to the idea of developing spatio-temporal methods to model the random wind field evolving through space and time.

Recall that we denote the wind speed data in Chapter 2 by V_t, which has only the time index. To model a spatio-temporal process, we expand the input variable set to include both the location variable, denoted by $\mathbf{s} \in \mathbb{R}^2$, and the time variable, denoted still by $t \in \mathbb{R}$, so that the spatio-temporal random wind field is represented by $V(\mathbf{s}, t)$. In this chapter, unless otherwise noted, N is used to denote the number of sites, whereas n is used to denote the number of time instances in the training set.

One of the key aspects in spatio-temporal modeling is to model the covariance structure of V through a positive-definite parametric covariance function, $Cov[V(\mathbf{s}, t), V(\mathbf{s}', t')]$.

3.1 COVARIANCE FUNCTIONS AND KRIGING

In this section, we focus on spatial covariance. For the time being, $V(\mathbf{s}, t)$ is simplified to be $V(\mathbf{s})$. Recall that the temporal covariance, also known as autocovariance, is discussed in Section 2.4.3.

We use $C(\mathbf{s}, \mathbf{s}'; t, t')$ to represent a covariance function, namely

$$C(\mathbf{s}, \mathbf{s}'; t, t') := Cov[V(\mathbf{s}, t), V(\mathbf{s}', t')].$$

When the time is held still and only the spatial covariance is concerned, the covariance function $C(\mathbf{s}, \mathbf{s}'; t, t')$ can be simplified to $C(\mathbf{s}, \mathbf{s}') := Cov[V(\mathbf{s}), V(\mathbf{s}')]$, after dropping the time index.

Given a set of N locations, $\mathbf{s}_1, \ldots, \mathbf{s}_N$, we can compute the corresponding covariance matrix $\mathbf{C}$, whose (i, j)-th entry is $C_{ij} = C(\mathbf{s}_i, \mathbf{s}_j)$. The covariance matrix is positive definite if all its eigenvalues are strictly positive, or positive semidefinite if some of its eigenvalues are zeros while the rest are positive. It is not difficult to notice that the covariance function is related to the kernel function mentioned in Section 2.5.2 and the covariance matrix is related to the Gram matrix (or the kernel matrix). A covariance function is referred to as a covariance kernel in a general machine learning context, and it can be shown that a positive definite kernel can be obtained as a covariance kernel in which the distribution has a particular form [94].

3.1.1 Properties of Covariance Functions

We start with the discussion of some general properties of the covariance functions.

Stationarity. A covariance function can be used to characterize both stationary and nonstationary stochastic processes. We primarily consider the stationary covariance function in this book, which has the property

$$C(\mathbf{s}, \mathbf{s}') = g(\mathbf{s} - \mathbf{s}'), \tag{3.1}$$

where $g(\cdot)$ is a function to be specified. The stationarity means that the covariance does not depend on the start location of a stochastic process but only depends on the distance and orientation between two points in that process. The variance of a stationary stochastic process can be expressed as

$$Var[V(\mathbf{s})] = g(\mathbf{0}) = \sigma_V^2. \tag{3.2}$$

For a stationary function, σ_V^2 is a constant, so that the stationary covariance matrix can be further factorized as

$$\mathbf{C} = \sigma_V^2 \cdot \mathbf{R}, \tag{3.3}$$

where $\mathbf{R}$ is a correlation matrix whose (i, j)-th entry is $\rho_{ij} = C_{ij}/\sigma_V^2$.

The concept of stationarity extends to the spatio-temporal covariance functions. By assuming stationarity, the covariance function only depends on the spatial lag, $\mathbf{u} = \mathbf{s} - \mathbf{s}'$, and the time lag, $h = t - t'$, such that the general function form $C(\mathbf{s}, \mathbf{s}'; t, t')$ can be expressed as $C(\mathbf{u}; h)$.

Isotropy: A stationary covariance function is isotropic, provided that

$$C(\mathbf{s}, \mathbf{s}') = g(\|\mathbf{s} - \mathbf{s}'\|_2), \tag{3.4}$$

where $\|\mathbf{s} - \mathbf{s}'\|_2$ is the Euclidean distance between the two locations $\mathbf{s}$ and $\mathbf{s}'$. When it does not cause any ambiguity, the subscript "2" is dropped hereinafter. Isotropy is to require invariance under rotation. This is to say, every pair of data points at $\mathbf{s}$ and $\mathbf{s}'$, respectively, having a common interpoint distance, must have the same covariance regardless of their orientation. Apparently, isotropy is a stronger condition than stationarity.

Smoothness: Smoothness (continuity and differentiability) is a property associated with sample functions, which are the realization of the stochastic process under a specified covariance function. The smoothness requirement is an important consideration in choosing a covariance function. The general relationship between the smoothness of sample functions and the covariance function is not straightforward. It is easier to talk about smoothness of sample functions when a specific covariance function is considered.

3.1.2 Power Exponential Covariance Function

A popular family of covariance functions is the power exponential function,

$$C(\mathbf{s}, \mathbf{s}') = \sigma_V^2 \exp\left\{ -\frac{1}{2} \sum_{j=1}^{d} \left| \frac{s_j - s_j'}{\theta_j} \right|^{p_j} \right\}, \tag{3.5}$$

where d is the dimension of $\mathbf{s}$, $0 < p_j \le 2$ is the shape parameter, and θ_j is the scale parameter. Usually $d = 2$ in spatial statistics.

A special form of the power exponential covariance function is the isotropic squared exponential (SE) covariance function (the phrase "isotropic" is often omitted), whose parameters are $\theta_1 = \cdots = \theta_d = \theta$, and $p_1 = \cdots = p_d = p = 2$, so that

$$C_{\text{SE}}(u) = \sigma_V^2 \exp\left\{ -\frac{u^2}{2\theta^2} \right\}, \tag{3.6}$$

where $u = \|\mathbf{u}\| = \|\mathbf{s} - \mathbf{s}'\| = \sqrt{\sum_{j=1}^{d}(s_j - s_j')^2}$. This function is also called the Gaussian covariance function. Recall the radial basis kernel in Eq. 2.48. The $C_{\text{SE}}(\cdot)$ is the same as $K(\cdot, \cdot)$ if $\phi = 1/2\theta^2$ and $\sigma_V^2 = 1$.

An anisotropic form of the squared exponential covariance function is where the scale parameters are different along different input directions while its shape parameter is fixed at 2, namely $p_1 = \cdots = p_d = p = 2$. This anisotropic form is also known as the *automatic relevance determination* (ARD). The corresponding covariance function reads as,

$$C_{\text{SE-ARD}}(\mathbf{s}, \mathbf{s}') = \sigma_V^2 \exp\left\{ -\frac{1}{2} \sum_{j=1}^{d} \left| \frac{s_j - s_j'}{\theta_j} \right|^2 \right\}. \tag{3.7}$$

The impact of the three types of parameters in the power exponential covariance function can be understood as follows, and Fig. 3.1 presents a few examples of the sample function under different parameter combinations.

- The variance term, σ_V^2, is referred to as the *amplitude*, because it is related to the amplitude of a sample function.

- The shape parameter, p, determines the smoothness of the sample functions. In the above two special cases of the power exponential function,

$p = 2$. Analysts like this choice because the corresponding sample functions are infinitely differentiable, meaning that the sample paths are smooth. For the power exponential family, $p = 2$ is the only shape parameter choice under which the sample functions are differentiable. When $p = 1$, the corresponding covariance function is known as the *exponential covariance function*. This choice is less popular because its sample functions are not smooth.

- The scale parameter, θ, referred to as the *length scale*, determines how quickly the correlation decays as the between-point distance increases. When θ decreases, the correlation between a pair of points of a fixed distance decreases, and thus, the sample functions have an increasing number of local optima. As a result, the sample function exhibits fast changing patterns and a short wavelength, where as θ increases, the correlation between a fixed pair of points increases, and the sample function hence exhibits slow changing patterns and a long wavelength.

Another popular family of the covariance function is the Matérn covariance function, which has a smoothness parameter, v, that can control the smoothness of sample functions more precisely. Specifically, the sample functions are almost surely continuously differentiable of order $\lceil v \rceil - 1$, where $\lceil \cdot \rceil$ rounds up to the next integer. We choose to omit the presentation of the Matérn covariance function because we do not use it in this book. Interested readers can refer to [173] for more information.

3.1.3 Kriging

Kriging is the method commonly used to make spatial predictions. The method is named after the South African mining engineer, D. G. Krige. In spatial statistics and machine learning, kriging is generally referred to as the Gaussian process regression [41, 173]. The problem setting is as follows. Consider sites, $\mathbf{s}_1, \ldots, \mathbf{s}_N$, and the wind speeds at these locations, denoted by $V(\mathbf{s}_1), \ldots, V(\mathbf{s}_N)$. The N sites can be the turbine sites in a wind farm, and the wind speeds at $\{\mathbf{s}_1, \ldots, \mathbf{s}_N\}$ can be the wind speed measurements obtained by the respective nacelle anemometers. Analysts express the sites and respective measurements as data pairs, such as $\{\mathbf{s}_i, V(\mathbf{s}_i)\}_{i=1}^N$. The objective is to make a prediction at a site, say, $\mathbf{s}_0$, where no measurements are taken.

Two popular versions of kriging are the ordinary kriging and universal kriging. The ordinary kriging uses the following model,

$$V(\mathbf{s}_i) = \beta_0 + \delta(\mathbf{s}_i) + \varepsilon_i, \quad i = 1, ..., N, \tag{3.8}$$

where β_0 is an unknown constant, $\delta(\cdot)$ is the term modeling the underlying random field via the spatial correlation among sites, and ε is the zero mean, i.i.d Gaussian noise, such that $\varepsilon_i \sim \mathcal{N}(0, \sigma_\varepsilon^2)$. The i.i.d Gaussian noise, ε, is also known as the *nugget effect*.

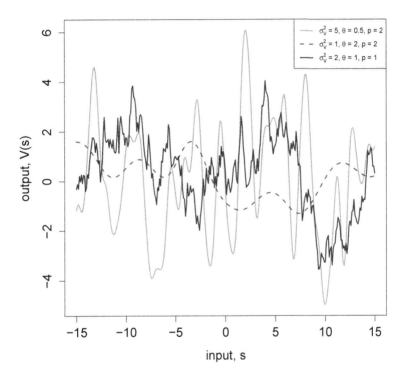

FIGURE 3.1   Three sample functions using a squared exponential covariance function with different parameter choices.

The random field term $\delta(\cdot)$ is assumed to be a zero-mean Gaussian process whose covariance structure is characterized, for instance, by a power exponential covariance function (other covariance functions can be used, too, but the power exponential family is a popular choice). Suppose that the squared exponential covariance function in Eq. 3.6 is used. It means that

$$C(\delta(\mathbf{s}), \delta(\mathbf{s}')) = \sigma_\delta^2 \exp\left\{-\frac{\|\mathbf{s} - \mathbf{s}'\|^2}{2\theta^2}\right\}, \tag{3.9}$$

where σ_δ^2 is the variance term associated with the random field function $\delta(\cdot)$. As such, the variance of wind speed is the summation of the variance associated with the random field and that of the i.i.d random noise, namely $\sigma_V^2 = \sigma_\delta^2 + \sigma_\varepsilon^2$.

What the ordinary kriging model implies is that the wind speed over a spatial field is centered around a grand average, β_0. The random fluctuation consists of two portions—the first depends on specific sites and is characterized by the spatial correlation between site $\mathbf{s}_0$ and the sites where observations are made or measurements are taken, and the second is the pure random noise, resulting from, for instance, the measurement errors.

Re-write Eq. 3.8 into a matrix form, i.e.,

$$\underbrace{\begin{pmatrix} V(\mathbf{s}_1) \\ V(\mathbf{s}_2) \\ \vdots \\ V(\mathbf{s}_N) \end{pmatrix}}_{\mathbf{V}} = \beta_0 \cdot \mathbf{1}_N + \underbrace{\begin{pmatrix} \delta(\mathbf{s}_1) \\ \delta(\mathbf{s}_2) \\ \vdots \\ \delta(\mathbf{s}_N) \end{pmatrix}}_{\delta} + \underbrace{\begin{pmatrix} \varepsilon_1 \\ \varepsilon_2 \\ \vdots \\ \varepsilon_N \end{pmatrix}}_{\varepsilon}, \tag{3.10}$$

where $\mathbf{1}_N$ is an $N \times 1$ vector of all ones. Denote the covariance matrix of δ by $\mathbf{C}_{NN} = (C_{ij})_{N \times N}$, where the subscript "$NN$" means that this is a covariance matrix for the N sites. Recall that δ and ε are two normal random variables having different covariance structures. This suggests that $\mathbf{V}$ follows a multivariate normal distribution, such as,

$$f(\mathbf{V}) = \mathcal{N}(\beta_0 \cdot \mathbf{1}_N, \mathbf{C}_{NN} + \sigma_\varepsilon^2 \mathbf{I}). \tag{3.11}$$

For the new site $\mathbf{s}_0$, the wind speed to be measured there, whose notation is simplified to V_0, has covariances with the existing N sites. The covariances can be characterized using the same covariance function, such as $C_{0j} = C(\mathbf{s}_0, \mathbf{s}_j)$, for $j = 1, \ldots, N$. Introduce a new $1 \times N$ row vector,

$$\mathbf{c}_{0N} := (C_{01}, \ldots, C_{0N}).$$

Then, the multivariate joint distribution of $(V_0, \mathbf{V}^T)^T$ is

$$f\left(\begin{bmatrix} V_0 \\ \mathbf{V} \end{bmatrix}\right) = \mathcal{N}\left(\beta_0 \cdot \mathbf{1}_{N+1}, \begin{bmatrix} \sigma_\delta^2 + \sigma_\varepsilon^2 & \mathbf{c}_{0N} \\ \mathbf{c}_{0N}^T & \mathbf{C}_{NN} + \sigma_\varepsilon^2 \mathbf{I} \end{bmatrix}\right), \tag{3.12}$$

where $\sigma_\delta^2 + \sigma_\varepsilon^2$ is the variance of V_0, namely σ_V^2, which is also known as the prior variance at the unseen site $\mathbf{s}_0$ before the prediction. Invoke the conditional Gaussian distribution formula, which says that if $\mathbf{x}$ and $\mathbf{y}$ are jointly Gaussian, i.e.,

$$\begin{pmatrix} \mathbf{x} \\ \mathbf{y} \end{pmatrix} \sim \mathcal{N}\left(\begin{bmatrix} \boldsymbol{\mu}_x \\ \boldsymbol{\mu}_y \end{bmatrix}, \begin{bmatrix} \mathbf{A} & \mathbf{D} \\ \mathbf{D}^T & \mathbf{B} \end{bmatrix} \right),$$

then, the condition distribution $f(\mathbf{x}|\mathbf{y})$ is

$$f(\mathbf{x}|\mathbf{y}) = \mathcal{N}(\boldsymbol{\mu}_x + \mathbf{D}\mathbf{B}^{-1}(\mathbf{y} - \boldsymbol{\mu}_y), \mathbf{A} - \mathbf{D}\mathbf{B}^{-1}\mathbf{D}^T). \qquad (3.13)$$

By using this conditional Gaussian distribution formula, we can express

$$f(V_0|\mathbf{V}) = \mathcal{N}(\beta_0 + \mathbf{c}_{0N}(\sigma_\varepsilon^2\mathbf{I} + \mathbf{C}_{NN})^{-1}(\mathbf{V} - \beta_0 \cdot \mathbf{1}),$$
$$\sigma_V^2 - \mathbf{c}_{0N}(\sigma_\varepsilon^2\mathbf{I} + \mathbf{C}_{NN})^{-1}\mathbf{c}_{0N}^T). \qquad (3.14)$$

This conditional distribution leads to the predictive distribution of V_0, once the observations on the existing N sites are obtained. We can write the predictive mean and predictive variance, respectively, as

$$\hat{V}_0 := \hat{\mu}_0 = \hat{\beta}_0 + \mathbf{c}_{0N}(\hat{\sigma}_\varepsilon^2\mathbf{I} + \mathbf{C}_{NN})^{-1}(\mathbf{V} - \hat{\beta}_0 \cdot \mathbf{1}),$$
$$Var(\hat{V}_0) := \hat{\sigma}_0^2 = \sigma_V^2 - \mathbf{c}_{0N}(\hat{\sigma}_\varepsilon^2\mathbf{I} + \mathbf{C}_{NN})^{-1}\mathbf{c}_{0N}^T. \qquad (3.15)$$

The first equation is the *kriging predictor*, which is a linear combination of the observed wind speeds in $\mathbf{V}$. The linear coefficients (the weights) depend on the correlation between the unseen site, $\mathbf{s}_0$, and the N training sites as well as the variance in the training data. The coefficients are bigger, namely the weights are greater, if the correlation is strong and the training data have small variances. The predictive variance is reduced from the prior variance σ_V^2 at the unseen site. The reduced amount depends also on the correlation between the unseen site and the training sites as well as the variance in the training data. The $100(1 - \alpha)\%$ confidence interval for the prediction at $\mathbf{s}_0$ can be obtained as

$$[\hat{V}_0 - z_{\alpha/2}\hat{\sigma}_0, \qquad \hat{V}_0 + z_{\alpha/2}\hat{\sigma}_0].$$

In the ordinary kriging model, Eq. 3.8, where an SE covariance function is used, there are four parameters, $\{\beta_0, \sigma_\delta^2, \theta, \sigma_\varepsilon^2\}$. These parameters can be estimated by maximizing a log-likelihood function, which is the density function in Eq. 3.11. Specifically, the log-likelihood function reads

$$\mathcal{L}(\mathbf{V}|\beta_0, \sigma_\delta^2, \theta, \sigma_\varepsilon^2) = -\frac{1}{2}(\mathbf{V} - \beta_0 \cdot \mathbf{1})^T \left(\sigma_\varepsilon^2\mathbf{I} + \mathbf{C}_{NN}\right)^{-1} (\mathbf{V} - \beta_0 \cdot \mathbf{1})$$
$$- \frac{1}{2}\log\left|\sigma_\varepsilon^2\mathbf{I} + \mathbf{C}_{NN}\right| - \frac{N}{2}\log(2\pi). \qquad (3.16)$$

Alternatively, one can first estimate β_0 by using the average of $\{V_i\}_{i=1}^N$ and then center the raw wind speed data by subtracting its average. After that, one can use the centered wind speed data and the maximum likelihood estimation to estimate the remaining three parameters (replace β_0 by $\bar{V}$ in Eq. 3.16).

Conceptually, the universal kriging is not much different from the ordinary kriging. The main extension is to make the mean component in Eq. 3.8 and Eq. 3.11 more flexible. In the ordinary kriging, the mean component is assumed a constant, β_0. In the universal kriging, the mean component is assumed as a polynomial model, $\beta_0 + \mathbf{g}^T(\mathbf{s})\boldsymbol{\beta}$, so that the universal kriging model can be expressed as

$$V(\mathbf{s}_i) = \beta_0 + \mathbf{g}^T(\mathbf{s}_i)\boldsymbol{\beta} + \delta(\mathbf{s}_i) + \varepsilon_i, \quad i = 1, ..., N, \tag{3.17}$$

where $\mathbf{g}(\cdot) = (g_1(\cdot), \ldots, g_q(\cdot))^T$ is a set of basis functions, $\boldsymbol{\beta} = (\beta_1, \ldots, \beta_q)^T$ is the coefficient vector, and q is the number of terms in the polynomial model in addition to the grand average, β_0. There are many different choices for the basis function $\mathbf{g}(\cdot)$ but it can be simply that $g_1(\mathbf{s}) = s_1$ and $g_2(\mathbf{s}) = s_2$ (in this case, $q = d = 2$). By expanding the mean component, the parameters in a universal kriging are $\{\beta_0, \beta_1, \ldots, \beta_q, \sigma_\delta^2, \theta, \sigma_\varepsilon^2\}$, and the number of parameters is $q + 4$, which is q more than that in the ordinary kriging. Nonetheless, the maximum likelihood estimation can still be used to estimate all the parameters after adjusting the log-likelihood function properly.

Kriging can be assisted by using the geoR package in R. Using the likfit function to estimate the parameters, such as

```
para <- likfit(spatialdata, ini.cov.pars = c(1,1), nugget =
                0.5).
```

Here, spatialdata is the wind spatial data object holding the $\{\mathbf{s}_i, V_i\}_{i=1}^N$ data pairs, ini.cov.pars provides the initial value for σ_δ^2 and θ, respectively, nugget provides the initial value for σ_ε^2. By default, the likfit uses the SE covariance function and estimates the parameters by the maximum likelihood estimation in an ordinary kriging. To make a prediction at $\mathbf{s}_0$, one can use

```
V0 <- krige.conv(spatialdata, locations = s0, krige =
                krige.control(obj.model = para)),
```

where obj.model = para in the krige.control function passes the parameters just estimated to the prediction function. The default setting is an ordinary kriging.

Fig. 3.2 presents an example of applying the ordinary kriging predictor to the wind speed data from ten turbines in the Wind Spatial Dataset. An ordinary kriging model is established based on the wind speed data collected in July at ten turbine sites. Then the kriging model is used to predict the wind speed at site #6 using the observed wind speed at the other nine sites for the month of August. Fig. 3.2, right panel, shows that the spatially predicted wind speed at site #6 closely matches the observed wind speed at the same site.

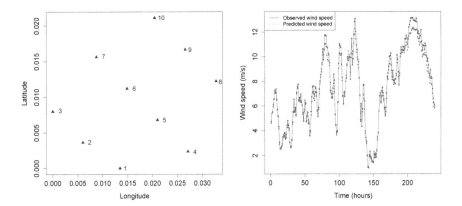

FIGURE 3.2 Left panel: the layout of the ten turbines. Right panel: the predicted and observed wind speeds of the first ten days in August at site #6.

3.2 SPATIO-TEMPORAL AUTOREGRESSIVE MODELS

The previous section considers purely the spatial correlation. This section presents a method that combines the spatial model feature and time series model feature in a method known as the Gaussian spatio-temporal autoregressive model (GSTAR) [166].

3.2.1 Gaussian Spatio-temporal Autoregressive Model

The wind speed in GSTAR model is assumed to follow a truncated normal distribution, a distribution choice as popular as Weibull used for modeling wind speed [75]. For notational simplicity, the site notation, $\mathbf{s}_i$, is shortened as site i, and consequently, $V(\mathbf{s}_i; t)$ is simplified to $V_i(t)$. To handle the wind speed nonstationarity over time, the time in a day is split into a number of *epochs*, during which the wind speed is assumed stationary [88]. For instance, 6 a.m. to 12 p.m. in a day can be treated as one epoch. With these notations, Pourhabib et al. [166] express $V_i(t) \sim \mathcal{N}^+(\mu_i(e_t), \sigma_i^2(e_t))$, where $i = 1, \ldots, N$, and e_t denotes the "epoch" at time t.

The GSTAR model assumes that the mean of wind speed at site i is a function of the past wind speeds at not only the target site but also other sites in its neighborhood, such that

$$\mu_i(e_t) = \beta_0 + \sum_{\ell=1}^{p} \sum_{j \in J_i} a_{ij\ell} V_j(t-\ell), \qquad \text{for} \quad i = 1, 2, \ldots, N, \qquad (3.18)$$

where β_0 is an unknown constant, p is the autoregressive model order, $J_i \subset$

$\{1, 2, \ldots, N\}$ denotes the set of neighborhood sites whose wind speeds have a strong enough correlation with the wind speed at the target site i, and $a_{ij\ell}$ are the coefficients that quantify the spatio-temporal dependency. Note that Eq. 3.18 is a model for the expectation, so that the zero-mean, i.i.d noise term, ε, disappears.

GSTAR relies on one important assumption, which is to assume the spatio-temporal parameters, $a_{ij\ell}$, can be factorized into the respective spatial and temporal parts, such that

$$a_{ij\ell} = a_{ij}^s a_{i\ell}^t \quad \text{for} \quad i = 1, 2, \ldots, N, \quad j \in J_i, \quad \ell = 1, 2, \ldots, p, \tag{3.19}$$

and GSTAR models the spatial part, a_{ij}^s, and the temporal part, $a_{i\ell}^t$, individually. GSTAR models its spatial dependency coefficient, a_{ij}^s, through a Gaussian kernel,

$$a_{ij}^s = \exp\left\{-(\mathbf{s}_i - \mathbf{s}_j)^T \mathbf{\Lambda}_i (\mathbf{s}_i - \mathbf{s}_j)\right\}, \quad i = 1, 2, \ldots, N, \quad j \in J_i, \tag{3.20}$$

where $\mathbf{\Lambda}_i = \text{diag}\{\lambda_{i1}, \lambda_{i2}\}$, and λ_{i1} and λ_{i2} characterize the spatial decay in the longitudinal and latitudinal directions, respectively. Differing from that in Eq. 2.48, the Gaussian kernel in Eq. 3.20 has different scale parameters along the two spatial directions, whereas Eq. 2.48 has a single scale parameter ϕ for all directions. In this sense, this Gaussian kernel is the counterpart of the $C_{\text{SE-ARD}}$ covariance function in Eq. 3.7, whereas Eq. 2.48 is the counterpart of the C_{SE} covariance function in Eq. 3.6.

GSTAR models its temporal dependency, $a_{i\ell}^t$, through an exponential decay in terms of time distance, such as

$$a_{i\ell}^t = \exp\left\{-\lambda_{i3}\ell\right\}, \quad i = 1, 2, \ldots, N, \quad \ell = 1, \ldots, p \tag{3.21}$$

where λ_{i3} characterizes the temporal decay. Using Eq. 3.19–3.21, the otherwise large number of spatio-temporal parameters for site i is reduced to the three parameters, λ_{i1}, λ_{i2}, and λ_{i3}.

Let $\mathbf{A}_i$ denote an $N \times p$ matrix of spatial dependency for site i, of which the (j, ℓ)-th entry, $(\mathbf{A}_i)_{j\ell}$, is a_{ij}^s. Because a_{ij}^s does not have the ℓ index, all the entries are the same for the j-th row. For instance, the elements in the first row are all a_{i1}^s. If $j \notin J_i$, the corresponding row of $\mathbf{A}_i$ is entirely zero. Let $\mathbf{D}_i$ denote a $p \times p$ diagonal matrix whose (ℓ, ℓ)-th entry is $a_{i\ell}^t$. Let $\mathbf{V}_i(t) = (V_i(t-1), \ldots, V_i(t-p))^T$ be the time series data vector at site i, and $\mathcal{V}(t)$ be the $N \times p$ time series data matrix for all sites, namely

$$\mathcal{V}(t) = \begin{pmatrix} \mathbf{V}_1^T(t) \\ \mathbf{V}_2^T(t) \\ \vdots \\ \mathbf{V}_N^T(t) \end{pmatrix}_{N \times p}. \tag{3.22}$$

With the above notations, Eq. 3.18 can be expressed in a matrix form as,

$$\mu_i(e_t) = \beta_0 + \text{tr}\left(\mathbf{A}_i \mathbf{D}_i \mathcal{V}^T(t)\right), \quad i = 1, 2, \ldots, N. \tag{3.23}$$

This model is referred to as the GSTAR of order p, or, simply GSTAR(p).

To estimate the parameters in Eq. 3.23, GSTAR uses a regularized least-squares estimation procedure as,

$$\min_{\lambda_{i1},\lambda_{i2},\lambda_{i3}} \sum_{\ell=1}^{n} \left\{ L\left[V_i(\ell+h) - \bar{V}_i, \text{tr}\left(\mathbf{A}_i\mathbf{D}_i\mathcal{V}^T(\ell)\right)\right] + \gamma \text{Pen}\left(\mathbf{A}_i\right) \right\}, \quad (3.24)$$

where h is the look-ahead time at which the GSTAR model is trained for making a forecast, n is the number of time stamps in the training set, $\bar{V}_i = \frac{1}{n}\sum_{\ell=1}^{n} V_i(\ell)$, $L[\cdot,\cdot]$ is a loss function (see Section 2.6 for various choices), γ is the penalty coefficient, and $\text{Pen}\left(\mathbf{A}_i\right)$ is the penalty term that controls the size of the neighborhood, to be discussed in the next section. This optimization problem is solved using the Broyden-Fletcher-Goldfarb-Shanno (BFGS) algorithm [64], which belongs to the class of quasi-Newton methods.

Following the approach in [75], GSTAR models the standard deviation of wind speed as a linear combination of volatility, which measures the magnitude of recent changes in wind speed, such as,

$$\hat{\sigma}_i(e_t) = b_0 + b_1\hat{\nu}_i(t), \qquad i = 1, 2, \ldots, N, \quad (3.25)$$

where

$$\hat{\nu}_i(t) = \left[\frac{1}{2|J_i|} \sum_{j \in J_i} \sum_{\ell=0}^{1} \left\{(V_j(t-\ell) - V_j(t-\ell-1))^2\right\}\right]^{\frac{1}{2}}, \quad (3.26)$$

and $|J_i|$ is the number of elements in J_i. In the above equation, only the immediately past two moving range values, i.e., the difference between wind speed at t and at $t-1$ and that between wind speed at $t-1$ and at $t-2$, are used to estimate the volatility, $\hat{\nu}$. The two coefficients, b_0 and b_1, can be estimated by regressing the sample standard deviation in the left-hand side of Eq. 3.25 on $\hat{\nu}_i(t)$.

GSTAR makes an h-step ahead forecast at site i based on the α-quantile of the truncated normal distribution, such as

$$\hat{V}_i(t+h) = \hat{\mu}_i(t+h) + \hat{\sigma}_i(t+h) \cdot \Phi^{-1}\left[\alpha + (1-\alpha)\Phi\left(-\frac{\hat{\mu}_i(t+h)}{\hat{\sigma}_i(t+h)}\right)\right], \quad (3.27)$$

where $\Phi(\cdot)$ is the cdf of the standard normal distribution, $\hat{\mu}_i(\cdot)$ is the estimated mean found through Eq. 3.23, in which $t+h$ denotes a forecasting time that falls in the epoch e_t, and $\hat{\sigma}_i(\cdot)$ is the estimated standard deviation, decided through Eq. 3.25. The value of α should be decided based on the choice of the loss function. Using MAE or RMSE, $\alpha = 0.5$. Using PCE, α should be chosen consistently with ξ in Eq. 2.58.

3.2.2 Informative Neighborhood

GSTAR only uses the sites within a neighborhood to make forecasts at the target site. This neighborhood of site i, denoted by J_i, is much smaller than the whole wind farm. The rationale of this treatment is that not every single site on the farm has strong enough correlation with the target site to provide meaningful information and hence facilitate forecasting. The use of the Gaussian kernel essentially means that when the distance grows to a certain extent, the turbine sites lying beyond would have very little impact. For this reason, this neighborhood is referred to as an *informative neighborhood* for the purpose of forecasting. An obvious benefit of using an informative neighborhood is the reduced computational burden in the solution procedure.

Unlike the traditional wisdom that uses a time-invariant distance-based criterion [88, 128], leading to a disc-like neighborhood with a fixed radius, GSTAR uses the correlation among the rate of change in wind speed to determine the spatial dependency. Pourhabib et al. [166] discover through their analysis that two locations are informative to each other if the two sites have similar rates of change in wind speed for a given period, which explains why a pure distance-based criterion alone could be ineffective. Employing this criterion to find the informative neighborhood is done by designing a special penalty term in Eq. 3.24.

Denote by $Z_i(t) = dV_i'(t)/dt \approx V_i'(t) - V_i'(t-1)$ the first derivative of wind speed (the change rate), where $V_i' = V_i/\max\{V_i(t)\}$ is the wind speed normalized by the maximum wind speed for the whole farm during the training period. Then, compute the $N \times N$ sample covariance matrix for $\mathbf{Z}(\ell) = [Z_1(\ell), Z_2(\ell), \ldots, Z_N(\ell)]^T$ as,

$$\mathbf{C}_Z = \frac{1}{n} \sum_{\ell=1}^{n} \left(\mathbf{Z}(\ell) - \bar{\mathbf{Z}}\right) \left(\mathbf{Z}(\ell) - \bar{\mathbf{Z}}\right)^T, \qquad (3.28)$$

where $\bar{\mathbf{Z}} = \sum_{\ell=1}^{n} \mathbf{Z}(\ell)/n$. To create the penalty term, Pen $(\mathbf{A}_i)$, it goes through three steps of action:

(a) Set an entry in $\mathbf{C}_Z$ to zero if its value is smaller than a prescribed threshold $\kappa \in [0, 1]$;

(b) Create a new matrix whose entries are the element-wise inverse of the entries of the matrix obtained in step (a) (with the convention that the inverse of zero is ∞); and

(c) Calculate the Frobenius norm of the product between the matrix obtained after step (b) and $\mathbf{A}_i$ in Eq. 3.24, with the convention that $0 \times \infty = 0$.

The specific mathematical steps are as follows. Let $\mathbf{C}_Z^\kappa$ denote the matrix after $\mathbf{C}_Z$ is truncated using κ, i.e.,

$$C_{Z,jk}^\kappa = C_{Z,jk} \quad \text{if} \quad C_{Z,jk} \geq \kappa, \quad \text{otherwise} \quad C_{Z,jk}^\kappa = 0, \qquad (3.29)$$

where $C_{Z,jk}^{\kappa}$ and $C_{Z,jk}$ are the (j,k)-th entry of $\mathbf{C}_Z^{\kappa}$ and $\mathbf{C}_Z$, respectively. Then, let $(\mathbf{C}_Z^{\kappa})^-$ denote the entry-wise inverse of $\mathbf{C}_Z^{\kappa}$. As such, the penalty term is defined as,

$$\text{Pen}(\mathbf{A}_i) = \|\mathbf{A}_i^T(\mathbf{C}_Z^{\kappa})^-\|_F, \tag{3.30}$$

where $\|\cdot\|_F$ represents the Frobenius norm and, inside $\text{Pen}(\mathbf{A}_i)$, one uses the notational convention that $0 \times \infty = 0$.

What this penalty term does is to associate each spatial dependency term, a_{ij}^s, with the inverse of a $C_{Z,jk}^{\kappa}$, in a fashion that can be loosely expressed as $a_{ij}^s/C_{Z,jk}^{\kappa}$. To reduce the cost resulting from the penalty term, one apparently wants to keep $a_{ij}^s/C_{Z,jk}^{\kappa}$ as small as possible. If $C_{Z,jk}^{\kappa} = 0$, meaning that the sample covariance of the first derivative of wind speed is smaller than the threshold, κ, then the corresponding a_{ij}^s is forced to zero. If $C_{Z,jk}^{\kappa}$ is not zero but small, indicating a weak correlation between the two first derivatives, then the corresponding a_{ij}^s is penalized more, whereas if $C_{Z,jk}^{\kappa}$ is large, indicating a strong correlation, then a_{ij}^s is penalized less. The informative neighborhood $J_i = \{j : C_{Z,ij}^{\kappa} \neq 0\}$ is then selected through this penalizing scheme.

Fig. 3.3 presents an example of the informative neighborhoods selected for three different target sites. Note that informative neighborhoods are irregularly shaped, rather than disc-like, and they are different when the target site is at a different location. The shape and size of the informative neighborhoods are time varying, and they will be updated through the learning process as the new wind data arrives. This informative neighborhood concept and method is more flexible and versatile in terms of capturing the spatial relevance.

Concerning the choice for the threshold, κ, the general understanding is that a smaller κ leads to a larger neighborhood, because it causes $\mathbf{C}_Z^{\kappa}$ to have fewer zero entries, whereas a large κ creates a smaller informative neighborhood, because the resulting $\mathbf{C}_Z^{\kappa}$ has more zero entries. Here GSTAR sets the κ value at 0.85 for all forecast horizons. Analysts can certainly conduct fine-scale adjustments by, say, setting a lower and an upper threshold for the size of an informative neighborhood. If the number of turbines in the neighborhood is below the lower threshold, the κ value is to be reduced, which in turn makes the neighborhood bigger to accommodate more turbines. If the number of turbines is above the upper threshold, then the κ is to be increased, to make the neighborhood smaller. In the numerical analysis in Section 3.2.3, the lower and upper bounds are set as 2 and 15, respectively.

3.2.3 Forecasting and Comparison

This section applies the GSTAR method to the Wind Spatial-Temporal Dataset1. In this application, GSTAR defines four epochs for each day in a calendar month: (1) 12:00 am to 6:00 am, (2) 6:00 am to 12:00 pm, (3) 12:00 pm to 6:00 pm, and (4) 6:00 pm to 12:00 am. Consequently, an individual GSTAR model for each epoch is fit, which is used to make forecasts for the horizon belonging to the same epoch. Each GSTAR model is trained using one month of data and then makes h-hour ahead forecasts for $h = 2, 3, 4,$ and 5.

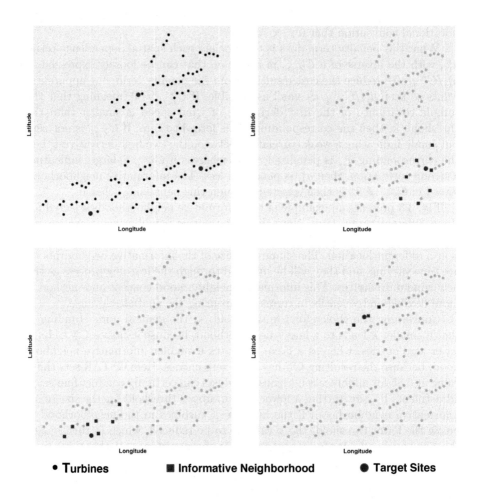

FIGURE 3.3 Neighborhoods selected by GSTAR based on one month of data in the `Wind Spatio-Temporal Dataset1`. Top-left: three turbine sites and the surrounding turbines; top-right, bottom-left and bottom-right: informative neighborhood selected for each site.

Pourhabib et al. [166] choose the PCE loss function as $L[\cdot,\cdot]$ in Eq. 3.24. When using PCE, a power curve function is needed as $g(\cdot)$ in Eq. 2.58. Using a nonlinear power curve function complicates the optimization in Eq. 3.24. Because of that, Pourhabib et al. simplify the power curve function to a piecewise linear function, such that

$$g(V) = \begin{cases} 0, & V \leq 3.5; \\ 0.1053(V - 3.5), & 3.5 < V \leq 13; \\ 1, & 13 < V. \end{cases}$$

This piecewise linear power curve function does not differ that much from the nonlinear power curve function. The ξ parameter used in PCE is set to 0.73. To ease the computational burden to go through the sizeable combinations of turbines, months, and epochs, each of the $N = 120$ turbine cases is randomly assigned to evaluate one of the epochs for a given month. The forecast error for a given month, averaged over roughly 30 evaluation cases, is then reported.

The competing models used in this comparison are ARMA(p, q), ARMA$^*(p, q)$, and the persistence model. ARMA$^*(p, q)$ is the same as ARMA(p, q), except that the analysis is performed on the residuals after removing a diurnal trend using Eq. 2.15. As seen in Chapter 2, a small time lag usually suffices to capture the temporal dependency. For the datasets used in this section, the partial autocorrelation of lag 1 is dominant, suggesting $p = 1$. Using BIC would select $p = 1$ and $q = 2$ for most of the cases. So the model order in the ARMA model is set as $p = 1$ and $q = 2$. When evaluating the ARMA models, another random sampling is applied to the 30 evaluation cases mentioned above, further reducing the number of runs to about 25% of what is used for GSTAR.

Table 3.1 presents the forecasting results of GSTAR and the comparison with the two versions of ARMA models and the persistence model. GSTAR, on average, outperforms the other three methods, indicating the benefit of incorporating the spatial dependency information. Interestingly, in this comparison, the persistence model wins over the ARMA models.

Table 3.2 shows some results using CRPS to give a sense of the quality of predictive distribution. Forty turbines are randomly chosen, to which the GSTAR and ARMA(1,2) are applied. Please note that here CRPS is computed for power response, meaning that the integration is conducted over y; please refer to Eq. 5.23.

In practice, the optimal value of ξ used in PCE may change over time and a variation of ξ around 0.73 can be expected. A sensitivity analysis is conducted, which is to change ξ between 0.6 and 0.8, and then average the PCE over this range. One hundred turbines are randomly chosen and the 2009 data are used in this analysis. Table 3.3 shows that the performance of the GSTAR model is reasonably robust when ξ is around 0.73.

TABLE 3.1 Forecasting results for 2009 and 2010 using PCE. The values in parentheses are the standard deviations of the corresponding forecasting. The row of "Imp. over PER" shows the improvement of GSTAR over PER in percentage.

	2-hour	3-hour	4-hour	5-hour
		2009		
PER	0.0614(0.0159)	0.0741(0.0184)	0.0857(0.0215)	0.0943(0.0212)
ARMA(1,2)	0.0663(0.0375)	0.0826(0.0386)	0.0844(0.0473)	0.0991(0.0463)
ARMA*(1,2)	0.0752(0.0366)	0.0917(0.0421)	0.1002(0.0485)	0.1038(0.0486)
GSTAR(1)	0.0608(0.0297)	0.0716(0.0318)	0.0816(0.0327)	0.0884(0.0321)
Imp. over PER	1.1%	3.3%	4.8%	6.3%
		2010		
PER	0.0484(0.0137)	0.0572(0.0160)	0.0644(0.0185)	0.0698(0.0208)
ARMA(1,2)	0.0650(0.0398)	0.0779(0.0437)	0.0783(0.0394)	0.0794(0.0400)
ARMA*(1,2)	0.0690(0.0386)	0.0823(0.0418)	0.0838(0.0460)	0.0857(0.0380)
GSTAR(1)	0.0477(0.0212)	0.0569(0.0231)	0.0630(0.0260)	0.0692(0.0277)
Imp. over PER	1.5%	0.5%	2.1%	0.8%

TABLE 3.2 CRPS values using forty randomly selected turbines and 2009 data.

	2-hour	3-hour	4-hour	5-hour
ARMA(1,2)	0.1538	0.1452	0.1496	0.1559
GSTAR	0.1243	0.1299	0.1378	0.1467

TABLE 3.3 Average PCE while ξ varying in $[0.6, 0.8]$ for 100 turbines using the data of 2009. The values in parentheses are the standard deviations.

	2-hour	3-hour	4-hour	5-hour
PER	0.0616(0.0122)	0.0731(0.0220)	0.0855(0.0327)	0.0937(0.0286)
GSTAR	0.0628(0.0235)	0.0723(0.0332)	0.0835(0.0364)	0.0900(0.0357)

3.3 SPATIO-TEMPORAL ASYMMETRY AND SEPARABILITY

3.3.1 Definition and Quantification

One of the key assumptions made in the GSTAR model is that the spatio-temporal dependency structure, $a_{ij\ell}$, can be expressed as the product of a spatial part and a temporal part; please refer to Eq. 3.19. Generally, a covariance structure is said to be *separable* if its covariance function can be factored into the product of purely spatial and purely temporal components such that $C(\mathbf{u}, h) = C^s(\mathbf{u}) \cdot C^t(h)$. This assumption of spatio-temporal separability is in fact rather common in spatio-temporal analysis [43] because separable spatio-temporal models are easier to be dealt with mathematically.

Assuming separability suggests the lack of interaction between the spatial and temporal components and implies full symmetry in the spatio-temporal covariance structure, which brings up the concept of *spatio-temporal symmetry*. A covariance structure is symmetric if

$$C(\mathbf{s}_1, \mathbf{s}_2; t_1, t_2) = C(\mathbf{s}_1, \mathbf{s}_2; t_2, t_1). \tag{3.31}$$

This is to say that the correlation between sites $\mathbf{s}_1$ and $\mathbf{s}_2$ at times t_1 and t_2 is the same as that between $\mathbf{s}_1$ and $\mathbf{s}_2$ at times t_2 and t_1. For a stationary covariance function, this can be written as $C(\mathbf{u}, h) = C(-\mathbf{u}, h) = C(\mathbf{u}, -h) = C(-\mathbf{u}, -h)$ [72]. Separability is a stronger condition. It can be shown that a separable spatio-temporal covariance structure must have symmetry but the converse is not necessarily true, meaning that a symmetric covariance structure may or may not be separable [74].

To quantify asymmetry, Stein [204] proposes a metric in terms of spatio-temporal semi-variograms. The spatio-temporal empirical semi-variogram of $V_i(t)$ between site $\mathbf{s}_1$ and site $\mathbf{s}_2$ at time lag h is defined as,

$$\varpi(\mathbf{s}_1, \mathbf{s}_2; h) = \frac{1}{2(n - h - 1)} \sum_{j=1}^{n-h-1} [V_1(t_j + h) - V_2(t_j)]^2. \tag{3.32}$$

Then, introduce two semi-variograms between $\mathbf{s}_1$ and $\mathbf{s}_2$: $\varpi(\mathbf{s}_1, \mathbf{s}_2, h)$ and $\varpi(\mathbf{s}_2, \mathbf{s}_1, h)$. Both of them represent the dissimilarity between the two spatial sites, but $\varpi(\mathbf{s}_1, \mathbf{s}_2, h)$ means that measurements taken at $\mathbf{s}_2$ are h time lag behind that at $\mathbf{s}_1$, whereas $\varpi(\mathbf{s}_2, \mathbf{s}_1, h)$ means that measurements at $\mathbf{s}_1$ are behind those at $\mathbf{s}_2$. A quantitative asymmetry metric can be thus defined as the difference between the two semi-variograms, namely

$$asym(\mathbf{s}_1, \mathbf{s}_2, h) := \varpi(\mathbf{s}_1, \mathbf{s}_2, h) - \varpi(\mathbf{s}_2, \mathbf{s}_1, h). \tag{3.33}$$

When the two semi-variograms are the same, the wind field is said to be symmetric. But when there is a dominant wind blowing from $\mathbf{s}_1$ towards $\mathbf{s}_2$, the propagation of wind from $\mathbf{s}_1$ towards $\mathbf{s}_2$ would generate a significantly positive value for $asym$, indicating a lack of symmetry. To signify the dominant wind direction, denoted by ϑ, the asymmetric metric is also expressed as $asym(\mathbf{s}_1, \mathbf{s}_2, h, \vartheta)$.

3.3.2 Asymmetry of Local Wind Field

The space-time symmetry assumption is not universally valid, and it is especially not true in many geophysical processes, such as wind fields, in which the prevailing air flow, if existing, causes the correlation in space and time stronger in one direction than other directions, thus invalidating the symmetry assumption. To see this, let us look at Fig. 3.4, left panel. Consider two sites and a wind flow primarily from s_1 to s_2. Let $t_1 = t$ and $t_2 = t+k$, $k > 0$. Were the assumption of symmetry true, it meant that $C(s_1, s_2; t, t+k) = C(s_1, s_2; t+k, t)$. The left-hand side covariance, $C(s_1, s_2; t, t + k)$, dictates how much information at s_1 and t is there to help make predictions at a down-wind site s_2 and a future time $t + k$. A significant $C(s_1, s_2; t, t + k)$ suggests that the upstream wind measurements at t help with the downstream wind prediction at $t + k$. This makes perfect sense, considering that wind goes from s_1 to s_2. The assumption of symmetry, were it true, says that the right-hand side covariance, $C(s_1, s_2; t + k, t)$, is equally significant, meaning that the downstream wind measurements at t could help with the upstream prediction at $t + k$, as much as the upstream helps the downstream. This no longer makes sense.

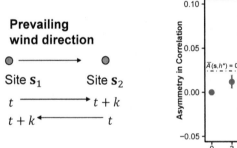
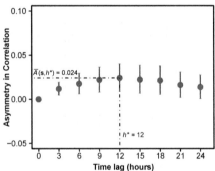

FIGURE 3.4 Under a dominant air flow, the covariance structure of the underlying wind field may become asymmetric. (Right panel reprinted with permission from Ezzat et al. [59].)

While studying large-scale atmospheric processes, analysts have in fact noted that when there exists a dominant air or water flow in the processes, the resulting random field does not have a symmetric covariance structure [42, 72, 114, 204, 225]. The question is—does this lack of symmetry phenomenon also take place on a small-scale wind field as compact as a wind farm?

Ezzat et al. [59] set out to investigate this question for the wind field on a farm. In their analysis, the diurnal trend for wind speed is first fitted using Eq. 2.15 to remove nonstationarity in the wind data. The fitted trend is then subtracted from the actual wind speed data and the residuals are subsequently used for quantifying asymmetry. Using the `Wind Spatio-Temporal Dataset2`, the yearly average wind direction is estimated as $\bar{\vartheta} = 264.24°$ (due west is $270°$). Because of this, for every pair of turbines i and j such that $\mathbf{s}_i$ is west of $\mathbf{s}_j$, Ezzat et al. compute $\varpi(\mathbf{s}_i, \mathbf{s}_j, h) - \varpi(\mathbf{s}_j, \mathbf{s}_i, h)$ using the residuals in place of V in Eq. 3.32. This computation is repeated for every pair of turbines and for different time lags ranging from 0 to 24 hours. All of the computed quantities are then transformed into the correlation scale. For the ℓ-th pair of turbines, the resulting quantity at each temporal lag h is the spatio-temporal asymmetry, $asym^\ell(\mathbf{s}_i, \mathbf{s}_j, h, \bar{\vartheta})$.

Denote the collection of asymmetry values at each temporal lag by $A(\mathbf{s}, h) = \{asym^\ell(\mathbf{s}_i, \mathbf{s}_j, h, \bar{\vartheta})\}_{\ell=1}^{\mathfrak{L}}$, where $\mathfrak{L}$ is the total number of turbine pairs. Represent by $\bar{A}(\mathbf{s}, h)$ the 50-th percentile of this collection. Fig. 3.4, right panel, presents the 25-th, 50-th and 75-th percentiles of $A(\mathbf{s}, h)$ for $h \in \{0, \dots, 24\}$ with a three-hour increment. On the one hand, all median asymmetry values in Fig. 3.4, right panel, are slightly positive, indicating a potential tendency towards spatio-temporal asymmetry. On the other hand, the largest median occurs at $h^* = 12$ and is approximately 0.024 on the correlation scale. To put this value in perspective, please note that Gneiting [72] reports a value of 0.12 for asymmetric large-scale wind flow over Ireland. The values of asymmetry reported in [74] range between 0.04 and 0.14, and are averaged at 0.11. Relative to those levels, an asymmetry of 0.024 appears to be rather weak to justify the existence of asymmetry in the local wind field. Analysts would understandably trade such weak asymmetry for computational efficiency and model simplicity gained by making the symmetry assumption. This may explain why separable, symmetric models are dominant in the wind application literature.

On the surface, the above analysis appears to indicate that there does not exist significant asymmetry in a local wind field within an area as compact as a wind farm. Ezzat et al. [59] believe that the weak asymmetry is due to the non-optimal handling of wind farm data, especially in terms of its temporal handling. When producing the right panel of Fig 3.4, the wind data is grouped for the whole year. Ezzat et al. test different temporal resolutions like monthly or weekly. Under the finer temporal resolutions, the asymmetric level indeed increases but still not much. Ezzat et al. hypothesize that a special spatio-temporal "lens" is needed to observe the wind data in order to detect strong degrees of asymmetry in a local wind field. This makes intuitive sense. In a large-scale atmospheric process, a dominant wind can persist for a sustained period of time and travel a substantial distance. These patterns can be pre-identified through climatological expertise over a region of interest, and as such, regular calendar decompositions, like weekly, monthly, seasonal, or yearly, appear to be reasonable choices. For a local wind field, however, observational data suggest that alternations in local winds occur at a relatively high rate, resulting in several distinct wind characteristics at each wind alternation. In such settings, regular calendar periods rarely contain a single dominant wind scenario. Rather, they contain various dominant winds that create multiple asymmetries having distinct directions and magnitudes. Consequently, aggregating the heterogeneous, and perhaps opposite, asymmetries leads to an underestimation of the true asymmetry level.

3.3.3 Asymmetry Quantification

The physical differences between local wind fields and large-scale atmospheric processes require special adjustments to the spatio-temporal resolution used to analyze wind measurements, in order to reveal the underlying asymmetry pattern. Ezzat et al. [59] devise a special lens consisting of two components—a temporal adjustment and a spatial adjustment.

The main reason that temporal aggregations based purely on calendar periods are not going to be effective is because such decomposition intervals are created arbitrarily. Hence, one key step for a successful temporal adjustment is to isolate the time intervals in which a unique dominant wind persists—such intervals are referred to as the *prevailing periods*, and detecting them is basically solving a change-point detection problem.

A binary segmentation version of the circular change-point detection [106] is used to detect the change points in wind direction. The R package `circular` is used to facilitate the task. The change-point detection method is applied to the wind direction data measured at one of the masts. Fig. 3.5 presents the detected change points for two weeks of the wind direction data, for the sake of illustration. For the whole year, a dominant wind direction lasts, on average, for 3.04 days with a standard deviation of 2.46 days. For 50% of the prevailing periods, the wind direction alternates in less than 2.27 days. The maximum interval of time in which a dominant wind direction is found to be persistent is 15.5 days, while the shortest length of a prevailing period is found to be 6 hours. These statistics indicate a fast dynamics and unpredictable nature in wind direction change, explaining why a typical calendar period-based approach is ineffective. A total of 119 change points are detected in the yearlong wind direction data, leading to 120 prevailing periods identified over the year. For the ℓ-th prevailing period, the dominant wind direction is denoted by ϑ_ℓ.

On the spatial level, the relative position of the turbines on a wind farm is another factor that affects the asymmetry level at a given time. Physically, asymmetry exists when wind propagates from an upstream turbine to a downstream one, implying that the latter is in the along-wind direction with respect to the former. Therefore, the spatial adjustment is to select only the along-wind turbines for asymmetry quantification.

A spatial bandwidth, denoted by b_ℓ, is to be selected for the ℓ-th prevailing period. The specific procedure is executed as follows: vary the bandwidth in the range $[2.5°, 45°]$ in increments of $2.5°$ and then select the bandwidth that maximizes the median asymmetry and denote that choice as the optimal bandwidth b_ℓ^*. With the spatial adjustment, the asymmetry metric, $asym(\cdot)$, is now denoted as $asym(\mathbf{s}_1, \mathbf{s}_2, h_\ell, \vartheta_\ell, b_\ell)$. Finally, an optimal time lag h_ℓ^* is

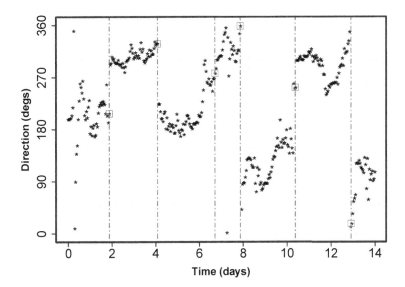

FIGURE 3.5 Change points detected in the first two weeks of wind direction data. The vertical dashed lines indicate the change points. (Reprinted with permission from Ezzat et al. [59].)

chosen to maximize the median asymmetry level in each prevailing period when spatial and temporal parameters are set at b_ℓ^* and ϑ_ℓ, respectively.

Under the parameter setting, $h_\ell^*, \vartheta_\ell, b_\ell^*$, the asymmetric metric $asym(\cdot)$ is computed using the Wind Spatio-Temporal Dataset2. Fig. 3.6 presents the 25-th, 50-th and 75-th percentiles of the asymmetry level versus the separating distance subgroups for the different scenarios thus considered: yearly, seasonal, monthly, weekly, temporal-only lens scenario, and spatio-temporal lens scenario. It is apparent that applying the spatio-temporal lens detects much higher asymmetry levels. For instance, at separating distances greater than 20 km, all of the turbine pairs exhibit positive asymmetry and 50% of them exhibit an asymmetry level higher than 0.2 on the correlation scale, a level considered significant in the past study [72] and nearly an order of magnitude greater than the median asymmetry of 0.024 detected earlier on the yearly data.

Table 3.4 classifies the median asymmetry values of all distance subgroups, where 93% of the prevailing periods exhibit positive median asymmetry, nearly a quarter of them exhibit a greater than 0.2 median asymmetry, and more than 41% of them exhibit a median asymmetry larger than 0.1, the level of asymmetry previously reported in [72, 74] for signaling the existence of ap-

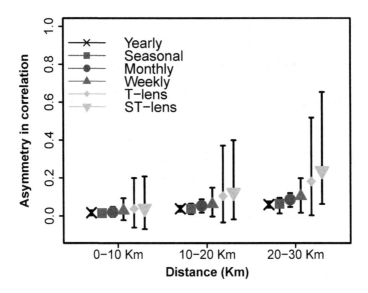

FIGURE 3.6 The 25-th, 50-th and 75-th percentiles of asymmetry values of different scenarios versus separating distance in kilometers. "T-lens" means temporal adjustment only, whereas "ST-lens" means spatio-temporal adjustments. (Reprinted with permission from Ezzat et al. [59].)

preciable asymmetric behavior in the large-scale atmospheric processes. The findings suggest that not only does strong asymmetry exist in local wind fields, but also the discovered asymmetry appears to fluctuate spatially and temporally in both magnitude and direction. Each prevailing period appears to have a unique asymmetry pattern, creating a temporal fluctuation of asymmetry throughout the year.

3.3.4 Asymmetry and Wake Effect

The implication of capturing the asymmetry in a local wind field can enrich the understanding of complex physical phenomena on a wind farm such as the wake effect. The spatio-temporal dynamics within a wind farm are affected by the wake effect because the rotating turbine blades cause changes in the speed, direction and turbulence intensity of the propagating wind [40]. For each prevailing period, Ezzat et al. [59] divide the whole farm, based on the wind direction, into two regions having approximately the same number of turbines. The first region is the set of wake-free wind turbines that receive

TABLE 3.4   Classification of prevailing periods according to
the median asymmetry level.

Group	Range	Percentage
1.	$\bar{A}(\mathbf{s}_1, \mathbf{s}_2, h_\ell^*, \vartheta_\ell, b_\ell^*) \leq 0$	7%
2.	$0 < \bar{A}(\mathbf{s}_1, \mathbf{s}_2, h_\ell^*, \vartheta_\ell, b_\ell^*) < 0.05$	27%
3.	$0.05 \leq \bar{A}(\mathbf{s}_1, \mathbf{s}_2, h_\ell^*, \vartheta_\ell, b_\ell^*) < 0.1$	25%
4.	$0.1 \leq \bar{A}(\mathbf{s}_1, \mathbf{s}_2, h_\ell^*, \vartheta_\ell, b_\ell^*) < 0.2$	20%
5.	$0.2 \leq \bar{A}(\mathbf{s}_1, \mathbf{s}_2, h_\ell^*, \vartheta_\ell, b_\ell^*)$	21%

Source: Ezzat et al. [59]. With permission.

less turbulent wind, whereas the second region is the set of wind turbines which are in the wake of other turbines and receive the disturbed, turbulent wind. Fig. 3.7 plots the medians of the asymmetry for each region. The wake-free region appears to exhibit stronger asymmetry, which is consistent with the physical understanding since the less-turbulent wind is the driving force creating the asymmetry. This analysis indicates that the asymmetry level spatially varies on a wind farm due to the wake effect. Incorporating such patterns in a spatio-temporal model could benefit modeling and prediction, as well as aid research in wake characterization.

3.4   ASYMMETRIC SPATIO-TEMPORAL MODELS

3.4.1   Asymmetric Non-separable Spatio-temporal Model

Consider a simple spatio-temporal model, the counterpart of the ordinary kriging in Eq. 3.8, such as

$$V_i(\ell) = \beta_0 + \delta_i(\ell), \quad i = 1, \ldots, N, \text{ and } \ell = t, t-1, \ldots, t-n, \qquad (3.34)$$

where β_0 is the unknown constant, like in Eq. 3.8. Unlike Eq. 3.8, which has two random terms, the i.i.d noise term ε is absorbed into the spatio-temporal random field term $\delta_i(\ell)$ here.

The key in spatio-temporal modeling, as mentioned at the beginning of this chapter, is to specify the covariance function for the spatio-temporal random field term, $\delta_i(\ell)$. The specific asymmetric non-separable spatio-temporal model presented here is a modified version of that proposed in [74], in which the asymmetric, non-separable covariance function is expressed as follows,

$$C_{\text{ASYM}}(\mathbf{u}, h) = \sigma_{\text{ST}}^2 \left\{ (1 - \varphi)\rho_{\text{NS}}(\mathbf{u}, h) + \varphi\rho_{\text{A}}(\mathbf{u}, h) \right\} + \eta \mathbb{1}_{\{\|\mathbf{u}\| = |h| = 0\}}, \quad (3.35)$$

where ρ_A is an asymmetric correlation function to be given below and ρ_{NS} is a non-separable symmetric correlation function such that

$$\rho_{\text{NS}}(\mathbf{u}, h) = \frac{1 - \tau}{1 + \zeta|h|^2} \left(\exp\left[-\frac{\phi\|\mathbf{u}\|}{(1 + \zeta|h|^2)^{\frac{\beta}{2}}} \right] + \frac{\tau}{1 - \tau} \mathbb{1}_{\{\|\mathbf{u}\| = 0\}} \right). \quad (3.36)$$

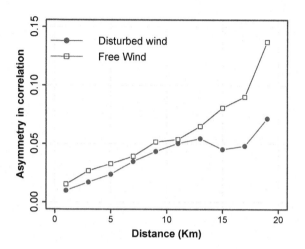

FIGURE 3.7 Wake effect and its implication on spatio-temporal asymmetry. (Reprinted with permission from Ezzat et al. [59].)

In Eq. 3.35 and Eq. 3.36,

- ζ and ϕ are, respectively, the temporal and spatial scale parameters,

- τ and η are, respectively, the spatial and spatio-temporal nugget effects (i.e., i.i.d random noise),

- σ_{ST}^2 is the spatio-temporal variance,

- β is the non-separability parameter, characterizing the strength of the spatio-temporal interaction, and

- φ is the asymmetry parameter, characterizing the lack of symmetry.

- The valid ranges of these parameters are: $\tau \in [0, 1)$, $\beta \in [0, 1]$, $\varphi \in [0, 1]$, $\sigma_{ST}^2 > 0$, and ϕ, ζ and η are all non-negative.

The $\rho_A(\cdot, \cdot)$ defined in [74] is a Lagrangian compactly supported function,

$$\rho_A(\mathbf{u}, h) = \left(1 - \frac{1}{2\|\mathbf{U}\|}\|\mathbf{u} - \mathbf{U}h\|\right)_+, \tag{3.37}$$

where $\mathbf{U} = (U_1, U_2)^T$ is the two-dimensional velocity vector having a longitudinal component and a latitudinal component and to be defined based on the knowledge of the weather system. For example, if the dominant wind is known to be strictly westerly, then $\mathbf{U}$ is chosen to be $(U_1, 0)^T$, namely a non-zero longitudinal wind velocity reflecting the traveling of the wind along the longitudinal axis. A generalized version of ρ_A is proposed by Schlater in [195]. Instead of using a constant vector, Schlater defines $\mathbf{U}$ as a random variable that follows a multivariate normal distribution, i.e., $\mathbf{U} \sim \mathcal{N}(\boldsymbol{\mu}, \frac{\mathbf{D}}{2})$. As such, ρ_A is defined as,

$$\rho_A(\mathbf{u}, h) = \frac{1}{\sqrt{|\mathbf{1}_{2\times2} + h^2\mathbf{D}|}} \exp\left\{-(\mathbf{u}-\boldsymbol{\mu}h)^T(\mathbf{1}_{2\times2}+h^2\mathbf{D})^{-1}(\mathbf{u}-\boldsymbol{\mu}h)\right\}, \quad (3.38)$$

where $|\cdot|$ denotes the matrix determinant.

The asymmetric non-separable model used by Ezzat et al. [59] consists of the modeling components in Eq. 3.35, Eq. 3.36, and Eq. 3.38, and it is referred to hereinafter as ASYM.

3.4.2 Separable Spatio-temporal Models

By setting $\beta = \varphi = 0$ in ASYM, the asymmetric, non-separable model is reduced to a symmetric, separable model. Analysts could entertain two variants of the symmetric, separable model. The first variant is to take the parameters of ASYM after all of them are estimated but simply reset $\beta = \varphi = 0$. The second variant is to first set $\beta = \varphi = 0$ before parameter estimation and then freely estimate the remaining parameters from the data. Understandably, the second variant generally works better and is what is used in Section 3.5. This symmetric, separable model is referred to as SEP.

3.4.3 Forecasting Using Spatio-temporal Model

The short-term wind forecasting may benefit from using an asymmetric, separable spatio-temporal covariance structure. Once the covariance function is specified, the forecasting is conducted similarly as in the kriging method of Section 3.1.3.

Let us arrange the spatio-temporal wind speed, $V_i(t)$, into an $Nn \times 1$ vector, such as

$$\mathbf{V} = (V_1(t), \cdots, V_N(t), V_1(t-1), \cdots, V_N(t-1), \cdots, V_1(t-n), \cdots, V_N(t-n))^T.$$

The objective is to make a forecast at site $\mathbf{s}_0$ and time $t + h$, denoted by $V_0(t + h)$, which is an h-hour ahead forecast at $\mathbf{s}_0$.

A covariance matrix corresponding to $\mathbf{V}$ can be constructed by using the covariance function C_{ASYM} and is hence denoted by $\mathbf{C}_{ASYM}$. A covariance row vector, $\mathbf{c}_0$, can be constructed by treating its i-th element $(\mathbf{c}_0)_i$ as the covariance between $V_0(t + h)$ with the i-th element in $\mathbf{V}$. The notation of $\mathbf{c}_0$ bears the same meaning as the notation of $\mathbf{c}_{0N}$ used earlier in Section 3.1.3. Here we drop the subscript "N" because the size of $\mathbf{V}$ for this spatio-temporal process is no longer $N \times 1$ but $Nn \times 1$. Denote by $\sigma_0^2 := C_{ASYM}(0,0)$ the prior variance of the underlying spatio-temporal process. Similar to the kriging forecasting in Eq. 3.15, the forecast of $V_0(t + h)$ can be obtained as

$$\hat{V}_0(t + h) = \hat{\beta}_0 + \mathbf{c}_0 \mathbf{C}_{\text{ASYM}}^{-1}(\mathbf{V} - \hat{\beta}_0 \cdot \mathbf{1}),$$
$$Var(\hat{V}_0(t + h)) = \sigma_0^2 - \mathbf{c}_0 \mathbf{C}_{\text{ASYM}}^{-1} \mathbf{c}_0^T. \tag{3.39}$$

The flowchart in Fig. 3.8 presents the steps of the forecasting procedure. To perform an h-hour ahead forecast, only the data in the preceding prevailing period that share similar wind asymmetry characteristics are used for model training. This implies that a small subset of data relevant to the current prevailing period is used in the model training stage. The benefit of such an approach is two-fold. First, it eliminates the computational burden in fitting a complicated asymmetric, non-separable spatio-temporal model, because the data in the preceding prevailing period are usually limited to from a few hours to a few tens of hours, rather than weeks or months. Second, this approach makes use of a local informative spatio-temporal neighborhood that is most relevant to the short-term forecasting horizon. In this sense, it bears the similarity with the spatial informative neighborhood discussed in Section 3.2.2 or the temporal neighborhood used in [231].

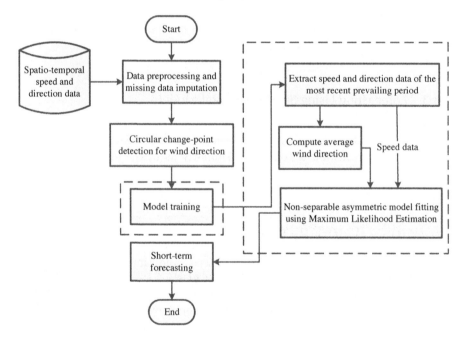

FIGURE 3.8 A flowchart that outlines the short-term forecasting based on a spatio-temporal model. (Reprinted with permission from Ezzat et al. [59].)

3.4.4 Hybrid of Asymmetric Model and SVM

A spatio-temporal model can be used together with some machine learning models to improve further the forecasting capability. Here, an ASYM is fit to the spatio-temporal wind training data, and then an SVM model is fit to the residuals obtained by ASYM, in order to capture any nonlinearities that are not covered by the base ASYM model. The final hybrid model has an additive form as

$$V_i(t) = V_i^{\mathrm{ASYM}}(t) + \mathcal{E}_i^{\mathrm{SVM}}(t) + \tilde{\varepsilon}_i(t), \tag{3.40}$$

where $V_i^{\mathrm{ASYM}}(t)$ is the ASYM model fit, $\mathcal{E}_i^{\mathrm{SVM}}(t)$ represents the SVM model fit to the spatio-temporal residuals after the ASYM model fit, and $\tilde{\varepsilon}_i(t)$ is the final residual term. This hybrid forecasting model is referred to as HYB.

3.5 CASE STUDY

In contrast to the situations where wind measurements come from a small number of locations spread over large areas, as in [75, 91, 208], the within-farm local wind field is much denser. Recall that the spatial and temporal resolutions of wind data in the `Wind Spatio-Temporal Dataset2` are one mile and one hour, respectively. The purpose of this case study is to demonstrate the existence of an asymmetric wind pattern in certain time periods and the benefit that a non-separable model may render in terms of short-term wind forecasting on such a compact wind field.

Four periods are chosen from different times in the `Wind Spatio-Temporal Dataset2`. For each of the four periods, six hours of data are used for model training. The choice for this short training period is motivated by observing that the shortest prevailing period length, as shown in Section 3.3.3, is about six hours. As such, a training period of six hours ensures temporal homogeneity and stationarity in the training data, allowing for reliable model estimation. Furthermore, for short-term wind forecasting, using a longer history of wind measurements is not necessarily helpful, as evident by the low time lag order used in the time series models in Chapter 2 or in the GSTAR model in Section 3.2.3.

In this study, forecasting is made for up to four hours ahead, i.e., $h=$ 1, 2, 3, or 4. A variety of forecasting models are studied and compared, including ASYM, SEP, the persistence model, a time-series model chosen as ARMA(1,1), an SVM model using a radial basis kernel function and the wind speeds measured at $t-1$, as well as an HYB that combines ASYM and SVM.

Although we by and large follow the numerical analysis conducted in Ezzat et al. [59], there are a couple of differences in treatment here leading to different numerical outcomes. But the main messages stay consistent with those advocated in Ezzat et al. [59].

This section employs a missing data imputation procedure, and as a result, the `Wind Spatio-Temporal Dataset2` does not have any missing data for wind speed. The power curve used here is a turbine-specific power curve,

TABLE 3.5 Log-likelihoods of asymmetric versus separable spatio-temporal models.

Period	ASYM	SEP
1.	−2090.900	−2091.294
2.	−2033.135	−2033.140
3.	−1800.352	−1800.702
4.	−2181.999	−2185.815

rather than a single power curve averaged for all the turbines. To specify $\boldsymbol{\mu}$ and $\mathbf{D}$ in ASYM, the most recent period is used as the training dataset. The speed and direction time-series data, recorded at one of the masts, is used to compute a time-series vector of wind velocities, along the longitudinal and latitudinal directions, respectively. The estimate of $\boldsymbol{\mu}$ is the sample average of the wind velocity vector, whereas the estimate of 2×2 matrix $\mathbf{D}$ is the sample covariance matrix of the horizontal and vertical velocities. This estimate of $\mathbf{D}$ is different from that in Ezzat et al. [59] but is the same as what is used in Chapter 4. This new estimate of $\mathbf{D}$ is used so that the ASYM model and its parameter estimation are consistent between Chapter 3 and Chapter 4.

The rest of the parameters in ASYM are estimated through a maximum likelihood estimation, implemented in R using the routine nlm. To appreciate the space-time coupling and asymmetry, take the prevailing period in January 2011 as an example. The non-separability parameter, $\hat{\beta} = 0.840$, and the asymmetry parameter, $\hat{\varphi} = 0.102$. These estimated values suggest that the underlying spatio-temporal process has space and time coupling and is asymmetric. When fitting the asymmetric model and its separable counterpart, i.e., ASYM and SEP, analysts can compare the respective log-likelihood values and observe which modeling option provides a better fit. Table 3.5 presents the log-likelihood values for ASYM and SEP model fits for all four periods. The numerical results show that ASYM has a higher log-likelihood value, albeit sometimes marginally so, than that of SEP.

In this study, two performance metrics are used—RMSE and MAE; for their definitions, please refer to Section 2.6. Tables 3.6 and 3.7 present the RMSE and MAE values for up to a 4-hour ahead forecast using the aforementioned temporal or spatio-temporal models. The aggregate measure reported is the average over all 4-hour ahead forecasts.

The results presented in Tables 3.6 and 3.7 show that the forecasts based on the asymmetric non-separable model outperform the competing methods considered in the study. The improvement of ASYM over the separable models is due to ASYM's capturing of the strong asymmetries, whereas its improvement over ARMA and SVM is mostly due to the characterization of spatial correlations as well as asymmetry, both of which the ARMA and SVM models fail to capture. Hybridizing ASYM with SVM (the HYB model) appears to achieve a further enhancement in forecasting accuracy over the ASYM only

TABLE 3.6   RMSE of wind speed forecasting. The percentage improvements are the error inflation rate relative to HYB.

Period	Method	$h = 1$	$h = 2$	$h = 3$	$h = 4$	Average	% Imp.
1	ASYM	0.993	1.441	2.853	3.122	2.289	3%
	SEP	1.070	1.727	3.242	3.469	2.582	14%
	PER	1.287	1.719	2.984	3.161	2.424	8%
	ARMA(1,1)	1.627	2.056	3.480	3.622	2.833	22%
	SVM	1.611	1.912	3.335	3.437	2.701	18%
	HYB	1.019	1.441	2.784	2.981	2.222	
2	ASYM	1.618	2.747	2.573	2.093	2.300	5%
	SEP	1.616	2.743	2.569	2.090	2.297	5%
	PER	1.832	2.877	2.569	2.075	2.374	8%
	ARMA(1,1)	1.986	3.054	2.781	2.222	2.547	14%
	SVM	2.543	3.777	3.531	2.977	3.243	33%
	HYB	1.585	2.667	2.438	1.874	2.184	
3	ASYM	0.897	0.946	1.078	1.390	1.095	0.2%
	SEP	0.900	1.184	1.269	1.654	1.281	15%
	PER	1.007	1.067	1.358	1.510	1.253	13%
	ARMA(1,1)	1.114	1.316	1.303	1.648	1.359	20%
	SVM	1.035	1.155	1.340	1.683	1.326	18%
	HYB	0.894	0.944	1.077	1.388	1.093	
4	ASYM	1.319	1.521	1.934	3.745	2.336	6%
	SEP	1.415	1.630	2.028	3.681	2.362	7%
	PER	1.880	2.096	2.526	5.281	3.248	33%
	ARMA(1,1)	2.070	1.769	2.144	3.809	2.575	15%
	SVM	1.806	1.859	2.392	4.375	2.810	22%
	HYB	1.239	1.422	1.942	3.446	2.191	

approach, demonstrating the additional benefit brought by the machine learning method. The improvements of HYB over ASYM for wind speed forecast range from 0.2% to 6%, and on average, 3.6%. Combining the strength of the asymmetrical modeling and machine learning, in terms of wind speed forecast, HYB improves, based on the average of the four periods, 10% in RMSE (12% in MAE, same below) over SEP, 16% (14%) over PER, 18% (20%) over ARMA(1,1), and 23% (24%) over SVM.

Measuring the performance metrics in terms of wind power, analysts can first make a wind speed forecast and then convert the wind speed to wind power, using the power curve as explained in Fig. 1.2. The nominal power curve is usually provided by the turbine manufacturer. To get more accurate representation of the actual power curve, the site-specific wind speed and wind power data can be used to estimate the turbine-specific power curve. The topic of estimating a power curve is the focus of Chapter 5. The specific procedure used here for power curve estimation is the binning method, the standard nonparametric method used in the wind industry [102]; for more details about the binning method, please refer to Chapter 5. Using the estimated power curves of individual turbines, analysts can predict the wind power generated at each turbine given the wind speed forecasts.

TABLE 3.7   MAE of wind speed forecasting. The percentage improvements are the error inflation rate relative to HYB.

Period	Method	$h = 1$	$h = 2$	$h = 3$	$h = 4$	Average	% Imp.
1	ASYM	0.846	1.248	2.636	2.912	1.911	4%
	SEP	0.919	1.540	3.046	3.273	2.194	16%
	PER	1.048	1.491	2.803	2.901	2.061	11%
	ARMA(1,1)	1.379	1.833	3.292	3.395	2.475	26%
	SVM	1.404	1.694	3.142	3.175	2.354	22%
	HYB	0.879	1.236	2.533	2.712	1.840	
2	ASYM	1.268	2.526	2.379	1.813	1.997	6%
	SEP	1.266	2.522	2.375	1.810	1.993	6%
	PER	1.489	2.552	2.265	1.749	2.013	7%
	ARMA(1,1)	1.615	2.806	2.520	1.894	2.209	15%
	SVM	2.308	3.485	3.211	2.610	2.904	35%
	HYB	1.232	2.442	2.240	1.599	1.878	
3	ASYM	0.729	0.773	0.906	1.224	0.908	0.4%
	SEP	0.736	1.017	1.110	1.476	1.085	17%
	PER	0.807	0.840	1.054	1.203	0.976	7%
	ARMA(1,1)	0.930	1.151	1.136	1.429	1.161	22%
	SVM	0.835	0.937	1.065	1.403	1.060	15%
	HYB	0.722	0.771	0.904	1.222	0.905	
4	ASYM	1.049	1.267	1.578	3.538	1.858	7%
	SEP	1.129	1.361	1.671	3.470	1.908	9%
	PER	1.488	1.711	2.060	4.782	2.510	31%
	ARMA(1,1)	1.668	1.437	1.757	3.503	2.091	17%
	SVM	1.469	1.525	1.934	3.968	2.224	22%
	HYB	0.968	1.180	1.566	3.226	1.735	

TABLE 3.8 RMSE of wind power forecasting. The percentage improvements are the error inflation rate relative to HYB.

Period	Method	$h = 1$	$h = 2$	$h = 3$	$h = 4$	Average	% Imp.
1	ASYM	0.090	0.140	0.326	0.383	0.265	3%
	SEP	0.092	0.171	0.372	0.415	0.295	13%
	PER	0.111	0.161	0.333	0.370	0.267	4%
	ARMA(1,1)	0.138	0.201	0.396	0.430	0.317	19%
	SVM	0.133	0.186	0.376	0.405	0.299	14%
	HYB	0.090	0.142	0.321	0.363	0.256	
2	ASYM	0.221	0.354	0.356	0.312	0.315	6%
	SEP	0.221	0.353	0.355	0.310	0.314	6%
	PER	0.252	0.368	0.346	0.293	0.318	7%
	ARMA(1,1)	0.282	0.404	0.387	0.325	0.353	16%
	SVM	0.389	0.525	0.509	0.450	0.471	37%
	HYB	0.216	0.341	0.332	0.276	0.295	
3	ASYM	0.116	0.094	0.105	0.146	0.117	1%
	SEP	0.102	0.116	0.124	0.177	0.133	13%
	PER	0.137	0.122	0.155	0.172	0.148	21%
	ARMA(1,1)	0.126	0.127	0.130	0.170	0.140	17%
	SVM	0.111	0.115	0.135	0.176	0.137	15%
	HYB	0.114	0.092	0.105	0.146	0.116	
4	ASYM	0.168	0.171	0.212	0.442	0.255	−0.3%
	SEP	0.170	0.175	0.226	0.431	0.256	0%
	PER	0.255	0.279	0.331	0.660	0.389	34%
	ARMA(1,1)	0.251	0.205	0.248	0.482	0.299	14%
	SVM	0.225	0.239	0.298	0.568	0.339	24%
	HYB	0.193	0.175	0.216	0.425	0.256	

Table 3.8 compares the competing models in terms of the RMSE of wind power prediction. Similar degrees of improvement of using the asymmetric, nonseparable model are observed in wind power prediction as in wind speed forecast. Specifically, the improvement of HYB over ASYM is up to 6%, and on average, 2.4%. Compared to other methods, HYB on average improves, in terms of reduction in RMSE, 8% over SEP, 17% over PER, 17% over ARMA(1,1), and 23% over SVM. These results are aligned with the findings made in Section 3.3 that local wind fields can be strongly asymmetric at the fine-scale spatio-temporal resolutions. Spatio-temporal models that capture such physical phenomena are expected to enhance short-term forecasting.

GLOSSARY

ARD: Automatic relevance determination

ARMA: Autoregressive moving average

ASYM: Asymmetric, non-separable spatio-temporal model

BFGS: Broyden-Fletcher-Goldfarb-Shanno optimization algorithm

cdf: Cumulative distribution function

CRPS: Continuous ranked probability score

GSTAR: Gaussian spatio-temporal autoregressive model

HYB: Hybrid model combining ASYM and support vector machine

i.i.d: Identically, independently distributed

MAE: Mean absolute error

PCE: Power curve error

PER: Persistence forecasting

RMSE: Root mean squared error

SE: Squared exponential covariance function

SEP: Separable spatio-temporal model

SVM: Support vector machine

EXERCISES

3.1 In the machine learning literature, if a prediction mechanism can be expressed as $\hat{\mathbf{V}} = \mathbf{S}\mathbf{V}$, it is called a linear smoother, where $\mathbf{S}$ is the smoother matrix. It is also established that the effective number of parameters in a linear smoother is $\text{tr}(\mathbf{S})$. In the following, to make things simpler, assume $\beta_0 = 0$. Consider a total of N data pairs in the training set:

   a. Show that the kriging predictor in Eq. 3.15 is a linear smoother.

   b. Show that the effective number of parameters in a kriging predictor is

   $$\sum_{i=1}^{N} \frac{\lambda_i}{\lambda_i + \hat{\sigma}_\varepsilon^2},$$

   where λ_i's, $i = 1, \ldots, N$, are the eigenvalues of $\mathbf{C}_{NN}$.

   c. Show that for a kriging predictor without the nugget effect, its effective number of parameters is N, the same as that of the data points in the training set. What does this tell you about the difference between a linear regression predictor and a kriging predictor (i.e., a Gaussian process regression)?

3.2 When we discuss the support vector machine formulation (2.47), we state (page 44) that "SVM regression can be made equivalent to Gaussian process regression, if (a) the loss function uses a squared error loss function, (b) $\gamma/2$ is set to σ_ε^2, which is the variance of the i.i.d noise term, (c) when the kernel function, $K(\cdot, \cdot)$, is set to be a covariance function." Please show that this is true.

3.3 When the kriging model in Eq. 3.8 has no nugget effect term, then it is said that the process has noise-free observations. Under that circumstance, the kriging predictor has an interpolation property, which means $\hat{V}(\mathbf{s}_i) = V(\mathbf{s}_i)$, if $\mathbf{s}_i$ is in the training set.

a. Prove the interpolation property.

b. Suppose that an underlying true function is $g(x) = e^{-1.4x}\cos(7\pi x/2)$, and seven training data pairs $\{x, y\}$ are taken from the curve, which are, respectively,

$$\{0.069, 0.659\}, \{0.212, -0.512\}, \{0.355, -0.440\}, \{0.498, 0.344\},$$
$$\{0.641, 0.294\}, \{0.783, -0.229\}, \{0.926, -0.199\}.$$

Please use this set of data and the ordinary kriging model without the nugget effect to construct the predictive function $\hat{g}(x)$. Plot both $g(x)$ and $\hat{g}(x)$ with the seven data points marked. Observe whether the kriging predictor interpolates the training data points.

3.4 Take one month of 10-min wind speed data and wind power data from the Wind Time Series Dataset. Treat the wind speed data as x and the wind power data as y. Fit an ordinary kriging model. Use the squared exponential covariance function. Please generate a plot with the original data points, the mean prediction line, and the two standard deviation lines.

3.5 Please generate one-dimensional sample functions using a power exponential function for the following parameter combinations:

a. $\theta = 5$, $\sigma_V^2 = 1$, $p = 2$.

b. $\theta = 1$, $\sigma_V^2 = 0.1$, $p = 2$.

c. $\theta = 5$, $\sigma_V^2 = 1$, $p = 1$.

d. $\theta = 1$, $\sigma_V^2 = 0.1$, $p = 1$.

e. $\theta = 5$, $\sigma_V^2 = 1$, $p = 1.5$.

f. $\theta = 1$, $\sigma_V^2 = 0.1$, $p = 1.5$.

3.6 Complete the following:

a. Derive Eq. 3.12.

b. Derive Eq. 3.14.

c. Derive the log-likelihood function in Eq. 3.16.

d. Given the universal kriging model in Eq. 3.17, find its log-likelihood function.

3.7 Use the 2009 data in the `Wind Spatio-Temporal Dataset1` and compute the pairwise sample correlation between any two turbines. Then plot the correlation against the distance between the two turbines, in which the horizontal axis is the between-turbine distance and the vertical axis is the correlation in its absolute value in $[0, 1]$.

3.8 Derive Eq. 3.23.

3.9 Use the data of January 2009 from the `Wind Spatio-Temporal Dataset1` and select a target site. Try different values of κ and see how it affects the resulting informative neighborhood.

3.10 Derive the α-quantile of the truncated normal distribution in Eq. 3.27.

3.11 Use the `Wind Spatio-Temporal Dataset2` and group the data for a month. Compute the asymmetry level for any pair of turbines for that month under its specific average wind direction. Repeat this for each month in the yearlong dataset and group the asymmetry values based on their corresponding time lags. Create a plot similar to the right panel of Fig. 3.4.

3.12 In Eq. 3.5, when $p = 1$, we say that the resulting covariance function is an exponential covariance function, which reads, if assuming isotropy,

$$C_{\text{Exp}}(\mathbf{u}) = \sigma_V^2 \exp\left\{-\frac{\|\mathbf{u}\|_1}{2\theta}\right\} = \sigma_V^2 \exp\left\{-\frac{|u_1| + \cdots + |u_d|}{2\theta}\right\},$$

where $\mathbf{u}$ is assumed to have d elements. But there is another definition of the exponential covariance function, which uses a 2-norm inside the exponential to measure distances, namely

$$C_{\text{Exp}}(\mathbf{u}) = \sigma_V^2 \exp\left\{-\frac{\|\mathbf{u}\|_2}{2\theta}\right\} = \sigma_V^2 \exp\left\{-\frac{\sqrt{u_1^2 + \cdots + u_d^2}}{2\theta}\right\}.$$

a. Explain under what condition the covariance function, $C_{\text{ASYM}}(\mathbf{u}, h)$, is the same as $C_{\text{Exp}}(\mathbf{u})$ with the 2-norm distance.

b. Consider a separable spatio-temporal covariance function, $C(\mathbf{u}, h)$, that is constructed by the product of exponential covariance functions for both the spatial and temporal components, i.e.,

$$C(\mathbf{u}, h) = C_{\text{Exp}}(\mathbf{u}) \cdot C_{\text{Exp}}(h).$$

How is this separable covariance function, $C(\mathbf{u}, h)$, different from $C_{\text{ASYM}}(\mathbf{u}, h)$ when β, φ, and τ are set to zero?

3.13 Use the `Wind Spatio-Temporal Dataset2` to conduct the following studies.

a. Use the `circular` package to conduct a change-point detection on the yearlong wind direction measured on one of the met masts, and see how many change points you detect. Suppose that k change points are detected, then it leads to $k+1$ prevailing periods.

b. Calculate the asymmetry level for each one of the prevailing periods and tabulate the results in a fashion similar to Table 3.4.

c. Select a period in which the asymmetry is weak (smaller than 0.05) and make sure that its overall duration is longer than ten hours. Then, fit an ASYM model and an SEP model using the first six hours of data. Compare the common model parameters and the log-likelihood of the two models.

d. Use the next four hours of data to conduct an h-hour ahead forecasting for $h = 1, 2, 3, 4$. Compare ASYM and SEP using both RMSE and MAE.

CHAPTER 4

Regime-switching Methods for Forecasting

DOI: 10.1201/9780429490972-4

O ne particular class of wind forecasting methods worth special attention is the regime-switching approach. We hence dedicate this chapter to the discussion of regime-switching methods.

The motivation behind the regime-switching approach is to deal with non-stationarity in wind dynamics—in wind speed, in wind direction, or in spatial correlation. Recall that the spatio-temporal covariance structures introduced in Chapter 3 are all stationary in nature. While nonstationary covariance structures do exist, using them is not easy. Analysts find that a simpler approach is to compartmentalize the nonstationary variables into a finite number of disjoint intervals, each of which is referred to as a regime. Within a regime, the underlying wind process is assumed stationary. To account for the overall nonstationarity, a mechanism is needed for the forecasting model to transition from one regime to another, as the underlying wind process is progressing. The resulting approach is called regime-switching. In essence, a regime-switching method is a collection of distinct, and most often linear, models.

The regime-switching mechanism can be used with a temporal only process, considering only nonstationarity in time, or with a spatio-temporal process, considering nonstationarity in both space and time.

4.1 REGIME-SWITCHING AUTOREGRESSIVE MODEL

Suppose that analysts pre-define a number of regimes, indexed from 1 to R, and denote the wind regime at time t by $r(t) \in \{1, ..., R\}$, which is known as the *regime variable*. The regime-switching autoregressive (RSAR) model [234] is a collection of R autoregressive models, each of which is associated with a wind regime and thus uses a set of parameters peculiar to that regime to produce regime-dependent forecasts.

In an RSAR, the wind speed, $V(t)$, at time t and in regime $r(t)$ is modeled as an AR model of order $p^{r(t)}$ using a set of regime-dependent parameters $\{a_0^{r(t)}, a_1^{r(t)}, \ldots, a_j^{r(t)}, \ldots\}$, such as

$$V(t) = a_0^{r(t)} + \sum_{j=1}^{p^{r(t)}} a_j^{r(t)} V(t-j) + \varepsilon(t), \qquad (4.1)$$

where $\varepsilon(t)$ is a zero-mean, normally distributed, i.i.d random noise whose variance can be regime-dependent. In this section, the value of regime variable, $r(t)$, is determined based on the observed values of wind speed. Be aware that $r(t)$ can be decided using other explanatory variables, including, but not limited to, wind direction or temperature [75, 176].

Estimating the parameters for a regime-switching autoregressive model is usually conducted for each individual AR model separately. The procedure, model selection criteria, and model diagnostics, as explained in Section 2.4, can be used here without much modification. Zwiers and von Storch [234] note a number of differences in handling a bunch of AR models, as opposed to handling a single AR model, summarized below.

- One word of caution is on ensuring that each regime should have a sufficient amount of data for parameter estimation. This aspect is less problematic nowadays with much advanced data collection capability in commercial wind operations. Data appear to be more than enough even after being divided into a number of regimes. The data amount sufficiency could have been an issue 30 years ago.

- An analyst can choose to use an aggregated AIC to decide the overall model order for the regime-switching method. This practice becomes less popular, as analysts nowadays rely more on computational procedures that split the data into training and test sets, like in cross validation, to test on a model's forecasting performance and to adjust respective modeling decisions.

- As mentioned above, $\varepsilon(t)$ could have different variances in different regimes. An implication is that analysts should pay attention to the heteroscedasticity issue (i.e., different variances) when devising a statistical test. For more discussion, please refer to page 1351 in [234].

The use of an RSAR for forecasting is fairly straightforward. Analysts first identify either the current wind regime, per definition given below, or the regime anticipated in the forecasting horizon, select the AR model corresponding to the target regime, and then make forecasts using the selected AR model, as one would have while using a single AR model.

4.1.1 Physically Motivated Regime Definition

In a regime-switching method, here as well as in the methods introduced in the sequel, one crucial question is how to decide the number of wind regimes

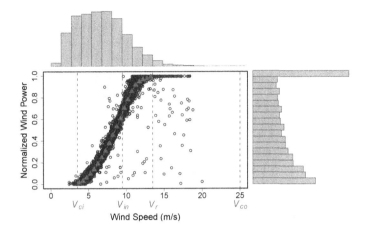

FIGURE 4.1 Normalized wind power versus wind speed. V_{ci}: cut-in speed, V_{in}: inflection point, V_r: rated speed and V_{co}: cut-out speed. On the top and right sides are the histograms of wind speed and power, respectively. The circle dots are raw wind data. (Reprinted with permission from Ezzat et al. [60].)

and the boundaries dividing these regimes. Consider R disjoint wind speed regimes, denoted by $\{r_1, r_2, \ldots, r_R\}$, such that $V(t)$ belongs to one and only one of the R wind regimes. Each regime, r_k, is defined by an interval $[u_k, v_k)$, such that u_k and v_k are the boundary values for r_k, with $u_1 = 0$ and $u_{k+1} = v_k$.

One approach is to pre-define the wind regimes based on physical understanding. We guide the selection of wind speed regimes in light of the regions associated with a wind power curve. Fig. 4.1 plots the wind speed against the normalized wind power recorded at one of the turbines for one year's worth of data in the `Wind Spatio-Temporal Dataset2`. The power curve is estimated by using the binning method [102] as mentioned in Chapter 3 and to be detailed in Chapter 5. The binning estimates are shown in Fig. 4.1 as the triangles.

Four physically meaningful values of wind speed are critical to defining a wind power curve, which are—the cut-in speed, V_{ci}, the inflection point, V_{in}, the rated speed, V_r, and the cut-out speed, V_{co}. We have explained in Section 1.1 the meanings of the cut-in speed, the rated speed, and the cut-out speed. A turbine manufacturer provides the values of V_{ci}, V_r, and V_{co} for a specific turbine. Their typical values are, respectively, 3.5, 13.5, and 25 m/s, although some turbines have their cut-out speed at 20 m/s. Between V_{ci} and V_r, the power curve follows a nonlinear relationship, with an inflection point separating the convex and concave regions. This inflection point, denoted by V_{in}, marks the start when the turbine control mechanism is used to regulate the power production. Hwangbo et al. [96] estimate V_{in} for modern wind turbines to be around 9.5 m/s. These physically meaningful values induced by the power curve motivate analysts to define a total of $R = 4$ regimes, with the regime boundaries set at V_{ci}, V_{in}, V_r, and V_{co}. We advocate using these values as a starting point and make necessary adjustment when needed.

For the Wind Spatio-Temporal Dataset2 specifically, only around 3% of wind speed data are higher than V_r. It makes sense to merge the last two wind speed regimes by eliminating the threshold at V_r. Moreover, V_{co} is in fact 20 m/s for the Wind Spatio-Temporal Dataset2, and adjusting the end point of the wind speed spectrum from 25 m/s to 20 m/s does not affect the above wind speed regime definition.

Wind regimes can also be defined by using wind direction, to be seen in Section 4.2, or by using the combination of wind speed regimes and wind direction regimes, to be seen in Section 4.4, where we define three wind speed regimes and two wind direction regimes, the combination of which produces a total of six wind regimes.

4.1.2 Data-driven Regime Determination

Another approach to identify the number of wind regimes is data-driven. Kazor and Hering [119] present a regime determination approach based on the Gaussian mixture model (GMM). The idea is to use a GMM to model the wind variable from the R regimes, each of which is treated as a stationary random process. Kazor and Hering use the 2×1 wind velocity vector, $\mathbf{v} = (V_1, V_2)^T$, where V_1 and V_2 are, respectively, the wind velocity along the longitudinal and latitudinal directions. Each regime is modeled as a bivariate normal density, i.e., $\mathbf{v} \sim \mathcal{N}(\boldsymbol{\mu}_k, \boldsymbol{\Sigma}_k)$, $k = 1, \ldots, R$. Denote by τ_k the proportion of observations available under the k-th regime. Then, the Gaussian mixture density function of the R regimes is expressed as

$$f(\mathbf{v}|\Theta) = \sum_{k=1}^{R} \tau_k \mathcal{N}(\mathbf{v}|\boldsymbol{\mu}_k, \boldsymbol{\Sigma}_k), \tag{4.2}$$

where $\Theta := \{\tau_1, \ldots, \tau_R; \boldsymbol{\mu}_1, \ldots, \boldsymbol{\mu}_R; \boldsymbol{\Sigma}_1, \ldots, \boldsymbol{\Sigma}_R\}$ is the set of parameters in this GMM. Kazor and Hering further simplify the covariance matrices by assuming their off-diagonal elements all zeros, leaving only two variance terms per covariance matrix to be estimated for this bivariate distribution. This assumption implies that the two wind velocity variables are uncorrelated. Under this assumption, there are five parameters per regime—one τ, two mean terms, and two variance terms—or a total of $5R$ parameters for R regimes. The parameters can be estimated by using a maximum likelihood estimation.

To determine the number of regimes, R, Kazor and Hering suggest computing the BIC for the GMM for a range of regime numbers. They specifically recommend computing the BIC for models with between one and five regimes. Recall the definition of BIC in Eq. 2.23, it can be expressed for this GMM model as

$$\mathrm{BIC}(R) = \ln(n) \cdot (5R) - 2\ln(\hat{f}(\mathbf{v}|\hat{\Theta})),$$

where n is the amount of data used to estimate the parameters, $5R$ is the number of parameters with the presence of R regimes, and $\ln(\hat{f}(\mathbf{v}|\hat{\Theta}))$ is the log-likelihood evaluated at the estimated parameters. For selecting the number of regimes, one can plot the BIC values against the number of regimes and then choose the elbow point and its corresponding number of regimes, similar to how analysts select the significant principal components using a scree plot [111].

This GMM approach does not need to define the boundaries of the regimes explicitly. Each regime is represented by its mean and variance parameters, which are in turn estimated from the data. Upon a new wind observation, $\mathbf{v}_{\mathrm{new}}$, analysts can compute the likelihood of each individual regime, which is $\hat{\tau}_k \mathcal{N}(\mathbf{v}_{\mathrm{new}}|\hat{\boldsymbol{\mu}}_k, \hat{\boldsymbol{\Sigma}}_k)$, for $k = 1, \ldots, R$, and then select the regime corresponding to the largest likelihood. This treatment is called *hard thresholding*, implying that one regime is chosen while all other regimes are discarded. By contrast, the *soft thresholding* treatment is to compute the normalized weighting to be given to each regime model as

$$w_k = \frac{\hat{\tau}_k \mathcal{N}(\mathbf{v}_{\mathrm{new}}|\hat{\boldsymbol{\mu}}_k, \hat{\boldsymbol{\Sigma}}_k)}{\sum_{i=1}^{R} \hat{\tau}_i \mathcal{N}(\mathbf{v}_{\mathrm{new}}|\hat{\boldsymbol{\mu}}_i, \hat{\boldsymbol{\Sigma}}_i)}, \quad k = 1, \ldots, R, \tag{4.3}$$

and then the forecasting is made by using all R models and by associating each model with the corresponding weight w_k.

4.1.3 Smooth Transition between Regimes

Analysts recognize that abrupt changes between regimes may not be desirable. The concept of smooth transition between regimes is therefore introduced. The soft-thresholding GMM is a type of smooth transition approach, as there are no rigid boundaries between regimes, and for each forecast, all regime-dependent models are used with their respective weights.

Pinson et al. [164] introduce another smooth transition autoregressive model (STAR, not to be confused with GSTAR in Section 3.2). The model takes the form of

$$
\begin{aligned}
V(t) = \sum_{i=1}^{R-1} \Bigg(&\left[a_0^{r_i} + \sum_{j=1}^{p^{r_i}} a_j^{r_i} V_{t-j} \right] \tilde{G}_i(\hat{V}^{r(t)}) \\
&+ \left[a_0^{r_{i+1}} + \sum_{j=1}^{p^{r_{i+1}}} a_j^{r_{i+1}} V_{t-j} \right] G_i(\hat{V}^{r(t)}) \Bigg) + \varepsilon(t),
\end{aligned}
\tag{4.4}
$$

where $\tilde{G}_i(\cdot) = 1 - G_i(\cdot)$ is the smooth transition function that assigns weights to the AR models associated with the i-th and $(i+1)$-th regimes, and $\hat{V}^{r(t)}$ is the estimated wind speed corresponding to the regime at time t. Pinson et al. suggest using the d-step lagged wind speed, $V(t-d)$, as $\hat{V}^{r(t)}$, and then, using a logistic function to create a soft-thresholding transition, such as

$$G_k(V(t-d)) = \frac{1}{1 + \exp\{-\varphi_k(V(t-d) - c_k)\}}, \quad k = 1, \ldots, R, \quad (4.5)$$

where $\varphi_k > 0$ and c_k are the parameters in the transition function. The set of parameters in the smooth transition model includes those of the AR models as well as these for the transition functions. Typically, the AR model parameters can be estimated separately for each regime, following the approach outlined for ARMA models in Section 2.4. Then, the parameters for the transition functions, $\{\varphi_k, c_k\}$, are decided by using a cross-validation approach.

4.1.4 Markov Switching between Regimes

A Markov-switching autoregressive (MSAR) model [6, 164, 201] uses a group of AR models, similar to those expressed in Eq. 4.1, but MSAR assumes that the switch between the regimes is triggered by a Markov chain and thus employs a transition probability matrix to govern regime changes.

The one-step ahead transition probability matrix, $\mathbf{\Pi}_{R \times R}$, is expressed as

$$\mathbf{\Pi}_{R \times R} = \begin{pmatrix} \pi_{11} & \pi_{12} & \cdots & \pi_{1R} \\ \pi_{21} & \pi_{22} & \cdots & \pi_{2R} \\ \vdots & \vdots & \ddots & \vdots \\ \pi_{R1} & \pi_{R2} & \cdots & \pi_{RR} \end{pmatrix}, \quad (4.6)$$

where the (i, j)-th element, π_{ij}, is defined as

$$\pi_{ij} = P[r(t+1) = r_j | r(t) = r_i].$$

In the above definition, the Markovian property is invoked, which says that the probability of a regime at time $t+1$ only depends on the regime status at the previous time, t, rather than on the entire history of regimes. Mathematically, what this means is

$$P[r(t+1)|r(t), r(t-1), \ldots, r(1)] = P[r(t+1)|r(t)]. \quad (4.7)$$

The transition matrix provides the probabilistic information for switching between regimes for one step ahead. The i-th row in $\mathbf{\Pi}$ represents the probabilities for the i-th regime to switch to other regimes, including itself (unchanged). The summation of all the probabilities per row should be one, i.e., $\sum_{j=1}^{R} \pi_{ij} = 1, \forall i$. The transition matrix can be estimated by using the data in a training period, namely that each π is estimated by the empirical probability based on the training data.

Once the one-step ahead transition matrix is estimated, its use mirrors that in the GMM-based approach described in Section 4.1.2. Individual AR models are fit using the data peculiar to specific regimes. The forecast at time $t+1$ is the weighted average of the forecasts made by individual AR models. Suppose that the forecast at $t+1$ made by the AR model in regime r_k is denoted by $\hat{V}^{(r_k)}(t+1)$. The weights to be used with $\hat{V}^{(r_k)}(t+1)$ come from the transition matrix. Here, again, analysts can use either the hard thresholding approach or the soft thresholding approach. Assuming the current regime is r_k, the final forecast while using the soft thresholding is

$$\hat{V}(t+1) = \sum_{j=1}^{R} \hat{\pi}_{kj} \hat{V}^{(r_j)}(t+1). \tag{4.8}$$

For the hard thresholding, one identifies the largest $\hat{\pi}_{kj}$ for $j = 1, \ldots, R$, and suppose it is $\hat{\pi}_{kj^*}$. Then the final forecast is simply to use the AR model corresponding to regime j^*, namely $\hat{V}(t+1) = \hat{V}^{(r_{j^*})}(t+1)$.

For h-step ahead forecasts, $h > 1$, a formula similar to Eq. 4.8 can be used, but one needs to replace $\hat{\pi}_{kj}$ with an h-step ahead transition probability and replace $\hat{V}^{(r_j)}(t+1)$ with the raw forecast at $t+h$, $\hat{V}^{(r_k)}(t+h)$, which can be made by the regime-specific AR model for h steps ahead. The h-step transition probability is denoted as

$$\pi_{ij}^{(h)} = P[r(t+h) = r_j | r(t) = r_i],$$

which can be recursively computed using the one-step ahead transition matrix, $\mathbf{\Pi}$, once per step. Apparently, $\pi_{ij}^{(1)} = \pi_{ij}$. Using the soft thresholding approach, the h-step ahead can be made by

$$\hat{V}(t+h) = \sum_{j=1}^{R} \hat{\pi}_{kj}^{(h)} \hat{V}^{(r_j)}(t+h).$$

The hard thresholding forecast can be attained similarly.

4.2 REGIME-SWITCHING SPACE-TIME MODEL

The previous section discusses how the regime-switching mechanism works with time series data or temporal only models. This section discusses the regime-switching space-time models, primarily based on the work reported in [75].

For a spatio-temporal wind process, the wind speed is denoted by $V_i(t)$, following the same notational convention used in Chapter 3, where the subscript i is the site index and t is the time index. Recall that we use n to indicate the data amount along the time axis and N to represent the number of sites. A generic spatio-temporal regime-dependent model [163] can be expressed as

$$V_*(t) = a_0^{r(t)} + \sum_{i=1}^{N} \sum_{\ell=1}^{p^{r(t)}} a_{i\ell}^{r(t)} V_i(t-\ell) + \varepsilon_*(t), \tag{4.9}$$

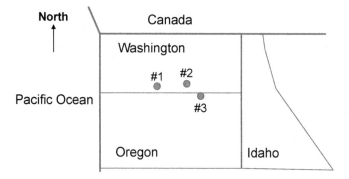

FIGURE 4.2 Geographic layout of the three sites in the border area of the states of Washington and Oregon. The actual state boundaries are not strictly parallel or vertical and the shapes of the states are approximated.

where $a_{i\ell}^{r(t)}$ is the spatio-temporal coefficient peculiar to the regime represented by $r(t)$ and '$*$' indicates the target site.

Gneiting et al. [75] consider a specific regime-switching spatio-temporal model. The setting in their study includes three geographical locations in the border area of the states of Washington and Oregon—see Fig. 4.2 for an illustration. The three sites are more or less on the same latitude but spread along a west-east line. The westernmost site is about 146 km from the middle site, which is in turn 39 km west of the easternmost site. The easternmost site is in the vicinity of the Stateline wind energy center, which is the target site for wind forecasting. The three sites are labeled as #1, #2, and #3, respectively, from the westernmost to the easternmost.

The regime is determined by the observed wind direction. The prevailing wind in that area, due to the pressure difference between the Pacific Ocean and the continental interior, is largely west-eastward. Gneiting et al. [75] pre-define their space-time regimes based on this physical understanding. They define two regimes—the westerly regime when the wind blows from the west and the easterly regime when the wind blows from the east, and then fit a space-time model for each regime.

The model used in [75] assumes a truncated normal predictive distribution at time $t + h$ and the target site, i.e., $\mathcal{N}^+(\mu_3(t + h), \sigma_3^2(t + h))$, where the subscript "3" indicates site #3, the target site for forecasting. This treatment resembles what is used in the GSTAR model in Section 3.2. In fact, the GSTAR model borrows this approach from [75], as [75] was published earlier, but their presentation order in this book may have left the readers with the opposite impression.

Gneiting et al. [75] propose a space-time model specific for each of the two regimes. For the westerly regime, the mean forecasting model is

$$\mu_3(t + h) = a_0^W + a_1^W V_3(t) + a_2^W V_3(t - 1) \\ + a_3^W V_2(t) + a_4^W V_2(t - 1) + a_5^W V_1(t), \tag{4.10}$$

where a_i^W, $i = 0, 1, \ldots, 5$, are the model coefficients to be estimated by using the data in the westerly regime. Note that in the above model, a low temporal order is used, only going back in history for two steps, i.e., t and $t - 1$. For the westernmost site (site #1), Gneiting et al. find that it is only beneficial enough to include the time history at t, not even that at $t - 1$.

For the easterly regime, the mean forecasting model is

$$\mu_3(t + h) = a_0^E + a_1^E V_3(t) + a_2^E V_2(t). \tag{4.11}$$

Here, Gneiting et al. [75] find that it is not beneficial to use the wind speed measurements at site #1 (westernmost) to make forecasts at site #3 (easternmost), because while the westerly wind creates a much stronger correlation between the two sites, the correlation is multi-fold weaker under the easterly wind. Another difference of the model in the easterly regime is that its temporal order is one lower than that used in the westerly regime.

The predictive standard deviation at $t + h$, $\sigma_3(t + h)$, is modeled similarly to that in Eq. 3.25, i.e.,

$$\sigma_3(t + h) = b_0 + b_1 \nu_3(t), \tag{4.12}$$

where in this specific case,

$$\nu_3(t) = \sqrt{\frac{1}{6} \sum_{i=1}^{3} \sum_{\ell=0}^{1} (V_i(t - \ell) - V_i(t - \ell - 1))^2},$$

and b_0, b_1 take different values in the two different regimes, although we drop the regime-indicating superscript for a clean presentation.

Gneiting et al. [75] further suggest removing the diurnal pattern from the data using Eq. 2.15 and then fitting the above space-time model to the residuals, corresponding to V'' in Eq. 2.16. But Gneiting et al. only recommend doing so for the westerly regime while leaving the easterly regime to use the original data. The dominant westerly wind, from the ocean to land, creates a special pattern causing all these differences in the above treatments.

The aforementioned models are supposed to be established for the respective regimes using the data collected in the corresponding regime. When making forecasts, the wind direction measured at site #1 is used to invoke one of the regimes and hence the corresponding AR model. In [75], Gneiting et al. are only concerned with making a forecast at $h = 2$, i.e., a two-hour ahead forecast, but the model above can be used for other h's in its current form. If the mean of the predictive distribution is used as the point forecast at $h = 2$ and site #3, then

$$\hat{V}_3(t + 2) = \hat{\mu}_3(t + 2) + \hat{\sigma}_3(t + 2) \frac{\phi\left(\frac{\hat{\mu}_3(t+2)}{\hat{\sigma}_3(t+2)}\right)}{\Phi\left(\frac{\hat{\mu}_3(t+2)}{\hat{\sigma}_3(t+2)}\right)},$$

where $\phi(\cdot)$ is the pdf of the standard normal distribution. If the median, or more generally, the α-quantile of the predictive distribution is used as the point forecast, then Eq. 3.27 is to be used; for median, i.e., the 0.5-quantile, set $\alpha = 0.5$.

For parameter estimation, Gneiting et al. [75] use the CRPS criterion, to be consistent with their probabilistic modeling approach. For a truncated normal distribution with its distribution parameter estimated as $\hat{\mu}$ and $\hat{\sigma}$, Gneiting et al. show that the CRPS can be expressed as

$$
\begin{aligned}
\text{CRPS}_{\text{TN}} = \frac{1}{n} \sum_{t=1}^{n} \hat{\sigma} \cdot \Phi\left(\frac{\hat{\mu}}{\hat{\sigma}}\right)^{-2} & \left\{ \frac{V_3(t) - \hat{\mu}}{\hat{\sigma}} \Phi\left(\frac{\hat{\mu}}{\hat{\sigma}}\right) \right. \\
& \times \left[2\Phi\left(\frac{V_3(t) - \hat{\mu}}{\hat{\sigma}}\right) + \Phi\left(\frac{\hat{\mu}}{\hat{\sigma}}\right) - 2 \right] \\
& \left. + 2\phi\left(\frac{V_3(t) - \hat{\mu}}{\hat{\sigma}}\right) \Phi\left(\frac{\hat{\mu}}{\hat{\sigma}}\right) - \frac{1}{\sqrt{\pi}} \Phi\left(\sqrt{2}\frac{\hat{\mu}}{\hat{\sigma}}\right) \right\},
\end{aligned}
\tag{4.13}
$$

where π is the circumference constant, not to be confused with the transition probability variable used in Eq. 4.6. The smaller the CRPS, the better. Minimizing the CRPS may run into numerical issues, especially as $\hat{\mu}/\hat{\sigma} \to -\infty$. Gneiting et al. recommend setting the CRPS to a large positive number when $\hat{\mu}/\hat{\sigma} \leq -4$ to resolve this issue.

Gneiting et al. [75] admit that the characteristics of this geographical area make the choice of regimes easier. Under other circumstances, the identification of forecast regimes may not be so obvious. Motivated to extend the regime-switching space-time model to a general setting, Hering and Genton [91] propose to include the wind direction as a circular variable in the model formulation to relax the model's dependence on arbitrary regime selections. Denote $\vartheta_i(t)$ as the wind direction measured at site i and time t, and the model in Eq. 4.10 now becomes

$$
\begin{aligned}
\mu_3(t + h) = a_0 &+ a_1 V_3(t) + a_2 V_3(t - 1) + a_3 V_2(t) + a_4 V_2(t - 1) + a_5 V_1(t) \\
&+ a_6 \sin(\vartheta_3(t)) + a_7 \cos(\vartheta_3(t)) + a_8 \sin(\vartheta_2(t)) + a_9 \cos(\vartheta_2(t)) \\
&+ a_{10} \sin(\vartheta_1(t)) + a_{11} \cos(\vartheta_1(t)).
\end{aligned}
\tag{4.14}
$$

Hering and Genton [91] recommend fitting the model in Eq. 4.14 to the residuals after removing the diurnal pattern using Eq. 2.15 and refer to the resulting

TABLE 4.1   RMSE for 2-hour ahead point forecasts for wind speed at site #3 in May to November 2003. Boldface values indicate the best performance.

	May	Jun	Jul	Aug	Sep	Oct	Nov
PER	2.14	1.97	2.37	2.27	2.17	2.38	2.11
AR-N	2.04	1.92	2.19	2.13	2.10	2.31	2.08
AR-D	2.01	1.85	2.00	2.03	2.03	2.30	2.08
RST-N	1.76	1.58	1.78	1.83	1.81	2.08	**1.87**
RST-D	**1.73**	**1.56**	**1.69**	**1.78**	**1.77**	**2.07**	**1.87**

Source: Gneiting et al. [75]. With permission.

TABLE 4.2   CRPS for probabilistic 2-hour ahead forecasts for wind speed at site #3 in May to November 2003. Boldface values indicate the best performance.

	May	Jun	Jul	Aug	Sep	Oct	Nov
AR-N	1.12	1.04	1.19	1.16	1.13	1.22	1.10
AR-D	1.11	1.01	1.10	1.11	1.10	1.22	1.10
RST-N	0.97	0.86	0.99	0.99	0.99	**1.08**	**1.00**
RST-D	**0.95**	**0.85**	**0.94**	**0.95**	**0.96**	**1.08**	**1.00**

Source: Gneiting et al. [75]. With permission.

method the trigonometric direction diurnal (TDD) model. For TDD, analysts do not need to estimate the model coefficients, $a_0, \ldots, a_{11}$, separately for the respective pre-defined regimes. The wind direction variable, ϑ, is supposed to adjust the model automatically based on the prevailing wind direction observed at the relevant sites. Pourhabib et al. [166] combine this regime switching idea with their GSTAR model and create a regime-switching version of the GSTAR model, which is called RSGSTAR. But the numerical results in [166] show that RSGSTAR produces only a marginal benefit as compared to the plain version of GSTAR.

Table 4.1 presents the comparison between the regime-switching space-time model with the AR model and the persistence model in terms of RMSE, whereas Table 4.2 presents the comparison in terms of CRPS. The persistence model is not included in Table 4.2 because it only provides point forecasts and no probabilistic forecasts. Here the regime-switching space-time model uses the pre-defined two regimes, i.e., the models in Eq. 4.10 and Eq. 4.11.

In the tables, the autoregressive model uses the acronym AR and the regime-switching space-time model uses the acronym RST. The suffix '-N' means that the respective model is fit to the original data, where the suffix '-D' means that the model is fit to the residual data after removing the diurnal pattern.

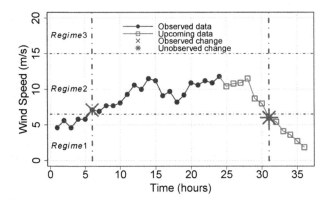

FIGURE 4.3 Wind speed at one of the turbines for a 36-hour duration. Two regime changes are identified: one in-sample and the other out-of-sample.

4.3 CALIBRATION IN REGIME-SWITCHING METHOD

The regime-switching autoregressive model and the regime-switching space-time method can be perceived as a "reactive" approach. Plainly speaking, a reactive model observes a regime change or a manifestation of it, and then adapts itself accordingly to accommodate it. In other words, the regime switching *reacts* to the regime observed and uses the forecasting model peculiar to the current wind regime to produce regime-dependent forecasts. The GMM-based approach, the smooth transition, and the Markov switching add flexibility to account for multiple possible wind regimes in the upcoming forecast period.

Ezzat et al. [60] argue that one key shortcoming of the reactive regime-switching approaches is their lack of anticipation of the upcoming regime changes in the forecast horizon. Fig. 4.3 plots the wind speeds recorded at one of the turbines in the `Wind Spatio-Temporal Dataset2` for a 36-hour duration. In practice, forecasting is often carried out in a rolling forward fashion. One could run into a situation where the goal is to obtain predictions for the next 12 hours, based on the past 24-hour data. Assume the number of regimes and regime boundaries have been pre-specified as shown in Fig. 4.3. Two regime changes are identified in the 36-hour duration, one of which takes place in the unobserved forecasting horizon. Reactive approaches may have the ability to deal with the in-sample change, but do not in their current treatment handle the unobserved, out-of-sample change. Extrapolating the characteristics learned from the training data, which are obviously not representative of the near future, could lead to negative learning and poor predictive performance. Note that the in-sample change in Fig. 4.3 is from Regime 1 to Regime 2, while the out-of-sample change is the opposite.

In the near ground wind fields like those on a wind farm, wind patterns can change rather frequently. Standing at any time point, an out-of-sample regime change could be imminent. Our analysis using the first 30 days of data in the `Wind Spatio-Temporal Dataset2` shows that the minimum-time-to-change and the median-time-to-change in wind speed are 5 hours and 15 hours, respectively, while those in wind direction are 11 hours and 33 hours, respectively. On average, a change in wind speed or wind direction takes place every 10 hours. Ignoring the occurrence of out-of-sample regime changes can seriously undermine the extrapolation ability of a regime-switching forecasting model.

Fig. 4.4 illustrates the change points detected in both wind speed and wind direction, using the first 30 days of data in the `Wind Spatio-Temporal Dataset2`. The wind direction data are from one of the met masts on the wind farm. The wind speed data are from the turbine anemometers but to facilitate a univariate detection, the wind speeds at all 200 turbines are spatially averaged to produce a single time series. Given that the hourly data are used, both wind speed and wind direction data vectors for one month are of the size 720×1. One may have noticed that the first half portion of the change points in the wind direction plot (bottom panel) is the same as that in Fig. 3.5. The specific change-point detection methods used are: for wind speed, a binary segmentation for multiple change detection based on the package `changepoint` in R [122], while for wind direction, a binary segmentation version of the circular change-point detection [106] based on the package `circular`. Recall that the circular change-point detection method is also used in Section 3.3.3 when producing Fig. 3.5.

Prompted by this observation, Ezzat et al. [60] contemplate a more proactive approach for short-term wind forecasting, which involves an action of wind speed calibration, referred to as the calibrated regime-switching (CRS) method. The CRS approach distinguishes between the in-sample regime changes taking place in the observed portion of the data and the out-of-sample regime changes occurring in the unobserved forecasting horizon. Next we take a closer look at the two types of changes. Hereinafter in this chapter, unless otherwise noted, the time index, t, is used to indicate the present time, while ℓ denotes an arbitrary time index. A forecast is to be made at $t + h$ for $h = 1, 2, \ldots, H$, i.e., the forecast horizon could be as far as H hours ahead of the present time.

4.3.1   Observed Regime Changes

An observed, in-sample regime change takes place in the observed portion of the data. Formally, an in-sample regime change occurs at time $\ell^* \in (1, t]$, when $r(\ell^* - 1) = r_k$, while $r(\ell^*) = r_{k'}$, such that $k \neq k'$ and $k, k' \in \{1, \cdots, R\}$. The CRS method signals an observed change in wind regimes by monitoring the most recent history of wind speed and wind direction. In practice, the retrospective searching for a regime change usually goes no further back than one month.

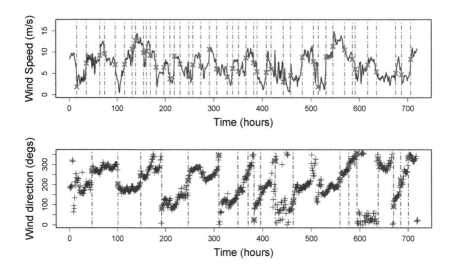

FIGURE 4.4 Top panel: change points in one month of spatially aggregated wind speed data. Bottom panel: change points in one month of wind direction data. The span of the x-axis is a month, or 720 hours. (Reprinted with permission from Ezzat et al. [60].)

4.3.2 Unobserved Regime Changes

An unobserved, out-of-sample regime change takes place in the forecasting horizon, $[t + 1, t + H]$. In other words, a future regime change may occur at $t + h$, where $r(t + h - 1) = r_k$, while $r(t + h) = r_{k'}$, such that $k \neq k'$ and $k, k' \in \{1, \cdots, R\}$.

Anticipating the out-of-sample regime changes is understandably much more challenging. It is important to identify certain *change indicator* variables that are thought to be able predictors of out-of-sample changes and whose values can be extracted from the observed data. Ezzat et al. [60] identify two principal change indicators: the current observed wind regime, i.e., $r(t)$, and the *runlength*, denoted by $x(t + h)$, which is to be explained below.

The current wind regime, $r(t)$, is naturally a useful indicator of upcoming wind regimes at $t + h$. For instance, in windy seasons, it is more likely to transit from low-speed to high-speed regimes, and the converse holds true for calmer seasons. This, in fact, is the essence of using Markov switching autoregressive models which translate the current regime information into transition probabilities for connections with the upcoming regimes.

Given the frequent changes in wind speed and direction as observed in Fig. 4.4, the current regime information alone is not sufficient to confidently inform about when and how out-of-sample changes occur. An additional input is required to make a good inference. Ezzat et al. [60] conclude that the runlength, which is the time elapsed since the most recent change point in the response of interest, is a far more potent indicator of upcoming changes than many other alternatives—other alternatives include the rate of change in wind speed, or that in wind direction, turbulence intensity and volatility measures. The use of runlength is first proposed in the online change-point detection literature [188].

The value of the runlength at any arbitrary time index ℓ is defined as $x(\ell) = \ell - \ell^*$, where ℓ^* is the time at which the most recent regime change is observed such that $\ell^* < \min(\ell, t)$. For a time point in the forecast horizon, i.e., $\ell = t + h$, Ezzat et al. [60] define the runlength in the forecast horizon as $x(t + h) = t + h - \ell^*$.

To appreciate the relevance of the runlength variable more intuitively, let us run a simple analysis using the change test results on the first 30 days of wind speed data, as shown in Fig. 4.4. Understandably, the change points in Fig. 4.4 are not exactly the regime change points, because the regime change points are defined using a set of prescribed wind speed or wind direction thresholds, whereas the change points in Fig. 4.4 are identified through a statistical significance test. Nevertheless, both types of changes serve a similar purpose, which is to identify a segment of time series data for which either the wind speed or the wind direction or both can be assumed relatively stationary. If the runlength is relevant to one, it ought to be relevant to the other.

The change test results in Fig. 4.4 suggest that there exist 43 change points in wind speed out of the 720 data points. For each of the 720 observations, one can compute the corresponding runlength, forming a 720×1 vector, namely $[x(1), \ldots, x(720)]^T$, where $x(1) = 0$. For instance, if the first change point was observed at $\ell = 16$, then $x(15) = 15$, $x(16) = 16$, but $x(17) = 1$, and so forth. Fig. 4.5, left panel, illustrates the runlength values for the first 100 points, where change points are marked by the crosses. Note how the runlength grows linearly with time, reaches its peak at change points, and then resets to one right after the change.

The 720 data points are subsequently grouped into two classes: the time points deemed as "not a change point," like at $\ell = 15$ and $\ell = 17$ as mentioned above, versus the "change points," like at $\ell = 16$. Fig. 4.5, right panel, presents the boxplots of the runlength values associated, respectively, with the two classes. The difference is remarkable: the median runlengths are 8.0 and 16.0 hours for the two classes, respectively. This means that for a given time point, which could be in the forecasting horizon, say at $t + h$, the larger its runlength $x(t+h)$, the more likely a change will occur. On the contrary, a small runlength makes it more likely that the wind follows the most recently observed pattern.

4.3.3   Framework of Calibrated Regime-switching

The basic idea of the CRS approach is as follows. Assume that a *base model*, $\mathcal{M}$, can produce a spatio-temporal forecast, $\hat{V}_i(t + h)$, at the i-th site and time $t + h$. This base model, $\mathcal{M}$, could be a spatio-temporal model yield-

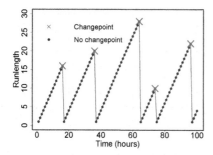 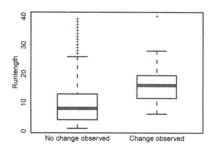

FIGURE 4.5 Left panel: runlength as a function of time. Right panel: boxplots of the runlengths for data points at which no change was observed versus those for data points at which a change was observed.

ing kriging-based forecasts, as we discuss in Chapter 3. Admittedly, this base model produces reactive, albeit regime-specific, forecasts. CRS seeks to calibrate the reactive forecasts to safeguard against upcoming, out-of-sample regime changes, by adding a regime-dependent term, $c_i^{r(t)}(t+h) \in \mathbb{R}$, to the raw forecast, $\hat{V}_i(t+h)$. This additional term, $c_i^{r(t)}(t+h)$, is referred to as the regime-dependent *forecast calibration*, and the quantity $\hat{V}_i(t+h)+c_i^{r(t)}(t+h)$ as the *calibrated forecast*. The idea behind CRS is illustrated in Fig. 4.6, where the goal of the calibration is to adjust the forecast at $t+h$ in anticipation of a regime change.

Determining the sign and magnitude of $c_i^{r(t)}(t+h)$ is arguably the most critical aspect of the CRS approach. Ezzat et al. [60] assume that the sign and magnitude of the forecasting calibration, $c_i^{r(t)}(t+h)$, can be informed by the observed data up to time t, denoted by $\mathbb{D}_t$. The dependence on $\mathbb{D}_t$ is signified by the notation, $c_i^{r(t)}(t+h|\mathbb{D}_t)$. For simplicity, $c_i^{r(t)}(t+h|\mathbb{D}_t)$ is assumed to only vary over time but be fixed across space, that is, $c_i^{r(t)}(t+h|\mathbb{D}_t) = c^{r(t)}(t+h|\mathbb{D}_t)$, for $i = 1, \cdots, N$. A general formulation to infer $c^{r(t)}(\cdot)$ can be expressed as

$$\min_{c^{r(t)} \in \mathcal{C}} \quad L\big[\hat{V}_i(t+h) + c^{r(t)}(t+h|\mathbb{D}_t), V_i(t+h)\big], \qquad (4.15)$$

where $\mathcal{C}$ is some class of functions to which $c^{r(t)}(\cdot)$ belongs, and $L[\cdot, \cdot]$ is a loss function that defines a discrepancy measure. To solve Eq. 4.15, $c^{r(t)}(\cdot|\mathbb{D}_t)$ ought to be parameterized.

Based on the discussion in Section 4.3.2, the sign and magnitude of a forecasting calibration is determined through the observed values of the two change indicators, $r(t)$ and $x(t+h)$. Ezzat et al. [60] further propose to use a log-normal cdf to characterize $c^{r(t)}(\cdot)$'s relationship with the two inputs. The choice of the lognormal cdf as a calibration function is motivated by its flexibility to model a wide spectrum of regime-switching behavior, ranging

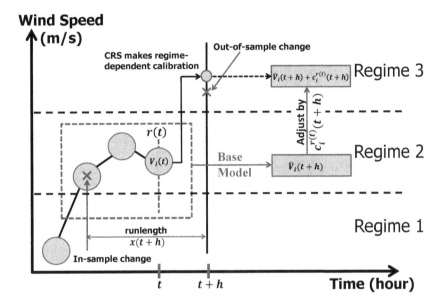

FIGURE 4.6 Illustration of the forecasting calibration for out-of-sample changes.

from abrupt shifts to gradual drifts, depending on the values of its parameters that are learned from the data.

Given R pre-defined wind regimes, $c^{r(t)}(\cdot)$ is modeled individually in each of them. The current regime information, $r(t)$, is then implicitly incorporated by the regime partition, as $c^{r(t)}(\cdot)$ uses the parameters specific to that particular regime. Consequently, the characterization of $c^{r(t)}(\cdot)$ has only the runlength variable, $x(t+h)$, as an explicit input. For the k-th regime, let us denote the regime-dependent parameters by $\Psi^k = \{\psi_1^k, \psi_2^k, \psi_3^k\}$, so that the regime-specific calibration function can be denoted as $c(x(t+h); \Psi^k | \mathbb{D}_t)$ and the superscript $r(t)$ is dropped. The log-normal cdf has the form of

$$c(x(t+h); \Psi^k) = \psi_1^k \Phi \left(\frac{\ln(x(t+h)) - \psi_2^k}{\psi_3^k} \right).$$

CRS aims to learn Ψ^k for each regime using the historical training data and continuously update them during the rolling forward forecasting.

The estimation procedure goes as follows. Assume that an analyst has at hand a sequence of forecasts obtained via a base model, $\mathcal{M}$, and their corresponding true observations. These forecasts are obtained in a rolling forward fashion, such that for the ℓ-th roll, the data observed up to time t^ℓ are used to obtain forecasts from $t^\ell + 1$ till $t^\ell + H$. Then, the window is slid by a specified interval, say s, so that the "present time" for the next forecasting roll is $t^{\ell+1} = t^\ell + s$. Suppose that there are $\mathfrak{L}$ forecasting rolls in the training set. For the ℓ-th forecasting roll, $\ell = 1, \ldots, \mathfrak{L}$, the following information is saved—the observed wind regime at the time of forecasting, $r(t)$, the associated runlength, $x^\ell(t+h)$, the raw forecast via $\mathcal{M}$, $\hat{V}_i^\ell(t+h)$, and the actual observation at $t+h$, $V_i^\ell(t+h)$. By employing a squared error loss function, the optimization problem of Eq. 4.15 can be re-written as,

$$\min_{\Psi^k} \frac{1}{\mathfrak{L}^k \times N \times H} \sum_{\ell=1}^{\mathfrak{L}^k} \sum_{i=1}^{N} \sum_{h=1}^{H} \left[\hat{V}_i^\ell(t+h) + c(x^\ell(t+h); \Psi^k) - V_i^\ell(t+h) \right]^2$$

(4.16)

where $\mathfrak{L}^k$ denotes the number of forecasting rolls relevant to regime k. Solving Eq. 4.16 for each regime individually, i.e., for $k = 1, \ldots, R$, gives the least-squared estimate of the parameters in $\{\Psi^k\}_{k=1}^R$.

Table 4.3 presents the features of various forecasting models. A checkmark "✓" means the presence of that feature, whereas a cross "X" means absence. The last column indicates the piece of information on which a method is actively invoked as a forecasting indicator. Please note that methods like ASYM, SEP and PER do not explicitly consider a wind regime and they are usually not included as a regime-switching approach. Nevertheless, they can be considered as a special case of reactive regime-switching, which has always a single regime and assume that the same regime continues in the forecast horizon. For this reason, ASYM, SEP, PER, RSAR, and RST are collectively referred to as the reactive methods.

TABLE 4.3   Features of various forecasting models.

Method	Temporal	Spatial	Asymmetry	In sample	Out of sample	Regime indicators
PER	X	X	X	X	X	X
SEP	✓	✓	X	X	X	X
ASYM	✓	✓	✓	X	X	X
RSAR	✓	X	X	✓	X	$r(t)$
MSAR	✓	X	X	✓	✓	$\{r(t), \Pi\}$
RST	✓	✓	✓	✓	X	$r(t)$
CRS	✓	✓	✓	✓	✓	$\{r(t), x(t+h)\}$

4.3.4   Implementation Procedure

To run a CRS comprises three sequential phases: (1) Phase I: generating the raw forecasts (via the base model $\mathcal{M}$) in the initialization period, (2) Phase II: learning the forecasting calibration function based on the raw forecasts and the actual observations solicited in the initialization period, (3) Phase III: making continuous rolling-forward forecasting and updating. Phases I and II use a subset of the data, say, the first month of data, to set up the CRS model. In Phase III, the actual forecasting and testing are carried out on the remaining months in the dataset. Fig. 4.7 presents a diagram for understanding the implementation of CRS.

Phases I and II are the training stage. Without loss of generality, the base spatio-temporal model, $\mathcal{M}$, is assumed to be parameterized by a set of parameters in Θ and thus denoted as $\mathcal{M}(\Theta)$.

The rolling mechanism in Phase I goes as follows. The first roll of training data is the first 12-hour data. Using the 12-hour data, the model parameters Θ are estimated and the raw forecasts from $t + 1$ till $t + H$ are made. The regime information, $r(t)$, and the forecasts, $\hat{V}_i(t + h)$, $h = 1, \ldots, H$, are saved for subsequent training. Then, the window is slid by a pre-specified interval s and all data points within that sliding interval are revealed, so that the runlength, $x(t + h)$, and the actual wind speed, $V_i(t + h)$, can be recorded and saved, too. Next, one is ready to make a new forecast, and for that, one needs to re-estimate Θ using the newly revealed data. One thing to bear in mind is that if the sliding interval contains any change points, one should use only the "relevant" data for estimating Θ. The "relevant" data refer to those from the most recent stationary data segment leading to the present time. For instance, Ezzat et al. [60] consider temporal lags for up to 4 hours into history. If the immediate past regime change happens within four time lags from the present time, Ezzat et al. use data with an even shorter time history, which is since the immediate past regime change. This rolling mechanism is continued until all data in the initialization period is exhausted, supposedly resulting in $\mathfrak{L}$ rolls.

Once Phase I is finished, the goal of Phase II is to learn the calibration

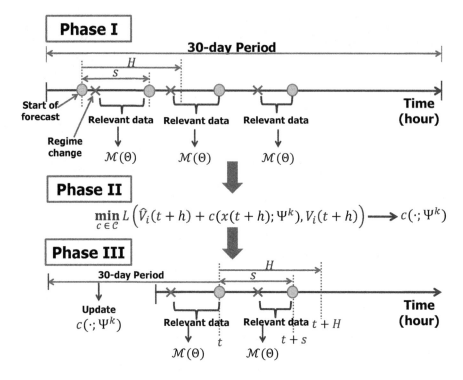

FIGURE 4.7 Steps and notations in the execution of the calibrated regime-switching approach.

function, $c(x(t+h); \Psi^k)$, using the Phase I data, where Eq. 4.16 is solved for each regime individually to estimate the regime-dependent parameters Ψ^k.

Then, proceed to Phase III, where rolling forecasts are performed. At the present time t, one should first look back and search for the most recent in-sample change point. Again, only the "relevant data," defined the same as before, are used to estimate the base model parameters in Θ. The base model is used to make the raw forecasts. The $c(x(t+h); \Psi^k)$, $h = \{1, \dots, H\}$, is calculated based on the knowledge of the current wind regime, the runlength, and Ψ^k. The resulting $c(x(t+h); \Psi^k)$ is used to calibrate the raw forecasts.

The window is then slid by s. At $t+s$, first use the last 30 days of data to update Ψ^k by re-solving Eq. 4.16 for $k = 1, \dots, R$, given the newly revealed observations, then estimate the base model parameters in Θ using the "relevant data," and finally, make forecasts for $t + s + h$, $h = 1, \dots, H$. The cycle is repeated until the forecasts for all the remaining months are produced.

4.4 CASE STUDY

This section applies the calibrated regime-switching method, together with a few alternatives, to the yearlong `Wind Spatio-Temporal Dataset2`. The performances of the respective methods are illustrated and compared.

4.4.1 Modeling Choices and Practical Considerations

In this analysis, the forecast horizon is up to $H = 12$. The sliding interval is set to $s = 6$ hours, meaning that after each roll, the first six hours of the forecast horizon are revealed, and the horizon is shifted by another six hours. This value appears reasonable considering the frequency at which forecasts are updated in practice.

The base model used in CRS is the non-separable, asymmetric spatio-temporal model presented in Section 3.4.1 and the corresponding forecasting model is the kriging method presented in Section 3.4.3. Same as in Section 3.4, by setting the asymmetry and separability parameters to zero, a separable version of the general spatio-temporal model can be obtained.

The base spatio-temporal model used is stationary, but wind fields have been reported to exhibit signs of nonstationarity [69, 166]. By considering only the most recent history of wind speed and direction for model training, it helps overcome the temporal nonstationarity, as the assumption of temporal stationarity is sufficiently reasonable in the short time window since the latest change point. Ezzat et al. [60] account for spatial nonstationarity by assuming local spatial stationarity within a subregion on the wind farm. Three subregions of wind turbines based on their proximity to the three masts are defined, and a region-specific stationary spatio-temporal model is fit and subsequently used for forecasting.

The physically motivated regime definition, as explained in Section 4.1.1, is used here for defining three wind speed regimes. Ezzat et al. [60] also define two wind direction regimes upon observing a dominant east-westward directional wind in the dataset. The combination of the wind speed regimes and wind direction regimes produces a total of $R = 6$ wind regimes.

A further fine-tuning is conducted to adjust the boundaries of the resulting regimes for boosting the performance of the CRS approach. Using the first month of data, the fine-tuning is conducted on 112 different combinations of regime thresholds, chosen as follows: $u_1 = 0$, vary v_1 from V_{ci} to $V_{ci} + 1.5$ with increments of 0.5 m/s, v_2 from $V_{in} - 1.5$ to V_{in} with increments of 0.5 m/s, D_1 from $180° - 45°$ to $180° + 45°$ with $15°$ increments, and set $D_2 = 360° - D_1$, where D_1 and D_2 are the wind direction thresholds. The fine-tuning based on the `Wind Spatio-Temporal Dataset2` yields the final regime thresholds at 4.5 and 9.0 m/s for wind speed and $45°$ and $225°$ for wind direction.

Fig. 4.8 illustrates the learned calibration functions for the six regimes as functions of the runlength. It appears that the wind speed variable is the main factor alluding to the upcoming out-of-sample changes. For instance,

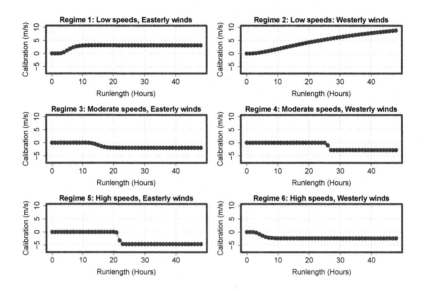

FIGURE 4.8 Learned forecasting calibration functions, $c(x(t + h); \Psi^k)$, using Phase I data for the six regimes. (Reprinted with permission from Ezzat et al. [60].)

the first two regimes (top row), which share the same wind speed profile (low wind speeds), both transit to higher wind speed regimes. Regimes 3 and 4, both with moderate wind speeds, and regimes 5 and 6, both with high wind speeds, likewise have a calibration function of the same pattern to their respective group. The wind direction appears to have a secondary, yet still important, relationship with the magnitude of the out-of-sample change, as well as its timing. For instance, it appears that the magnitude of change is larger in regime 2 (westerly) than in regime 1 (easterly), and larger in regime 4 (westerly) than in regime 3 (easterly). The opposite happens in regimes 5 (easterly) and 6 (westerly). The switching behavior difference between gradual shifts like in regimes 1, 2, 3, and 6 and abrupt shifts like in regimes 4 and 5 also implies a certain degree of interaction between the two factors. These functions may change with time and they are continuously re-estimated in Phase III.

Finally, during the actual testing in Phase III, Ezzat et al. [60] decide to impose maximal and minimal thresholds on the magnitude of forecast calibration to avoid over-calibrating the forecasts when extrapolating. Some numerical experiments indicate that restricting the magnitude of the calibration quantities to the range $(-3, 3)$ m/s yields satisfactory performance. Empirical evidence also suggests that, on average, forecast calibration does not offer much benefit in the very short-term horizon, like less than three hours ahead. For this reason, CRS only calibrates the forecasting for more than three hours ahead (three hours ahead included). This is understandable, since at very short time horizons, wind conditions are more likely to persist than to change drastically.

4.4.2 Forecasting Results

This subsection presents the numerical results comparing CRS with the following approaches: persistence forecast, the asymmetric model, the separable model, the regime-switching autoregressive model, a soft-thresholding Markov-switching model, and a Markov-switch vector autoregressive model (MSVAR) [119]. MSVAR generalizes the MSAR model to further account for the spatial dependence. The temporal order used in RSAR and MSAR (both versions) is one, i.e., $p = 1$.

The aforementioned models are compared in terms of both wind speed and wind power forecasting performances. The forecast accuracy is evaluated using MAE for each h. Specifically, the MAE used in this comparison study is expressed as

$$\text{MAE}(h) = \frac{1}{\mathfrak{L} \times N} \sum_{\ell=1}^{\mathfrak{L}} \sum_{i=1}^{N} \left| \hat{V}_i^\ell(t+h) - V_i^\ell(t+h) \right|, \qquad (4.17)$$

where $V_i^\ell(t+h)$ and $\hat{V}_i^\ell(t+h)$ are, respectively, the observed and forecasted responses from a forecasting model, obtained at the i-th site and for h-hour ahead forecasting during the ℓ-th forecasting roll, $\ell = 1, ..., \mathfrak{L}$. For each h, MAE is computed as an average over all turbines and forecasting rolls for the eleven-month test data. The MAE values are presented in Tables 4.4 and 4.5, for wind speed and power, respectively. Please note that when computing the MAE for CRS (as well as the PCE below), $\hat{V}_i(t+h)$ is substituted by the calibrated forecast, i.e., $\hat{V}_i(t+h) + c(x(t+h); \Psi^k)$.

The results in Table 4.4 demonstrate that, in terms of wind speed, CRS outperforms the competing models in most forecasting horizons. For $h \geq 2$, the CRS approach renders the best performance among all competing models. This improvement is mainly due to the use of regime-specific calibration functions, which help anticipate the out-of-sample regime changes hinted by run-length. Additional benefits over temporal-only and separable spatio-temporal models come from the incorporation of comprehensive spatio-temporal correlations and flow-dependent asymmetries. For the very short-term horizon, $h = 1$, PER offers the best performance, with CRS slightly behind, but still enjoying a competitive performance.

Fig. 4.9, upper panel, presents the percentage improvements, in terms of MAE and wind speed forecast, that the CRS approach has over the competing models at different forecast horizons. The percentage improvement over reactive methods such as ASYM, SEP, RSAR and PER is more substantial as the look-ahead horizon increases. This does not come as a surprise since the farther the look-ahead horizon is, the more likely a change will occur in that horizon, and hence, the benefit of using CRS is more pronounced.

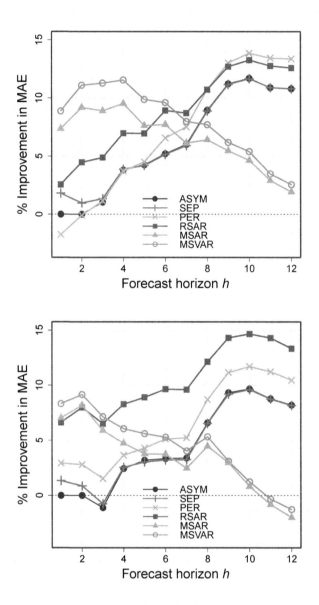

FIGURE 4.9 Percentage improvements in terms of MAE that CRS has over the competing approaches in wind speed (upper panel) and in wind power (lower panel). (Reprinted with permission from Ezzat et al. [60].)

TABLE 4.4   MAE for wind speed forecasting for h-hour ahead, $h = 1, 2, \ldots, 12$. Bold-faced values indicate best performance.

Method	1	2	3	4	5	6
ASYM	1.12	**1.45**	1.72	1.96	2.15	2.27
SEP	1.15	1.47	1.74	1.97	2.15	2.27
PER	**1.11**	1.46	1.73	1.97	2.16	2.31
RSAR	1.16	1.53	1.79	2.03	2.21	2.36
MSAR	1.23	1.64	1.92	2.14	2.28	2.38
MSVAR	1.21	1.60	1.87	2.09	2.23	2.33
CRS	1.12	**1.45**	**1.71**	**1.89**	**2.06**	**2.15**

	7	8	9	10	11	12
ASYM	2.39	2.51	2.68	2.77	2.83	2.87
SEP	2.40	2.52	2.68	2.77	2.84	2.87
PER	2.44	2.57	2.74	2.84	2.92	2.96
RSAR	2.46	2.56	2.73	2.82	2.89	2.93
MSAR	2.45	2.48	2.54	2.59	2.62	2.63
MSVAR	2.40	2.45	2.52	2.57	2.60	2.61
CRS	**2.25**	**2.29**	**2.37**	**2.44**	**2.52**	**2.56**

Source: Ezzat et al. [60]. With permission.

The trend of the improvement of CRS over the Markov-switching approaches, i.e., MSAR and MSVAR, is different. The Markov-switching approaches anticipate regime changes in the look-ahead forecast horizon, too, but use a different mechanism (the transition probabilities). For short-term horizons, the performance of CRS is remarkably better than the Markov-switching approaches. As the look-ahead horizon increases, the advantage of CRS over the Markov-switching models reaches a peak around $h = 4$ hours, and after that, the performance of the Markov-switching approaches gradually catches up with that of CRS. The difference between CRS and the Markov-switching approaches highlights the merit of using the runlength to anticipate the out-of-sample changes. The inclusion of runlength and regime information in CRS appears to offer higher sensitivity, and thus more proactivity, to out-of-sample changes than that offered by the transition probabilities in the Markov-switching approaches.

Similar findings are extended to the power prediction results in Table 4.5, in which CRS is shown to outperform the competing models for most forecasting horizons. Its improvement over the reactive methods is also higher as the look-ahead horizon increases, whereas its improvement over the Markov-switching approaches is best in the shorter forecast horizons. The percentage improvements shown in Fig. 4.9, lower panel, are somewhat different from

TABLE 4.5   MAE values for wind power forecasting for h-hour ahead, $h = 1, 2, \ldots, 12$. Bold-faced values indicate best performance.

Method	1	2	3	4	5	6
ASYM	**0.121**	**0.156**	**0.184**	0.212	0.227	0.236
SEP	0.123	0.158	0.185	0.212	0.227	0.236
PER	0.125	0.161	0.189	0.215	0.230	0.241
RSAR	0.129	0.169	0.199	0.226	0.241	0.253
MSAR	0.132	0.171	0.200	0.220	0.233	0.242
MSVAR	0.131	0.170	0.198	0.217	0.228	0.238
CRS	**0.121**	**0.156**	0.186	**0.207**	**0.220**	**0.229**

	7	8	9	10	11	12
ASYM	0.247	0.261	0.280	0.291	0.294	0.296
SEP	0.247	0.261	0.280	0.292	0.295	0.296
PER	0.253	0.268	0.286	0.299	0.303	0.304
RSAR	0.264	0.278	0.297	0.309	0.314	0.314
MSAR	0.249	0.258	0.263	0.267	0.268	.269
MSVAR	0.245	0.256	0.262	0.266	**0.267**	**0.267**
CRS	**0.239**	**0.244**	**0.254**	**0.263**	0.268	0.271

Source: Ezzat et al. [60]. With permission.

their counterparts in the upper panel. The difference is mainly due to the nonlinear speed-power conversion used in computing wind power.

In addition to MAE, Table 4.6 presents the average PCE errors across all forecasting horizons, for values of ξ ranging between 0.5 and 0.8 with a 0.1 increment, as well as $\xi = 0.73$, which is the value recommended in [91]. It appears that the improvement of CRS over the competing models is also realizable in terms of PCE. The CRS approach performs well when under-estimation is penalized more severely than over-estimation (namely $\xi > 0.5$), which describes the more realistic cost trade-off in power systems.

TABLE 4.6   Average PCE for competing models across all horizons. Bold-faced values indicate best performance.

Method	$\xi = 0.5$	$\xi = 0.6$	$\xi = 0.7$	$\xi = 0.73^*$	$\xi = 0.8$
ASYM	0.116	0.117	0.114	0.111	0.104
SEP	0.116	0.118	0.114	0.112	0.105
PER	0.118	0.121	0.124	0.125	0.127
RSAR	0.123	0.123	0.120	0.117	0.110
MSAR	0.113	0.123	0.127	0.124	0.126
MSVAR	0.112	0.118	0.122	0.118	0.119
CRS	**0.109**	**0.110**	**0.107**	**0.105**	**0.097**

Source: Ezzat et al. [60]. With permission.

GLOSSARY

AR: Autoregressive model

AR-D: Autoregressive model fit after the diurnal pattern is removed

AR-N: Autoregressive model fit to the original data

ARMA: Autoregressive moving average

BIC: Bayesian information criterion

cdf: Cumulative distribution function

CRPS: Continuous ranked probability score

CRS: Calibrated regime switching

GMM: Gaussian mixture model

GSTAR: Gaussian spatio-temporal autoregressive model

MAE: Mean absolute error

MSAR: Markov-switching autoregressive model

MSVAR: Markov-switching vector autoregressive model

PCE: Power curve error

pdf: Probability density function

PER: Persistence forecasting

RMSE: Root mean squared error

RSAR: Regime-switching autoregressive model

RSGSTAR: Regime-switching Gaussian spatio-temporal autoregressive model

RST: Regime-switching space time model

RST-D: Regime-switching space time model fit after the diurnal pattern is removed

RST-N: Regime-switching space time model fit to the original data

SEP: Separable spatio-temporal model

STAR: Smooth transition autoregressive model

TDD: Trigonometric direction diurnal model

EXERCISES

4.1 Use the `Wind Time Series Dataset` and conduct the following exercise.

    a. Use the three pre-defined wind speed regimes, $[0, 4.5)$, $[4.5, 9.0)$ and $[9.0, 20)$, and fit three AR models to the hourly data of April and May. To select the model order for the AR models, please use BIC.

    b. Use the hourly data of April and May to fit a single AR model. Still use BIC to decide the model order. Compare the AR model in (b) with the three AR models in (a).

    c. Use the AR models in (a) to make one-hour ahead rolling forward forecasts for the next ten hours. The regime for one hour ahead is assumed the same as the current regime. Compute the MAE of the ten one-hour ahead forecasts.

    d. Use the AR models in (b) to make one-hour ahead rolling forward forecasts for the next ten hours. Compute the MAE of the ten one-hour ahead forecasts. Compare the MAEs obtained in (c) and (d). What do you observe?

4.2 Use the `Wind Time Series Dataset` and fit a Gaussian mixture model to the yearlong hourly data. Here you do not have the wind direction data. So instead of fitting a bivariate Gaussian distribution, like in Eq. 4.2, you will fit a univariate Gaussian distribution.

    a. Explore the number of regimes between one and five. Use the BIC to decide the best number of regimes.

    b. Using the R decided in (a) and the associated GMM parameters, compute the weight w_k in Eq. 4.3 for wind speed between 0 m/s and 20 m/s with an increment of 1 m/s. Do this for $k = 1, \ldots, R$ and make a plot of w_k to demonstrate how each regime model is weighted differently as the wind speed changes.

4.3 Use the hourly data in `Wind Time Series Dataset` and assume three pre-defined wind speed regimes, $[0, 4.5)$, $[4.5, 9.0)$ and $[9.0, 20)$. Conduct the following exercise.

    a. Go through the first half year's data, i.e., January through June. At any data point, label the wind speed's current regime (namely, at t) as well as the regime at the next hour (namely, at $t+1$). For the entire half year of data, count the regime switching numbers between the three regimes, including the case of remaining in the same regime. Note that the regime switching from 1 to 2 and that from 2 to 1 are counted as different regime switchings. Then, divide each count by the total number of switchings. The relative frequency provides the empirical estimate of π_{ij}. Please write down the 3×3 transition probability matrix $\mathbf{\Pi}$. Verify if each row sums to one.

b. Do the same for the second half year's data, i.e., July through December. Compare the new $\mathbf{\Pi}$ with that obtained in (a). Do you find any noticeable difference between the two $\mathbf{\Pi}$'s?

4.4 If $F(\cdot)$ is the predictive cdf and V is the materialized wind speed, the continuous ranked probability score is defined as

$$\mathrm{crps}(F, V) = \int_{-\infty}^{\infty} \left(F(x) - \mathbb{1}(x \geq V) \right)^2 dx.$$

The expression in Eq. 2.60 is the sample average based on n_{test} observations, namely

$$\mathrm{CRPS} = \frac{1}{n_{\text{test}}} \sum_{i=1}^{n_{\text{test}}} \mathrm{crps}(\hat{F}, V_i).$$

Please derive the closed-form expression of $\mathrm{crps}(F, V)$ when $F(\cdot)$ is a normal distribution.

4.5 The cdf of the truncated normal distribution, $\mathcal{N}^+(\mu, \sigma^2)$, is

$$F(x) = \frac{\Phi(\frac{x-\mu}{\sigma}) - \Phi(-\frac{\mu}{\sigma})}{1 - \Phi(-\frac{\mu}{\sigma})} \tag{P4.1}$$

when $x \geq 0$, and $F(x) = 0$ when $x < 0$. Please drive the closed-form expression of $\mathrm{crps}(F, V)$ for the truncated normal distribution, which is the expression inside the summation in Eq. 4.13.

4.6 Use the wind speed data in `Wind Spatio-Temporal Dataset2`. Select three turbines from the wind farm, the west-most turbine, the east-most turbine, and a turbine roughly halfway from the two turbines on the periphery. If possible, try to select the turbines on a similar latitude. Use the average of the wind directions measured on the three met masts to represent the wind direction for the wind farm. Create four wind regimes—the easterly, southerly, westerly, northerly regimes of which the wind direction ranges are, respectively, $(45°, 135°)$, $(135°, 225°)$, $(225°, 315°)$, and $(315°, 45°)$. Use the first two months of data associated with the three turbines to fit four separate AR models, each of which has the same structure as in Eq. 4.10. Doing this yields a four-regime RST method. Use this RST method to make forecasts at the east-most turbine for $h = 2$. Shift the data by one month and repeat the above actions, and then, repeat for the whole year. One gets eleven 2-hour ahead forecasts. Compute the MAE and RMSE for these $h = 2$ forecasts.

4.7 Take the first month of wind direction data from a met mast and implement the circular variable detection algorithm to detect the change points. How many change points are there? Are the minimum-time-to-change and median-time-to-change different from those values reported on page 105?

4.8 Use the change-point detection results from the previous problem and produce boxplots similar to that in Fig. 4.5, right panel. Is there a noticeable difference between the two resulting boxplots? How do you feel using the runlength as a change indicator for a wind direction-based regime-switching method?

4.9 Test the sensitivity of the CRS approach by comparing the following competing alternatives:

a. No forecasting calibration for $h = 1$ and $h = 2$ versus conducting calibration for $h = 1$ and $h = 2$.

b. Cap the magnitude of the calibration quantities to the range $[-3, 3]$ versus $[-2, 2]$, or $[-5, 5]$, or no restriction at all.

c. Three wind speed regimes, with boundary values at 4.5 and 9.0 m/s, versus four wind speed regimes, with boundary values at 3.5, 9.5, and 13.5 m/s.

II

Wind Turbine Performance Analysis

CHAPTER 5

Power Curve Modeling and Analysis

DOI: 10.1201/9780429490972-5

P art II of this book focuses on modeling $f_t(y|\boldsymbol{x})$, where y is the wind power output and $\boldsymbol{x}$ is the vector of environmental variables, including wind speed. A common list of elements in $\boldsymbol{x}$ is given in Section 1.1. The variables in $\boldsymbol{x}$ are also called *covariates*, especially in the statistical literature. The dynamics of $f_t(y|\boldsymbol{x})$, which speaks to the innate change in a turbine's aerodynamic characteristics, is much slower as compared with the dynamics of the environmental covariates, particularly that of wind speed. For this reason, analysts often drop the subscript t and express the aforementioned conditional density as $f(y|\boldsymbol{x})$. This notation does not mean that $f(y|\boldsymbol{x})$ is a constant function over time; rather, it means that modeling the dynamics is not the focus here.

Modeling $f(y|\boldsymbol{x})$ embodies the power curve analysis. As explained in Section 1.2, $f(y|\boldsymbol{x})$ depicts a probabilistic power response surface, and the corresponding conditional expectation, $\mathbb{E}(y|\boldsymbol{x})$, is the power curve, when $\boldsymbol{x}$ is reduced to wind speed, V. In practice, $\mathbb{E}(y|\boldsymbol{x})$ is used more frequently than $f(y|\boldsymbol{x})$, as it is easier to model and to use. But modeling the conditional density, $f(y|\boldsymbol{x})$, is also beneficial, as doing so lays the basis for uncertainty quantification.

The importance of power curve modeling goes without saying. Two primary areas of impact are, respectively, for wind power forecasting and for turbine performance assessment. In Chapter 2, we note that wind power forecasting can be done by forecasting wind speed first and then converting a speed forecast to a power forecast through the use of a power curve. In the three chapters comprising Part I of this book, the use of the power curve is repeatedly mentioned. The second principal use of the power curve is for turbine performance assessment and turbine health monitoring [133, 205, 216], in which a power curve is used to characterize a turbine's power production efficiency. In both applications, accurate modeling of the power curve is essential, as it entrusts subsequent decision making.

In Chapter 5, we dedicate ourselves to various methods for power curve modeling and analysis. Chapter 6 discusses the relevance of power curves in turbine efficiency quantification. Chapter 7 focuses on one particular type of turbine change, known as turbine upgrade via retrofitting, and shows how data science methods can help quantify the change in power production due to an upgrade. Chapter 8 presents a study concerning how the wake effect affects a turbine's production performance.

5.1 IEC BINNING: SINGLE-DIMENSIONAL POWER CURVE

The current industrial practice of estimating the power curve relies on a nonparametric approach, known as the binning method, recommended by the International Electrotechnical Commission (IEC) [102]. The basic idea of the binning method is to discretize the domain of wind speed into a finite number of bins, say, using a bin width of 0.5 m/s. Then, the value to be used for representing the power output for a given bin is simply the sample average of all the data points falling within that specific bin, i.e.,

$$y_i = \frac{1}{n_i} \sum_{j=1}^{n_i} y_{i,j}, \tag{5.1}$$

where $y_{i,j}$ is the power output of the jth data point in bin i, and n_i is the number of data points in bin i.

The physical law of wind power generation [1, 15] states that:

$$y = \frac{1}{2} \cdot C_p \cdot \rho \cdot \pi R^2 \cdot V^3, \tag{5.2}$$

where R is the radius of the rotor and C_p is the power coefficient, which is believed to be a function of (at least) the blade pitch angle and the turbine's tip speed ratio. What else might affect C_p is still a matter under debate. Currently no formula exists to express C_p analytically in terms of its influencing factors. C_p is therefore empirically estimated. Turbine manufacturers provide for a specific turbine its nominal power curve with the corresponding C_p values under different combinations of wind speed, V, and air density, ρ. The above expression also provides the rationale why temperature, T, and air pressure, P, are converted into air density, ρ, to explain wind power, rather than used individually.

Even though the expression in Eq. 5.2 on the surface suggests that the electrical power that a wind turbine extracts from the wind is proportional to V^3, an actual power curve may exhibit a different nonlinear relationship. This happens because the tip speed ratio is a function of wind speed, V, making C_p also a function of V and adding complexity to the functional relationship between wind speed and wind power. Another complexity is brought by turbine controls. The power law in Eq. 5.2 governs the wind power generation before the rated wind speed, V_r. The use of the pitch control mechanism levels off, and ultimately caps, the power output when it reaches the rated power output, y_r. Recall the shape of the power curve shown in Fig. 1.2. The power curve has an inflection point somewhere near the rated wind speed, so that the whole curve consists of a convex segment, between V_{ci} and the inflection point and a concave segment, between the inflection point and V_{co}.

Given the physical relation expressed in Eq. 5.2, the wind industry recognizes the need to include air density as a factor in calculating the power output, and does so through a formula known as the air density correction. If V is the raw average wind speed measured in a 10-minute duration, the air density correction is to adjust the wind speed based on the measured average air density, ρ, in the same 10-minute duration, namely

$$V' = V \left(\frac{\rho}{\rho_0} \right)^{\frac{1}{3}}, \tag{5.3}$$

where ρ_0 is the sea-level dry air density ($=1.225 \text{ kg/m}^3$) per the International Organization for Standardization's atmosphere standard. The binning method with air density correction uses this corrected wind speed, V', and the power output, y, to establish a power curve. In the subsequent numerical analysis, by "binning method" we refer to this air density corrected version, unless otherwise noted. To make the notation simpler, we continue using V to denote the wind speed even after the air density adjustment.

Another adjustment analysts practice in the wind industry is to identify the free sectors for a wind turbine. A free sector is a subset of wind directions under which a wind turbine is supposedly free of wake effect from its neighboring turbines. The use of a free sector is effectively a filtering action, which often removes as many as two-thirds of the raw data.

Please note that the IEC binning method only provides the estimation of the average power curve, not that of the conditional density. One simple way to get the conditional density is to assume a distribution type for a given bin, say, Gaussian, and use the data in that bin to estimate the parameters in the assumed distribution. In this way, the density estimation for the whole input region is the collection of a bunch of bin-based individual density estimations.

5.2 KERNEL-BASED MULTI-DIMENSIONAL POWER CURVE

Wind power production is apparently affected by more than just wind speed. The current IEC method, explained above, primarily considers wind speed, while using the air density information in an *ad hoc* manner. The IEC method does not actually use the wind direction information—it simply controls for that condition. The power curve established under the free sector has a poor predictive capability for wind power production under general wind conditions. In this sense, the IEC method is not created for power prediction or turbine performance assessment purposes. Rather, the IEC's intention is to create a standardized condition when a turbine's power production can be compared and verified. Accomplishing this is important for activities like contracting, in which a manufacturer's claim of its wind turbine's production ability ought to be verified at the time of a transaction, under a condition agreed upon by both parties.

For the purposes of wind power prediction, turbine control, and turbine performance assessment, all under general wind directions, it is more desirable to have a multi-dimensional power curve that can account for the effects of as many environmental variables as possible. Some works [17, 109, 165, 191] study the impact of having wind direction incorporated as one of the input variables, or as an additional covariate, and find that much can be gained by this inclusion. Bessa et al. [17] also include in their power curve a third covariate, in addition to wind speed and wind direction, which is either the time of the day or a look-ahead time step. Lee et al. [132] present one of the first truly multi-dimensional power curve models, referred to as the additive-multiplicative kernel (AMK) method, for both mean estimation, namely $\mathbb{E}(y|\boldsymbol{x})$, and for the density estimation, namely $f(y|\boldsymbol{x})$. The AMK power curve model can, in principle, take as many inputs as possible, although the test cases included in [132] use up to seven covariates.

5.2.1   Need for Nonparametric Modeling Approach

The underlying physics of wind power generation expressed in Eq. 5.2 provides some clues concerning a preferable power curve model. The following summarizes the observations:

1. There appear to be at least three important factors that affect wind power generation: wind speed, V, wind direction, D, and air density, ρ. This does not exclude the possibility that other environmental factors may also influence the power output.

2. The functional relationships between the environmental factors and the power response are generally nonlinear. The complexity partially comes from the lack of understanding of C_p, which is affected by many environmental factors (V, D, and ρ included). As there is no analytical expression linking C_p to any of the influencing factors, the functional form of the power curve is unknown.

3. The environmental factors appear in a multiplicative relationship in the power law equation, Eq. 5.2, indicating interactions among the factors.

The lack of precise physical understanding in power curve modeling presents an opportunity for data science methods. While developing data-driven power curve models, the second observation above makes it a compelling case for needing a nonparametric modeling approach to model a power curve, because the specific functional form of power curves is not known and

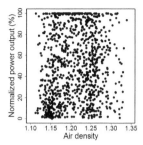 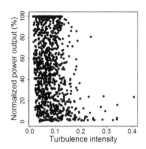 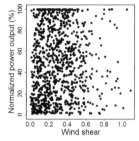

FIGURE 5.1 Scatter plots of the power output versus three environmental factors for $3.5 < V < 20, 0° < D < 360°$. Left panel: power versus air density; middle panel: power versus turbulence intensity; right panel: power versus wind shear. (Reprinted with permission from Lee et al. [132].)

can be rather different under various circumstances. A parametric approach, by contrast, is to assume a function of a known form with a set of unknown parameters, say, a polynomial function to some degree, and then estimate the unknown parameters using the data. The major shortcoming of the parametric approaches is its lack of flexibility, as there is no guarantee that the assumed functional form captures the true relationship between the power and the environmental inputs. The nonparametric approach, on the other hand, follows the philosophy of *"let the data speak for itself"* and can be much more adaptive without making too many assumptions. The IEC binning method is in and by itself a nonparametric method.

The third observation above touches upon the issue of factor interactions. To see this aspect more pointedly, consider the scatter plots in Figs. 5.1 and 5.2. Fig. 5.1 presents the scatter plots between wind power and three environmental variables. These scatter plots are unconditional on wind speed and wind direction. Under this setting, these environmental factors show no obvious effect on the power output. Fig. 5.2 presents the scatter plots between the same variables but instead under different wind speeds and wind directions. One does observe nonlinear relationships in the conditional plots, and the relationships appear to be different depending on the wind conditions. This implies that interaction effects do exist among wind speed, wind direction, and other environmental factors. A power curve model should hence characterize not only the nonlinear effects of wind speed and wind direction, but also the interaction effects among the environmental factors. The existence of interaction effects suggests that purely additive models or generalized additive models (GAM) are unlikely to work well in modeling a power curve.

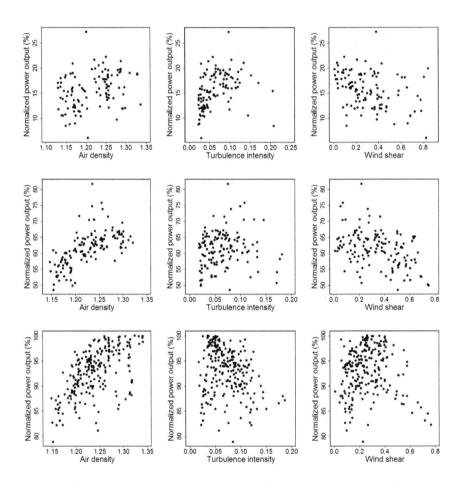

FIGURE 5.2 Scatter plots of power output versus environmental factors under specific wind speeds and wind directions. Top panel: $6.1 < V < 6.2, 270° < D < 300°$; middle panel: $9.1 < V < 9.2, 270° < D < 300°$; and bottom panel: $11.1 < V < 11.2, 270° < D < 300°$. (Reprinted with permission from Lee et al. [132].)

5.2.2 Kernel Regression and Kernel Density Estimation

Kernel regression or kernel density estimation methods have been used for modeling power curves [17, 109, 132]. The kernel-based methods appear to be a capable statistical modeling tool, not only capturing the complicated higher-order interaction effects but also avoiding the need to specify a functional form of the power curve relationship.

A kernel regression [86] is a type of localized regression method, which is to make an estimation, $\hat{y}$, at a target input value, x_0, by using observed data points close to x_0. This can be accomplished by a weighted average of the data points falling into a local neighborhood, such as

$$\hat{y}(x_0) = \sum_{x_i \in \mathfrak{N}(x_0)} w_i y_i,$$

$$\text{and} \quad \sum_i w_i = 1, \tag{5.4}$$

where $\mathfrak{N}(x_0)$ is the neighborhood of x_0, however it is defined, y_i is the observed response corresponding to input x_i, w_i is the weighting coefficient associated with y_i, and the constraint, $\sum_i w_i = 1$, is to ensure that the magnitude of $\hat{y}$ is consistent with that of y.

In the kernel regression, the localization is achieved by employing a weighting function symmetric with respect to x_0, known as the kernel function and denoted by $K(x_0, x_i)$. A kernel function is supposed to be integrable to one, following the same rationale above of requiring the summation of w_i's to be one. The kernel function has a bandwidth parameter λ that controls how fast the function decays from its peak towards zero and effectively defines the neighborhood, $\mathfrak{N}(x_0)$. Consider the one-dimensional Gaussian kernel function, taking the form of a normal probability density function,

$$K_\lambda(x_0, x_i) = \frac{1}{\sqrt{2\pi\lambda^2}} \exp\left(-\frac{\|x_0 - x_i\|^2}{2\lambda^2}\right), \tag{5.5}$$

where λ is equivalent to the standard deviation in a normal pdf. This kernel function is mathematically equivalent to the kernel function used in the support vector machine in Section 2.5.2. The term, $1/\sqrt{2\pi\lambda^2}$, in the above equation is the normalization constant to ensure that $K_\lambda(x_0, x)$ is integrable to one. When $K(\cdot, \cdot)$ is used to define the weighting coefficient, w_i, the normalization constant appears in both the numerator and denominator, so that it is cancelled out. For this reason, analysts have practically omitted the normalization constant and simply write the Gaussian kernel as

$$K_\lambda(x_0, x_i) = \exp\left(-\frac{\|x_0 - x_i\|^2}{2\lambda^2}\right). \tag{5.6}$$

This expression looks the same as that in Eq. 2.48 if one lets $\phi = 1/(2\lambda^2)$.

What matters in the kernel function is the difference, or the distance, between x_0 and x_i, just like in a stationary covariance function. Therefore, analysts choose to simplify the input arguments in $K(\cdot, \cdot)$ to be a single variable, say, $u = x_0 - x_i$. As such, the kernel function in kernel regression is often denoted by one of the following interchangeable expressions: $K_\lambda(u)$ or $K_\lambda(\|x_0 - x_i\|)$ or $K_\lambda(x_0, x_i)$.

Another way, arguably more common, to write the Gaussian kernel function is to first express it with a unit bandwidth, namely $\lambda = 1$, as

$$K(u) = \frac{1}{\sqrt{2\pi}} \exp\left(-\frac{\|u\|^2}{2}\right). \tag{5.7}$$

This $K(u)$ is integrable to one. It is then used as the building block for kernel function, $K_\lambda(u)$, with an arbitrary bandwidth λ—$K_\lambda(u)$ is referred to as the scaled kernel. Using $K(u)$, $K_\lambda(u)$ is written as

$$K_\lambda(u) = \frac{1}{\lambda} K\left(\frac{u}{\lambda}\right), \tag{5.8}$$

which gives back the expression in Eq. 5.5. The Gaussian kernel expression in Eq. 5.7 and Eq. 5.8 is used throughout the book when a kernel regression or a kernel density estimation is concerned.

Fig. 5.3, left panel, presents an illustration of a Gaussian kernel function. Suppose that one only has three data points, marked as #1, #2, and #3, respectively, and the corresponding data pair is $\{x_i, y_i\}$, $i = 1, 2, 3$. One wants to assess the response, $\hat{y}(x_0)$, at x_0. The weighting coefficient associated with each one of the data points is decided through the kernel function. Specifically,

$$w_i(x_0) = \frac{K_\lambda(\|x_0 - x_i\|)}{\sum_{j=1}^{3} K_\lambda(\|x_0 - x_j\|)}, \quad i = 1, 2, 3,$$

and using this weighting coefficient function, one has

$$\hat{y}(x_0) = w_1(x_0)y_1 + w_2(x_0)y_2 + w_3(x_0)y_3.$$

In this example, as illustrated in Fig. 5.3, left panel, points #1 and #2 have positive weights associated with them, whereas point #3 has a virtually zero weight, so that $\hat{y}(x_0)$ is effectively the weighted average of y_1 and y_2 at points #1 and #2, respectively. Other data points that are even farther away from x_0 than #3 hardly affect the estimation of $\hat{y}(x_0)$ at all. One may consider that the neighborhood of x_0, $\mathfrak{N}(x_0)$, covers a certain distance from x_0 on either side and contains #1 and #2 but not #3. Eq. 5.4, once factoring in the neighborhood constraint for this three-point case, can be expressed as

$$\hat{y}(x_0) = w_1(x_0)y_1 + w_2(x_0)y_2, \quad w_1(x_0) + w_2(x_0) = 1.$$

If one moves x_0 continuously from one end of the input domain to the other end and estimates $\hat{y}(x_0)$ at every x_0 using the same kernel-based localized

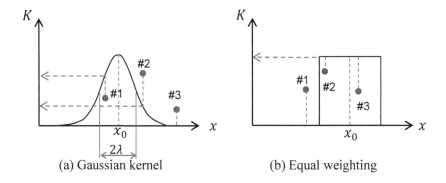

(a) Gaussian kernel (b) Equal weighting

FIGURE 5.3 Gaussian kernel function versus rectangular window function used in IEC binning. (Reprinted with permission from Ding et al. [50].)

regression, one practically estimates the input-output functional relationship between x and y. Let us drop the subscript "0" from the input variable, as it is now a generic input variable. The estimated relationship is denoted by $\hat{y}(x)$. The kernel regression leading to $\hat{y}(x)$, based on n pairs of data points, $\{(x_1, y_1), \ldots, (x_n, y_n)\}$, is

$$\hat{y}(x) = \sum_{i=1}^{n} w_i(x) y_i,$$

$$\text{where} \quad w_i(x) = \frac{K_\lambda(\|x - x_i\|)}{\sum_{j=1}^{n} K_\lambda(\|x - x_j\|)}. \tag{5.9}$$

This estimator in Eq. 5.9 is in fact the Nadaraya-Watson kernel regression estimator [152, 226]. The kernel function used therein does not have to be always the Gaussian kernel. The Gaussian kernel function does not go exactly to zero unless $\|x - x_i\| \to \infty$. There are other kernel functions, for instance, the Epanechnikov kernel function, that defines a window, so that the kernel function value is precisely zero outside the window.

The kernel regression is considered a nonparametric approach not because the kernel function does not have parameters or not use a functional form; in fact it does have a parameter, which is the bandwidth. Being a nonparametric approach, the kernel function is different from the target function, $y(x)$, that it aims at estimating. While the target functions may vary drastically from one application to another, the kernel function used in the estimation can remain more or less the same. The non-changing kernel regression is able to adapt to the ever-changing target functions, as long as there are enough data. The parameter in the kernel function serves a role differing substantially from the coefficients in a linear regression. The coefficients in a linear regression play a direct role in connecting an input x to the response y, while the bandwidth parameter in kernel functions, by contrast, defines the neighborhood and plays a rather indirect role in connecting x to y.

Comparing the kernel regression with the binning method, one notices that the binning method can be considered as a special kernel regression that uses a uniform kernel function; see the illustration in Fig. 5.3, right panel. A uniform kernel function is a rectangular window function, giving equal weights to all data points within the window, regardless of how far away they are from x_0. Once a data point is outside the function window (point #1), its weight is zero. The final estimate at x_0 is a simple average of all y's associated with the data points within the window (points #2 and #3 in this case). Of course, the IEC binning method is not really a kernel model due to another important reason: in the kernel regression, the kernel function moves continuously along the x-axis, producing a continuous, smooth curve, while the window functions in the binning method are disjoint, so that the resulting function response from the binning method, if magnified enough, is discretized.

When analysts use the weighted average of y values of the data points falling under the kernel function, the resulting outcome is a point estimate of the function value at a given input x. The kernel method, nonetheless, is capable of producing the estimate of a probability density function of y, conditioned on x, namely $f(y|x)$. The way of doing this is very similar to that in Eq. 5.9 but notice a different requirement here—for a density estimation, the left-hand side is supposed to be a density function, rather than a point estimate. One idea is to replace y_i with a density function, centered at y_i, so that the weighted averaging acts now on a series of density functions and thus results in a density function as well. Recall that the Gaussian kernel function is in fact a density function, and as such, a conditional density estimation [99, 184] can be obtained through a formula like

$$\hat{f}(y|x) = \sum_{i=1}^{n} w_i(x) K_{\lambda_y}(y - y_i), \tag{5.10}$$

where $w_i(x)$ is likewise defined as in Eq. 5.9, but λ is to be replaced with λ_x, as the bandwidth parameters associated with x may differ from that with y.

5.2.3 Additive Multiplicative Kernel Model

The AMK method [132] is a kernel-based approach. Prior to AMK, a bivariate conditional kernel density including wind speed and direction is used in [109] and a trivariate kernel density is used in [17]. AMK goes beyond the plain version of a kernel density estimation or a kernel regression and employs a special model structure that allows it to handle the multi-dimensional inputs in power curve modeling.

Recall the conditional density estimate $\hat{f}(y|x)$ in Eq. 5.10. For a multivariate input, denoted by $\boldsymbol{x}$, let us re-write Eq. 5.10 as

$$\hat{f}(y|\boldsymbol{x}) = \sum_{i=1}^{n} w_i(\boldsymbol{x}) K_{\lambda_y}(y - y_i), \tag{5.11}$$

where

$$w_i(\boldsymbol{x}) = \frac{K_{\lambda_x}(\|\boldsymbol{x} - \boldsymbol{x}_i\|)}{\sum_{j=1}^{n} K_{\lambda_x}(\|\boldsymbol{x} - \boldsymbol{x}_j\|)}, \tag{5.12}$$

$\boldsymbol{\lambda}_x = (\lambda_1, \ldots, \lambda_q)$ is the vector of the bandwidth parameters associated with the environmental factors in $\boldsymbol{x}$, and q is the number of explanatory variables in $\boldsymbol{x}$. In the above formulation, $K_{\lambda_x}(\|\boldsymbol{u}\|)$, where $\boldsymbol{u} = \boldsymbol{x} - \boldsymbol{x}_i$, is a multivariate kernel function and is composed of a product kernel that is a multiplication of univariate kernel functions, such as

$$K_{\lambda_x}(\|\boldsymbol{u}\|) := K_{\lambda_1}(u_1) K_{\lambda_2}(u_2) \cdots K_{\lambda_q}(u_q). \tag{5.13}$$

Here $K_{\lambda_j}(u_j)$ is generally a univariate Gaussian kernel, except for wind direction, D. The kernel function for D is chosen to be the von Mises kernel [212], because D is a circular variable that may cause trouble in numerical computation, had a Gaussian kernel been used. For more discussion regarding the handling of circular variables, please refer to [143, 144, 145]. The von Mises kernel function can characterize the directionality of a circular variable and takes the form of

$$K_\nu(D - D_i) = \frac{\exp\{\nu \cos(D - D_i)\}}{2\pi I_0(\nu)}, \tag{5.14}$$

where $I_0(\cdot)$ is the modified Bessel function of order 0, and ν is the concentration parameter of the von Mises kernel, which has now taken the role of the inverse of the bandwidth parameter λ_D.

In addition, the mean of the conditional density estimator in Eq. 5.11 provides an estimation of the conditional expectation, $\hat{y}(\boldsymbol{x}) := \mathbb{E}(y|\boldsymbol{x})$, as

$$\hat{y}(\boldsymbol{x}) = \int y \hat{f}(y|\boldsymbol{x}) dy. \tag{5.15}$$

Hydman et al. [99] note that the estimator in Eq. 5.15 is equivalent to the Nadaraya-Watson regression estimator in Eq. 5.9 with the input variable in Eq. 5.9 replaced by its multivariate counterpart, $\boldsymbol{x}$.

The AMK method [132] does not simply use the multivariate kernel as is. The reason is two-fold. One is concerning data scarcity in a multi-dimensional space. With wind data arranged in 10-minute blocks, one year's worth of data translates to slightly over 52,000 data pairs, which could still become scarce in a multi-dimensional factor space. The dimensionality of the input space, if using the list of elements in $\boldsymbol{x}$ given in Section 1.1, is seven. When 52,000 data points are dispersed into the seven-dimensional space, certain combinations of environmental conditions could have very little data or even no data at all, thereby deteriorating the performance of the resulting multivariate kernel model. In the future, technology innovation will almost surely make additional measurements available, so that a truly multi-dimensional power curve model should be able to entertain as many input variables as realistically possible. The scalability issue is also the reason why the IEC binning is not used for multi-dimensional cases. The second difficulty is that running a multi-dimensional kernel-based conditional density estimation takes longer computational times than analysts typically prefer. It is thus desirable to use fewer input variables to form the multivariate product kernels if possible.

Lee et al. [132] tailor the power curve modeling to an additive-multivariate model structure, which gives their kernel model the name AMK. The idea is to form a series of product kernel functions taking three input variables each, allowing up to three-factor interactions to be modeled. The use of trivariate kernels helps alleviate the data scarcity concern, as a trivariate kernel only needs to handle a three-dimensional space, low enough to avoid the curse of dimensionality. For high-dimensional covariates (more than three), AMK pools multiple trivariate product kernels together in an additive structure.

For notation simplicity, let us designate the first two elements of x, namely x_1 and x_2, as V and D, respectively. Other environmental variables are denoted by x_j, $j = 3, ..., q$. AMK employs the following model structure,

$$\hat{f}(y|x) = \sum_{i=1}^{n} \frac{1}{(q-2)} \left[w_i(x_1, x_2, x_3) + \cdots + w_i(x_1, x_2, x_q) \right] K_{\lambda_y}(y - y_i),$$

$$\hat{y}(x) = \frac{1}{(q-2)} \left[\hat{y}(x_1, x_2, x_3) + \cdots + \hat{y}(x_1, x_2, x_q) \right].$$

(5.16)

In the above expression, AMK keeps the multivariate kernels but limits them to be product kernels of three inputs. Based on the observations from Fig. 5.2, it is believed that it is important to include V and D, incorporating wind speed and direction information, in every multivariate kernel so that the trivarite kernels can capture the interaction effect between the third environmental factor with wind speed and wind direction. AMK can be used for high-dimensional data without causing computational or data sparsity problems. When additional explanatory variables become available, AMK would include extra additive terms, each of which has the same structure as the current terms, namely a trivariate kernel having inputs of V, D, and a third explanatory variable.

5.2.4 Bandwidth Selection

The key parameters in AMK are the bandwidth parameters, λ_y and $\boldsymbol{\lambda}_x$. Lee et al. [132] employ a data-driven selection criterion, known as the integrated squared error (ISE) criterion [61, 83], as follows,

$$
\begin{aligned}
\mathrm{ISE}(\boldsymbol{\lambda}_x, \boldsymbol{\lambda}_y) &= \int \int \left(f(y|\boldsymbol{x}) - \hat{f}(y|\boldsymbol{x}) \right)^2 f(\boldsymbol{x}) dy d\boldsymbol{x} \\
&= \int \int \hat{f}(y|\boldsymbol{x})^2 f(\boldsymbol{x}) dy d\boldsymbol{x} - 2 \int \int \hat{f}(y|\boldsymbol{x}) f(y|\boldsymbol{x}) f(\boldsymbol{x}) dy d\boldsymbol{x} \\
&\quad + \int \int f(y|\boldsymbol{x})^2 f(\boldsymbol{x}) dy d\boldsymbol{x} \\
&= I_1 - 2I_2 + I_3.
\end{aligned}
$$

(5.17)

With this criterion, one would choose the bandwidths that minimize the ISE. Because I_3 in the ISE expression does not depend on the bandwidth selection, it can be omitted during the minimization of ISE. For I_1 and I_2, Fan and Yim [61] suggest using a leave-one-out cross-validation estimator as

$$
\begin{aligned}
\hat{I}_1 &= \frac{1}{n} \sum_{i=1}^{n} \int \left(\hat{f}_{-i}(y|\boldsymbol{x}_i) \right)^2 dy, \quad \text{and} \\
\hat{I}_2 &= \frac{1}{n} \sum_{i=1}^{n} \hat{f}_{-i}(y_i|\boldsymbol{x}_i),
\end{aligned}
$$

(5.18)

where $\hat{f}_{-i}(y|\boldsymbol{x}_i)$ is the estimator $\hat{f}(y|\boldsymbol{x}_i)$ with the i-th data pair $\{\boldsymbol{x}_i, y_i\}$ omitted. The data-driven bandwidth selection is simply to choose the bandwidths $\boldsymbol{\lambda}_x$ and $\boldsymbol{\lambda}_y$ that minimize $\hat{I}_1 - 2\hat{I}_2$.

Using this cross-validation algorithm could, however, take a long time. In order to have a faster bandwidth selection for practical purposes, Lee et al. choose to employ a simpler, greedy procedure to select the bandwidth parameters one at a time, as described in Algorithm 5.1.

In the algorithm, to handle the von Mises kernel, Lee et al. [132] follow an approach suggested in [212] that ties the concentration parameter ν to the bandwidth parameter λ_2 as $\nu = 1/\lambda_2^2$. Then, λ_2 can be selected together with other bandwidth parameters for the Gaussian kernels.

In R, the package kernplus implements the kernel regression, i.e., the mean function estimation $\hat{y}(\boldsymbol{x})$ in Eq. 5.16. Suppose that wind data is stored in the matrix windpw. The syntax to fit a multi-dimensional power curve is

```
pc.est <- kp.pwcurv(windpw$y, windpw[, c('V', 'D', 'rho', 'I',
              'Sb')], id.spd = 1, id.dir = 2),
```

where the two arguments, id.spd=1 and id.dir=2, indicate the first two columns in the data matrix are, respectively, the wind speed and wind direction data. Five covariates are included in this example, i.e., $\boldsymbol{x} = (V, D, \rho, I, S_b)$.

5.3   OTHER DATA SCIENCE METHODS

Addressing the multi-dimensional power curve problem is essentially a regression problem. For this matter, other data science methods, especially those

---

Algorithm 5.1 Greedy kernel bandwidth selection.

---

1. Consider only a simple univariate kernel regression corresponding to individual environmental variables in x.

2. Calculate the bandwidth for each univariate kernel following the direct plug-in (DPI) approach suggested in [187]. This DPI estimator can be obtained by using the `dpill` function in the `KernSmooth` package and performing it on one input variable at a time, such as

$$\hat{\lambda}_j \text{ <- dpill}(x_j, \ y).$$

3. Denote the resulting bandwidths as $(\hat{\lambda}_1, \hat{\lambda}_2, \ldots, \hat{\lambda}_q)$;

4. Use a basic power curve model that includes only the wind speed, V and wind direction, D as inputs, and fix the bandwidths for the two univariate kernels corresponding to V and D as $\hat{\lambda}_1$ and $\hat{\lambda}_2$, respectively. Then, estimate the bandwidth $\hat{\lambda}_y$ that minimizes $\hat{I}_1 - 2\hat{I}_2$.

---

of semi-parametric or nonparametric nature, can be employed as well. Two methods introduced previously, the support vector machine in Section 2.5.2 and artificial neural network in Section 2.5.3, can certainly be applicable. As argued in Section 5.2.1, parametric regression methods are less effective and not robust in power curve modeling.

In this section, we would like to introduce three more data science methods: k-nearest neighborhood (kNN), tree-based methods, and spline-based methods; please also refer to [86] for the basics about these methods. Most of the methods produce only the mean estimation, $\hat{y}(x)$, but Bayesian additive regression trees (BART) [36], being a Bayesian method, naturally produces the posterior distribution, leading to the density estimation, $\hat{f}(y|x)$.

5.3.1 k-Nearest Neighborhood Regression

The idea of kNN is fairly simple. For a prediction at any target site, the method uses the average of the closest k data points. Suppose that we have n data points in the training set and want to make a prediction at x_0. Then, the kNN regression at x_0 is

$$\hat{y}(x_0) = \frac{1}{k} \sum_{x_i \in \mathfrak{N}_k(x_0)} y_i, \tag{5.19}$$

where the subscript k in the neighborhood notation, $\mathfrak{N}_k$, signifies that this neighborhood contains exactly k data points. The parameter in kNN is the neighborhood size, k, which needs to be selected *a priori*, usually through cross validations. In the power curve modeling, kNN is used for regression.

It can also be used for classification in other applications. For classification, a data instance will be assigned to the class to which the majority of data instances belong in the neighborhood.

One may notice that the kNN above looks awfully similar to the kernel regression in Eq. 5.4, especially once letting $w_i = 1/k$ for all i's. But there are a couple of differences. The kNN regression uses a simple average of all data points in its neighborhood, whereas the kernel regression uses a weighted average. In the terms of averaging, kNN is the same as the binning method that uses the uniform kernel.

A more important difference between kNN and kernel regression is in the definition of the neighborhood. The neighborhood in the kernel regression is decided through the use of a specific kernel function and its bandwidth parameter λ. Kernel regression does not directly control the number of data points in its neighborhood but once λ is chosen, the size of the neighborhood is more or less decided. By contrast, the kNN regression decides its neighborhood through setting a specific amount on the data points closest to the target site. The closeness metric used in kNN is usually based on the Euclidean distance, although it could be other distance alternatives.

To appreciate the difference of neighborhood definition between kNN and kernel regression, consider the following analogy. Pretend that two kids in a kindergarten want to decide who can be their friends. Alexandra declares whoever is similar enough to her (based on her definition of similarity) is her friends, and that it does not matter how many friends she ends up with. Nicholas says that he wants precisely k friends and those k kids who live closest in distance to him are his friends, regardless of how different they may be from each other. Alexandra uses the same approach as in the kernel regression, whereas Nicholas uses the kNN approach.

To use kNN for regression, analysts can call the `knn.reg` function in the FNN package. The syntax is

$$\text{knn.reg(training inputs, test inputs, } y, \ k).$$

If k is not specified, the default value is three.

5.3.2 Tree-based Regression

A tree-based model is commonly known as the Classification and Regression Trees (CART) [86]. Consider the objective of regression. Recall the learning formulation of SVM in Eq. 2.47. In fact, that formulation is extendable to a general class of learning problems, known as machine learning through regularization, which entails three key components: a loss function, $L(\cdot, \cdot)$, a penalty function, Penalty$(\cdot)$, and the structure of the hypothesis space, $\mathcal{H}$, in which the optimization is conducted. Given a set of training data, $\{(\boldsymbol{x}_1, y_1), (\boldsymbol{x}_2, y_2), \ldots, (\boldsymbol{x}_n, y_n)\}$, the learning problem can be loosely formulated as,

$$\hat{g} = \underset{g \in \mathcal{H}}{\text{argmin}} \left[\sum_{i=1}^{n} L(y_i, g(\boldsymbol{x}_i)) + \gamma \cdot \text{Penalty}(g) \right], \qquad (5.20)$$

where γ is the penalty coefficient, trading off between the loss function and the penalty function. Note that Eq. 2.47 uses $\gamma/2$ as the penalty coefficient but the inclusion of the constant, "2," is for mathematical convenience and does not fundamentally change the learning outcome.

The learned function, $\hat{g}(\cdot)$, is used to make a prediction at a target site, say, $\boldsymbol{x}_*$, such that $\hat{y}(\boldsymbol{x}_*) = \hat{g}(\boldsymbol{x}_*)$. In the SVM regression, the loss function is the ϵ-sensitive error loss function, the penalty is a squared function, and $\mathcal{H}$ is the reproducing kernel Hilbert space, so that $g(\boldsymbol{x}) = \sum_{i=1}^{n} \alpha_i K(\boldsymbol{x}, \boldsymbol{x}_i)$. In CART, the loss function is a squared error loss function, the penalty function is the tree size, and the most important difference is that the hypothesis space, $\mathcal{H}$, is piecewise constant.

Basically, a CART partitions the input data space into J regions, each of them denoted by R_j, $j = 1, \dots, J$. For each region, CART uses a single constant, c_j, to represent it. As such, a CART model is represented by the following parameters: $\{J; R_1, \dots, R_J; c_1, \dots, c_J\}$. Then, the prediction using CART can be expressed as

$$\hat{y}(\boldsymbol{x}) = \sum_{j=1}^{J} c_j \cdot \mathbb{1}(\boldsymbol{x} \in R_j).$$

Since CART uses a squared error loss function, the representation for each region is the sample average of all data points falling into that region, meaning that the estimate of c_j is $\bar{y}_j$ for R_j.

Practically, a CART is built through a greedy algorithm, which is to perform a binary splitting on a variable, one at a time. The action every time splits an input domain into subregions. Once a splitting is carried out, it will not be revisited, even if it may not be the optimal splitting in hindsight. This greedy algorithm runs efficiently, in the complexity of $O(pn \log n)$, where p is the dimension of the input space and n is the data amount.

If one carries out this binary partition process, the process can be visualized through the growing of a binary tree—this is how the method gets its name. One such example in a two-dimensional space is presented in Fig. 5.4. In the tree, the whole region corresponds to the root node and the final subregions correspond to the terminal nodes. When one region is split to two subregions, the two subregions correspond to the two children nodes of the same parent node. Two nodes sharing the same parent node are called sibling nodes. The tree size is decided by the number of terminal nodes. Once a tree is established, the data points scattered in the original space, i.e., in the root node, are now dispersed into the respective terminal nodes.

The tree growing process starts with all the data. At each step, it considers a splitting variable j and split point s and defines the pair of the half-planes

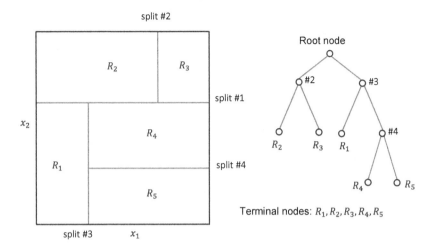

FIGURE 5.4 The two-dimensional example of binary splitting in the creation of a tree model.

as,

$$R_1(j, s) = \{x | x_j \leq s\} \text{ and } R_2(j, s) = \{x | x_j > s\}.$$

Note that R_1 and R_2 are the generic notations for any half planes at a splitting action and do not correspond to those same symbols used in Fig. 5.4. The tree model is built by searching all p elements in x, one at a time, and decide the best splitting variable and the corresponding split point at each step, by minimizing the following objective function,

$$\min_{j,s} \left[\sum_{x_i \in R_1(j,s)} (y_i - \hat{c}_1)^2 + \sum_{x_i \in R_2(j,s)} (y_i - \hat{c}_2)^2 \right] \qquad (5.21)$$

where $\hat{c}_j$ is the sample average of all data points in R_j.

The above is done by treating that the tree size, J, is fixed. The tree size represents the model complexity of CART. If the resulting tree is too small (too simple a model), the piecewise constant approximation is crude, causing a high bias in prediction, whereas if the resulting tree is too large (too complicated a model), then it fits the training data too hard, causing overfitting and again leading to poor prediction. Selecting an appropriate tree size is therefore important. A practical procedure of deciding the tree size is the bottom-up pruning. The basic idea is to build the largest possible tree first and then prune the large tree to an appropriate size. Pruning is the action of merging terminal nodes to reduce the size of the tree, as outlined in Algorithm 5.2.

Algorithm 5.2 Tree pruning procedure.

---

1. Start with the largest possible tree (split until some pre-specified minimum node size is reached);

2. Identify terminal siblings, which are sibling terminal nodes having the same parent node;

3. *Provisionally* merge each terminal sibling pair making the respective parent a terminal node (i.e., remove split);

4. Find the pair that, when merged, increases the fitting error of tree on training data the least;

5. Remove the corresponding split, making the parent node terminal, and creating a tree with one fewer terminal node;

6. Go to Step 2 until no splits left (i.e., only one root node).

The final tree size is chosen as J^* that minimizes the estimate of the test error using either an independent validation dataset or the cross-validation method.

Suppose that a tree method is applied to wind power data with a single input (wind speed). It could produce an outcome that looks like from the binning method. Using a tree, the final multi-bin result comes out of the iterations of binary splitting. One thing different, though, is that unlike the IEC binning using bins of equal width, the tree-based method less likely produces bins of equal size, because the actual split points depend on the solution to the optimization problem in Eq. 5.21.

According to Hastie et al. [86], Table 10.1, despite many appealing characteristics, CART has a relatively poor predictive power. This understanding motivates analysts to enhance the predictive power of a tree-based method through ensembling a set of trees. The general thought process is that a weak base learner like CART can be made much capable, or appreciably stronger, when many weak learners are made to work together. Specific ensembling mechanisms used include bagging, leading to bagged trees or the random forest (RF), or boosting, leading to the multiple additive regression trees (MART). Bayesian additive regression trees [36], or BART, a Bayesian version of sum of trees, is also an ensemble of trees, each of which explains a small and different portion of the predictive function. Conceptually, BART is closer to boosting than to bagging.

The technical details in BART are rather involved. For practitioners, it is advised to use the `bart` function in the `BayesTree` package. The syntax of using BART is

```
output<-bart(x.train, y.train, x.test),
```

where `output` is an R object and `output$yhat.test` contains the samples from the estimated conditional density function, $\hat{f}(y|\boldsymbol{x})$. Each column is a vector of samples drawn from $\hat{f}(y|\boldsymbol{x})$ for a specific $\boldsymbol{x}$ corresponding to a row in the input argument `x.test`, and the average of all the samples in that column is the corresponding conditional expectation, namely the point estimation, $\hat{y}(\boldsymbol{x})$.

5.3.3 Spline-based Regression

The spline-based regression is to use piecewise polynomials to model a nonlinear response. One of the popular spline functions used is the cubic spline [86]; see Fig. 5.5, middle panel, for an illustration. A cubic spline partitions the input domain into a few segments, which is in fact an action of binning, and models each segment using a cubic polynomial. In order to produce a smooth, coherent model for the whole domain, a cubic spline imposes continuity and smoothness constraints at the partition points, known as *knots*. In Fig. 5.5, two knots are used and denoted as ξ_1 and ξ_2, respectively. Although ξ_1 and ξ_2 partition the input domain in Fig. 5.5 into three roughly equal parts, knots in general do not have to be evenly spaced. Each cubic polynomial is specified by four parameters, producing a total of 12 parameters for the three piecewise cubic polynomials. The constraints imposed at the partition points, however, reduce the number of actual parameters that need to be estimated. For the cubic spline in Fig. 5.5, there are three constraints at each knot, which require, respectively, the equality of the function value, that of its first-order derivative and that of the second-order derivative, at each of the partition points. With the six constraints considered, the number of actual parameters to be estimated for the cubic spline is six.

One may have noticed that the spline method in fact injects the idea of binning into its action of modeling. If only using the idea of binning without the boundary constraints, however, the response looks like the plot in the right-most panel of Fig. 5.5. The three unconstrained piecewise cubic polynomials need a total of 12 parameters to specify. When a single global cubic polynomial is used to model a response, it uses four parameters, but its modeling adaptivity to local features is far worse than the other two alternatives. With only a slight increase in model complexity (measured by the number of parameters), the cubic spline is endowed with the level of modeling adaptivity as a binning method allows.

Analysts may argue that the binning method can use a single constant for each bin, so that the number of parameters for the right-most example in Fig. 5.5 can be three, instead of 12. The problem of this argument is that when using a constant to model a bin, the bin width needs to be much smaller, or equivalently, the number of bins needs to be much greater, so that a piecewise constant function can approximate a nonlinear response with sufficient accuracy. It is not unusual that with one single input variable such as wind speed, analysts need to use 20 bins to model the whole response. With 20 bins, the

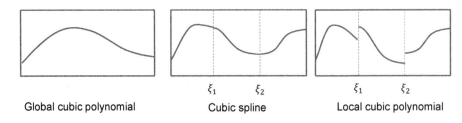

Global cubic polynomial   Cubic spline   Local cubic polynomial

FIGURE 5.5 Global cubic polynomial, cubic spline, and local cubic polynomials. (Reprinted with permission from Ding et al. [50].)

number of parameters cannot be fewer than 20, already producing a model that is unnecessarily complicated.

To use spline-based regression on actual applications, there are a few technical problems to be resolved. One of the problems is that the behavior of polynomial fit to data tends to be erratic near the boundaries. To fix the problem, analysts introduce the *natural cubic spline*, which postulates that the response function be linear beyond the boundary knots. This constraint translates to the continuous and continuous first derivative requirements at each boundary knot but it no longer requires continuous second derivatives, because the linear function outside the boundary knot does not have a second derivative.

Another technical problem is how many knots one should choose and where to position them. For addressing the knot selection problem, analysts introduce the *smoothing spline*. The smoothing spline finds a function that minimizes the following objective function,

$$\hat{g} = \operatorname{argmin}\left\{\sum_{i=1}^{n}[y_i - g(x_i)]^2 + \gamma \int_t \{g''(t)\}^2 dt\right\}. \tag{5.22}$$

This optimization formulation is consistent with the general regularization learning problem expressed in Eq. 5.20. Here the loss function is a squared error loss, the penalty function is the integration of the second derivative on $g(\cdot)$, and the hypothesis space $\mathcal{H}$ is that $g(\cdot)$ should have two continuous derivatives. When $\gamma = 0$, then g can be any function that interpolates the training data points, making the loss function zero, while when $\gamma = \infty$, it forces $g''(t) = 0 \ \forall t$, resulting a simple least squares line fit.

The solution of the above optimization formulation is that the smoothing spline is a natural cubic spline with knots at the unique values of every x_i in the training dataset. At the face value, a natural cubic spline with knots at each and every x_i implies that the spline may have as many as n knots and may cause over-parametrization and overfitting. In actuality, when $\gamma > 0$, the number of *effective* knot positions, or the effective number of parameters in the resulting model, can be much smaller than n. Which one of the x_i's to be selected as a knot, or which other to stay away, depends on the outcome of the optimization in Eq. 5.22.

In Exercise 3.1, we mention that for a linear smoother with a smoother matrix $\mathbf{S}$, the effective number of parameters is $\text{tr}(\mathbf{S})$. The spline regression can in fact be expressed as a linear smoother, so that the effective number of knots or the effective number of parameters can be decided by the trace of the smoother matrix (see Exercise 5.4). The resulting effective number of parameters apparently depends on the choice of γ, the cost coefficient trading off between the loss function and the penalty function.

One may have noticed that we use a scalar x_i in Eq. 5.22, implying a univariate regression and smoothing. This is because the spline-based regression, in its most general form, is not particularly scalable. When it comes to handling multivariate covariates like in the circumstance of building a multi-dimensional power curve, the plain version of smoothing splines is difficult to use. There are two popular multivariate extensions of the spline methods. One is the smoothing spline ANOVA (SSANOVA) [79] and the other one is the multivariate adaptive regression splines (MARS) [68]. MARS is used in Chapter 10 for extreme load analysis. The smoothing spline ANOVA is used here to be one of the alternatives for multi-dimensional power curve modeling.

SSANOVA employs a functional ANOVA (analysis of variance) decomposition that limits the interactions to the two-way interactions and ignores the higher-order terms, so as to help rein in the curse of dimensionality. A higher-order interaction can be included but through a recursive two-way interaction mechanism. For instance, a four-way interaction is modeled as adding one extra factor to an already existing three-way interaction term. This strategy is in fact similar to what is used in MARS, as MARS accomplishes the scalability also through a hierarchical inclusion of interaction terms.

Analysts can use the `ssanova` function from the `gss` package in R for implementing the smoothing spline ANOVA method.

5.4   CASE STUDY

Two datasets are used in this case study, the `Inland Wind Farm Dataset1` and `Offshore Wind Farm Dataset1`. Table 5.1 summarizes the specifications of the datasets; for certain entries an approximation rather than the accurate value is given for the protection of the identities of the turbine manufacturers and wind farms. Fig. 5.6 presents the turbines/masts layout and turbine-to-mast distances.

5.4.1   Model Parameter Estimation

The quality of point estimation is to be evaluated using RMSE, while that of the density estimation is to be evaluated using CRPS. An out-of-sample test is to be conducted based on a five-fold cross validation. In each iteration of the five-fold cross validation, a dataset is divided into a partition of 80%

TABLE 5.1   Specifications of turbines in the two wind farms.

Wind farm	Inland	Offshore
Number of met masts	Multiple	Single
Number of wind turbines	200+	30+
Hub height (m)	80	70
Rotor diameter (m)	about 80	about 90
Cut-in wind speed (m/s)	3.5	3.5
Cut-out wind speed (m/s)	20	25
Rated wind speed (m/s)	around 13	around 15
Rated power (MW)	1.5–2.0	around 3
Location	Inland, U.S.	Offshore, Europe

Source: Lee et al. [132]. With permission.

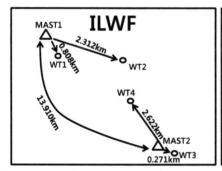 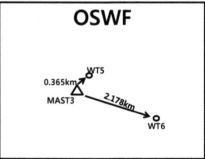

FIGURE 5.6 Layout of the turbines and masts and turbine-to-mast distances. ILWF: inland wind farm; OSWF: offshore wind farm. (Reprinted with permission from Lee et al. [132].)

for training and 20% for testing. Then, the average of the five error values is reported to represent a method's performance. Note that in Part II of the book, the performance metrics are computed in terms of wind power, requiring a change in notation to be made to the formulas of the performance metrics presented in Section 2.6. Specifically, we use

$$\text{RMSE} = \sqrt{\frac{1}{n_{\text{test}}} \sum_{i=1}^{n_{\text{test}}} \left(\hat{y}(\boldsymbol{x}_i) - y_i\right)^2}, \qquad \text{and}$$

$$\text{CRPS} = \frac{1}{n_{\text{test}}} \sum_{i=1}^{n_{\text{test}}} \int \left(\hat{F}(y|\boldsymbol{x}_i) - \mathbb{1}(y > y_i)\right)^2 dy. \tag{5.23}$$

Algorithm 5.1 works well for estimating the bandwidth parameters. For point estimation, Lee et al. [132] are able to use all the training data for bandwidth selection and the computational time is of no concern at all. But for density estimation, even with the greedy algorithm, the last step (Step 4 in Algorithm 5.1) that finds the bandwidth for y would still take a long running time, had all the training data been used. In the end, Lee et al. decide to randomly select 25% of the training data for estimating the bandwidth used in density estimation.

For the out-of-sample testing, Lee et al. are able to use all the testing data points for computing the out-of-sample RMSEs. But for computing the CRPS values, using all the testing data again requires a long time. Lee et al. find that using 1,000 randomly sampled data points to calculate the CRPS values remains reasonably stable over different random sampling. As such, the CRPS values reported here are based on 1,000 test data points.

Note that the RMSE values of the same method in this section may differ slightly in different studies because of the random split of the training and test datasets. The RMSE values here see a more noticeable difference from those presented in [132] because of two reasons: (a) the RMSE values associated with AMK are computed using the `kernplus` package. Although based on the algorithm originally developed in [132], the implementation in the `kernplus` package made small changes to deal with certain application complexities. (b) What is reported here is the average of a five-fold cross validation, whereas what was reported in [132] was based on a one-time 80-20 random split. Despite the numerical differences, the main message from the numerical studies stays the same as in [132].

5.4.2 Important Environmental Factors Affecting Power Output

Based on the physical understanding hinted by the power generation law in Eq. 5.2, it is apparent that wind speed, direction, and air density are important factors to be included in a power curve model. The question is what else may also need to be included.

To address that question, the first set of results is to show the RMSEs when

AMK includes a single additive term from $\boldsymbol{x} = (x_1, x_2, x_3)$ to $\boldsymbol{x} = (x_1, x_2, x_q)$, where $q = 5$ in the `Inland Wind Farm Dataset1` and $q = 7$ in the `Offshore Wind Farm Dataset1`. Recall that each additive term is a trivariate kernel with the first two of the variables always being the wind speed, V, and wind direction, D.

The baseline model for comparison is chosen to be the kernel model that has only the wind speed and wind direction (V, D) in a product kernel, which is in fact the same as the one used in [109]. This bivariate kernel model is referred to as BVK.

The results are shown in Table 5.2, in which the notation of $(\cdot, \cdot, \rho)$ means that the additive term included in the model has the wind speed, V, and wind direction, D, and air density, ρ, as its inputs, where the wind speed and wind direction are shorthanded as two dots. Other notations follow the same convention. These results lead to the following observations:

1. In both the inland wind farm and offshore wind farm, air density, ρ, is indeed, after the wind speed and wind direction, the most significant factor in wind power generation. Including ρ in the model delivers reductions in RMSE from 9% to 17% across the board. This outcome is consistent with the physical understanding expressed earlier.

2. For the offshore wind turbines, humidity, H, appears to be another important factor in explaining variations in power outputs. Because humidity measurements are not available in the `Inland Wind Farm Dataset1`, it is unknown whether humidity is also a significant factor there.

3. The remaining three factors, namely turbulence intensity and the two wind shears, which each represents some other aspects of wind dynamics, show also positive impact, except for the case of WT5. The impact of turbulence intensity is rather pronounced for the inland turbines, nearly as significant as humidity on the offshore turbines. The impact of the below-hub wind shear is noticeably positive, although not as much as turbulence intensity. Both factors have shown more obvious effects for the inland turbines than for the offshore ones, but the significance of their impact is definitely after that of ρ.

The next step undertaken is to determine which other factors may impact the power output when AMK includes more than one additive term, conditional on the factors that have already been included. Based on the observations expressed above, for both inland and offshore turbines, the first additive term included is always (V, D, ρ). For the inland turbines, in addition to this first term, there are two more terms that have either turbulence intensity, I, or the below-hub wind shear, S_b. For the offshore turbines, a second additive term, (V, D, H), is also always included. In addition to the first two terms, there are three more terms that have either the two wind shears, S_a, S_b, or turbulence intensity, I. The two wind shears are always included or excluded together in the numerical analysis to keep the total number of model

TABLE 5.2 Impact on RMSE when including different environmental factors. The percentages in the parentheses are the reduction in terms of RMSE when the corresponding model's point estimation is compared with that of BVK.

WT	BVK	$(\cdot,\cdot,\rho)$	$(\cdot,\cdot,I)$	$(\cdot,\cdot,S_b)$	$(\cdot,\cdot,S_a)$	$(\cdot,\cdot,H)$
WT1	0.0884	0.0748 (15.3%)	0.0856 (3.1%)	0.0869 (1.7%)	·	·
WT2	0.0921	0.0814 (11.6%)	0.0887 (3.8%)	0.0894 (3.0%)	·	·
WT3	0.0817	0.0681 (16.7%)	0.0755 (7.6%)	0.0747 (8.6%)	·	·
WT4	0.1163	0.1030 (11.4%)	0.1093 (6.0%)	0.1109 (4.6%)	·	·
WT5	0.0907	0.0824 (9.1%)	0.0928 (−2.2%)	0.0917 (−1.1%)	0.0922 (−1.6%)	0.0858 (5.4%)
WT6	0.0944	0.0815 (13.6%)	0.0939 (0.5%)	0.0918 (2.7%)	0.0927 (1.7%)	0.0873 (7.5%)

TABLE 5.3 Model comparison using data in the Inland Wind Farm Dataset1. RMSE values are reported in the table. Boldface values are the smallest RSME in the row.

WT	$(\cdot,\cdot,\rho)$	$(\cdot,\cdot,\rho,I)$	$(\cdot,\cdot,\rho,S_b)$	$(\cdot,\cdot,\rho,I,S_b)$
WT1	0.0747	0.0743	**0.0742**	0.0751
WT2	0.0816	**0.0800**	0.0802	0.0802
WT3	0.0680	0.0651	**0.0645**	0.0646
WT4	0.1028	**0.1001**	0.1010	0.1004

comparisons manageable. Tables 5.3 and 5.4 present the model comparison results.

For some of the inland turbines, the best AMK explaining their power output includes the input factors of the wind speed and wind direction (V and D), air density (ρ), and turbulence intensity (I), while some others include the wind speed and wind direction (V and D), air density (ρ), and wind shear (S_b). These versions differ marginally. For the offshore turbines, it is rather clear that the model with the wind speed (V), wind direction (D), air density (ρ), and humidity (H) produces the lowest RMSE. Including other environmental factors in the model could instead increase the RMSE. The increase in RMSE is consistent and can be as much as 5.1% for WT6. If the above analysis is repeated using CRPS, the insights remain the same, but the presentation of the CRPS results is omitted.

TABLE 5.4 Model comparison using data in the Offshore Wind Farm Dataset1. RMSE values are reported in the table. Boldface values are the smallest RSME in the row.

WT	$(\cdot, \cdot, \rho, H)$	$(\cdot, \cdot, \rho, H, I)$	$(\cdot, \cdot, \rho, H, S_a, S_b)$	$(\cdot, \cdot, \rho, H, I, S_a, S_b)$
WT5	**0.0790**	0.0801	0.0810	0.0818
WT6	**0.0800**	0.0816	0.0822	0.0832

5.4.3 Estimation Accuracy of Different Models

In this subsection, we compare various power curve methods. In the comparisons, AMK is selected based on the best subset of variables revealed in the previous section. Other methods use the same subset of variables to level the playground.

Let us first take a look at the prediction errors of IEC binning method and AMK. The binning method used here is the air density corrected version. Only the RMSE values are presented, because the IEC binning does not produce a density estimation.

Table 5.5 presents the comparison. The reduction in terms of RMSE made by AMK over IEC is astonishing, but it may not be that surprising. Recall that we mention earlier in this chapter that the IEC method's intention is to provide a benchmark for verification purpose, rather than providing a method for real-life performance assessment or wind power prediction. Consider the following analogy in the context of vehicle fuel economy. At the time of sale, a new car is displayed with a published fuel economy, in the unit of miles per gallon. The published fuel economy value is obtained under a standardized, ideal testing condition, which cannot be replicated in real-life driving. A car's real-life fuel economy based on someone's actual driving is almost always worse than the published one. In the wind power production, engineers observe something similar—using the IEC binning power curve often leads to a conclusion of under performance, which is to say that the actual power production falls short of the prediction.

Still, car buyers and car manufacturers feel that the fuel economy obtained under the ideal condition provides a reasonable benchmark, offering some ballpark ideas of how fast a car consumes its fuel. However, for consumers who care very much about the real-life fuel economy, such as in commercial driving, they are not advised to use the published fuel economy value, as using the published value will surely lead to biased estimations. The same conclusion should have been extended to the IEC method, but in actuality, in the vacuum of robust, reliable, and capable power curve models, the IEC binning method is routinely used in the tasks or for the missions it is not designed for.

We would also like to articulate one important limitation of the IEC binning method. From the wind power production law in Eq. 5.2, we understand that the inclusion of air density is important. That is the reason why the IEC binning uses the air density correction. The same piece of information (air

TABLE 5.5   Compare the RMSE between the IEC binning method and AMK.

Wind Farm	Turbine	IEC	AMK	Error reduction rate over IEC
Inland	WT1	0.1305	0.0742	43%
	WT2	0.1158	0.0800	31%
	WT3	0.1217	0.0645	47%
	WT4	0.1567	0.1001	36%
Offshore	WT5	0.0970	0.0790	19%
	WT6	0.1089	0.0800	27%

density) can be included in AMK as well. We wonder—by making use of the same covariate, which method will benefit more? The benefit of including air density can be discerned by comparing the same method with and without using air density while making wind power prediction. For the IEC binning, this is between the plain version of binning and the air density adjusted binning. For AMK, this is between the AMK with only wind speed and wind direction (which is in fact BVK) and the AMK with wind speed, wind direction, and air density.

Table 5.6 presents the comparison using the four inland turbines, but the same conclusion is extendable to the offshore turbine data as well. The percentage values reported in parentheses are the reductions in terms of RMSE between the two versions of the same method, rather than the reduction between the two different methods. Take WT1 as an example. The -0.1% in the fourth column means a very slight increase in RMSE when using the air density adjusted binning method, as compared to the plain version of the binning method, whereas the 15.3% in the sixth column is the reduction in RMSE when using the AMK with inputs (V, D, ρ), as compared to AMK with inputs (V, D).

The comparison makes it clear that while air density is an important factor to be included in a power curve model, the air density adjustment used in the IEC binning is not optimal. It does help reduce $1 - 2\%$ in RMSE in fitting the wind power. But on the other hand, the potential benefit of having air density in a power curve model is much greater. When used in AMK, it can help reduce RMSE as much as 17%. This example demonstrates the power of data science methods over a pure engineering heuristics.

Table 5.7 further compares the point estimation, in terms of RMSE, among the four data science-based multi-dimensional power curve methods: kNN, SSANOVA, BART, and AMK. In this comparative study, kNN uses the normalized covariates, i.e., each covariate is normalized by its standard deviation, while the other methods use the raw measurements. The action of normalization has a profound impact on kNN but not so much on other methods. When

TABLE 5.6 Impact of air density on IEC binning and on AMK. Reported below are the RMSE values.

Wind Farm	Turbine	IEC Binning		AMK	
		Plain	Adjusted	BVK	BVK + air density
Inland	WT1	0.1303	0.1305 (−0.1%)	0.0884	0.0748 (15.3%)
	WT2	0.1180	0.1158 (1.9%)	0.0921	0.0814 (11.6%)
	WT3	0.1237	0.1217 (1.6%)	0.0817	0.0681 (16.7%)
	WT4	0.1592	0.1567 (1.6%)	0.1163	0.1030 (11.4%)

using SSANOVA, a full model considering all possible interactions takes too long to run. Instead, the main effects and selected interactions are included. For WT5 and WT6, V, D, ρ, H, $V \times D$, $V \times \rho$, $V \times H$, $V \times D \times \rho$, and $V \times D \times H$ are included in the SSANOVA model. For WT1 and WT3, H in the aforementioned terms is replaced by S_b, whereas for WT2 and WT4, H in the aforementioned terms is replaced by I.

The comparison shows that AMK overall performs the best, with kNN as a close second. BART performs slightly better than kNN on WT1–WT3 cases and slightly worse than AMK, but it does slightly worse than kNN for WT4 and noticeably worse for WT5–WT6. SSANOVA is the worst performer among the four. In fact, SSANOVA performs closer to what BVK does, as shown in Table 5.6 (the fifth column). One closer look reveals that the data associated with WT4 have the largest variation among the four inland turbine datasets, as evident by its large RMSE values. This large variation could be due to the fact that WT4 is located the farthest away from its companion mast so that the wind measurements taken at the mast are less representative of the wind condition at the turbine site. For WT5 and WT6, the RMSE in terms of the normalized power is slightly higher than that of WT1 to WT3, but because WT5 and WT6 have a higher rated power, almost double that of the inland turbines, the absolute value of the noises are greater. This observation appears to suggest that BART is sensitive to the noise level in a dataset and its performance suffers when the data noise level is elevated.

Next, let us take a look at the density estimation based on CRPS. Note that the IEC binning method and kNN can produce only point estimation, while BVK, BART and AMK produce both point and density estimations. SSANOVA is supposed to produce density estimation as well, but doing so takes way too long. Therefore, in the CRPS-based comparison, only BVK, AMK and BART are included, and the baseline model is BVK.

The CRPS-based comparison is presented in Table 5.8, in which the per-

TABLE 5.7 Comparing various data science-based power curve methods. Reported below are the RMSE values. The boldface font indicates the best performance.

| | kNN | SSANOVA | BART | AMK | AMK improvement over | | |
					kNN	SSANOVA	BART
WT1	0.0766	0.0870	0.0762	**0.0741**	3%	15%	3%
WT2	0.0828	0.0907	0.0817	**0.0799**	4%	12%	2%
WT3	0.0669	0.0766	0.0667	**0.0645**	4%	16%	3%
WT4	0.1035	0.1118	0.1039	**0.1000**	3%	11%	4%
WT5	0.0810	0.0947	0.0876	**0.0792**	2%	16%	10%
WT6	0.0830	0.1039	0.0906	**0.0804**	2%	23%	11%

TABLE 5.8 Comparing CRPS among BVK, AMK, and BART. Boldface font indicates the best performance.

| Turbine | BVK | AMK | | BART | |
		three inputs	four inputs	three inputs	four inputs
WT1	0.0432	0.0377 (12.7%)	**0.0370** (14.3%)	0.0487 (−12.7%)	0.0475 (−9.9%)
WT2	0.0456	0.0413 (9.4%)	**0.0400** (14.0%)	0.0539 (−18.2%)	0.0518 (−13.6%)
WT3	0.0378	0.0337 (10.8%)	**0.0311** (17.7%)	0.0419 (−10.8%)	0.0385 (−1.8%)
WT4	0.0571	0.0498 (12.8%)	**0.0473** (17.2%)	0.0693 (−21.4%)	0.0631 (−10.5%)
WT5	0.0461	0.0408 (11.5%)	**0.0388** (15.8%)	0.0565 (−22.5%)	0.0553 (−19.9%)
WT6	0.0462	0.0378 (18.2%)	**0.0375** (18.8%)	0.0561 (−21.4%)	0.0550 (−19.0%)

centage values in the parentheses are the reductions in CRPS a method makes relative to BVK. A negative value suggests an increase, rather than a reduction, in the respective CRPS. There are two versions of AMK and BART that are included: the three-input version uses (V, D, ρ) for both inland and off-shore turbines, and the four-input version uses (V, D, ρ, I) for inland turbines but (V, D, ρ, H) for offshore turbines.

BART turns out to be the worst performer for predicting the conditional density among the three models and AMK the best. AMK is 14%–18% better than BVK, which is in turn 2%–20% better than BART. AMK appears to exhibit competitiveness and robustness, thanks in part to its model structure being advised by the physical understanding of wind power production.

To facilitate an intuitive understanding how AMK improves the density estimation, we present in Fig. 5.7 an illustration of density estimations using BVK and AMK. To produce the result in Fig. 5.7, WT5 data are used. A

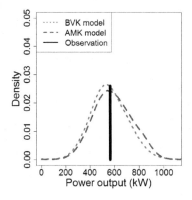 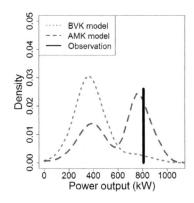

(a) Two models produce similar CRPS (b) Two models produce different CRPS

FIGURE 5.7 Comparison of the predictive distributions of power output when BVK and AMK produce similar CRPS values versus when they produce different CRPS values.

four-input AMK is employed, with input variables as (V, D, ρ, H). The left panel of Fig. 5.7 shows the predictive distributions of the power output from the two models, when their CRPS values are not much different. The two distributions are similar and either model produces a good estimate. The right panel of Fig. 5.7 presents the predictive distributions of the two models, when their CRPS values differ considerably. One can see that the distribution from the BVK model is centered incorrectly.

GLOSSARY

AMK: Additive multiplicative kernel method

ANOVA: Analysis of variance

BART: Bayesian additive regression trees

BVK: Bivariate kernel method

CART: Classification and regression tree

CKD: Conditional kernel density

CRPS: Continuous ranked probability score

DPI: Direct plug-in estimator

GAM: Generalized additive model

IEC: International Electrotechnical Commission

ILWF: Inland wind farm

ISE: Integrated squared error criterion

kNN: k nearest neighborhood

MARS: Multivariate adaptive regression splines

MART: Multiple additive regression trees

OSWF: Offshore wind farm

pdf: Probability density function

RF: Random forest

RMSE: Root mean squared error

SSANOVA: Smoothing spline ANOVA

SVM: Support vector machine

WT: Wind turbine

EXERCISES

5.1 Use the 10-min data in the `Wind Time Series Dataset` and treat the wind power as y and the wind speed as x. Conduct the following exercise.

    a. Random split the data into 80% for training and 20% for testing. Use the training data to build a power curve model, following the IEC binning method. Select the bin width as 0.5 m/s.

    b. Use the 20% test data to calculate the RMSE.

    c. Change the bin width to 1 m/s, 1.5 m/s, and 2 m/s, respectively, and for each one of them, build a respective power curve model and calculate its corresponding RMSE. Observe how the bin width affects the quality of the power curve method.

5.2 Suppose that the number of bins used in Exercise 5.1 under different bin widths are denoted as $B_{(0.5)}$, $B_{(1)}$, $B_{(1.5)}$, and $B_{(2)}$, respectively. Still use the 10-min data in the `Wind Time Series Dataset`.

    a. Build a CART model with the number of terminal nodes set to be $B_{(0.5)}$, $B_{(1)}$, $B_{(1.5)}$, and $B_{(2)}$, respectively.

    b. Visualize the partition on wind speed by the CART model for each of the terminal node choices, and compare the partition outcome with the respective partition used in the IEC binning.

c. Conduct an out-of-sample test through, again, the 80-20 random split. Compute the RMSE for each of the choices and compare with the respective binning outcome.

5.3 Again use the 10-min data in the `Wind Time Series Dataset`.

a. Build a one-dimensional kernel regression using the Gaussian kernel function. Set the bandwidth parameter λ to 0.5 m/s, 1 m/s, 1.5 m/s, and 2 m/s, respectively. Use the 80-20 random split, conduct the out-of-sample test, and report the corresponding RMSE.

b. Use five-fold cross validations to search for the optimal λ. Is the optimal λ different from the above prescribed choices?

c. Compare the RMSE of the kernel regression under the optimal λ, the best binning outcome in Exercise 5.1, and the best CART outcome in Exercise 5.2. How much are they different? If they do not differ a lot, does that surprise you? Why or why not? If they do differ a lot, can you explain why?

5.4 Because the smoothing spline is a natural cubic spline with knots at every data point x_j, $j = 1, \ldots, n$, we can write the smoothing spline function as

$$g(x) = \sum_{j=1}^{n} h_j(x)\beta_j,$$

where $h_j(x)$'s are the basis functions used in the natural cubic spline and n is the number of data points in the training set. For the natural cubic splines, the first two basis functions are $h_1(x) = 1$ and $h_2(x) = x$. The other basis functions take the form of a third-order polynomial function but the detailed expressions are omitted here. Please derive the smoother matrix $\mathbf{S}$ in terms of $h_j(\cdot)$ and β_j. Show that the degree of freedom of the smoothing splines, or the effective number of knots, equals n when $\gamma = 0$ and equals two when $\gamma \to \infty$. Do the two extreme values make intuitive sense? What this means is that the degree of freedom of the smoothing splines, or its effective number of knots, is between two and n, as γ goes from 0 to infinity.

5.5 Rasmussen and Williams [173, pages 138-141] state that if one chooses a particular type of covariance function (i.e., a kernel function), the smoothing spline and the one-dimensional Gaussian process regression (namely kriging) can be made equivalent. To appreciate this understanding, please generate a set of one-dimensional data and do a simple numerical test.

a. Let $y = e^{-1.4x} \cos(7\pi x/2) + \varepsilon$ and $\varepsilon \sim \mathcal{N}(0, 0.5)$. Use this function to simulate 200 data pairs, i.e., $\{(x_1, y_1), (x_2, y_2), \ldots, (x_{200}, y_{200})\}$.

b. First fit an ordinary kriging model to the simulated one-dimensional data, and then, fit a smoothing spline using the R function smoothing.spline in the stat package. Adjust the penalty coefficient used in the smoothing spline fit and see if you could produce a spline fit close to the kriging fit.

5.6 Use the 10-min data in the Wind Time Series Dataset, and build a kNN-based power curve model. Test different choices of k, the neighborhood size in kNN. Use the 80-20 random split and conduct an out-of-sample test. Observe what choice of k produces a model whose RMSE is close to, respectively, that of the best binning outcome in Exercise 5.1, that of the best CART outcome in Exercise 5.2, and that of the best kernel regression in Exercise 5.3.

5.7 Use the WT5 data in the Offshore Wind Farm Dataset1, and build a CART and a BART, respectively, using all seven covariates. Conduct an out-of-sample test based on a 80-20 random split and compare their RMSEs. Does BART outperform CART? Is that what you anticipated?

5.8 Use the WT1 data in the Inland Wind Farm Dataset1, and build an SVM, an ANN, and an AMK, respectively, using all five covariates. Conduct an out-of-sample test based on a 80-20 random split and compare their RMSEs. How do their performances compare to each other?

5.9 To select the best subset of variables to be included in the final model, two versions of a greedy strategy are used and referred to, respectively, as the *forward stepwise selection* and *backward stepwise selection* [86, Section 3.3.2].

- The forward stepwise selection is to screen through all the candidate variables, one at a time, and select the one whose addition to the model reduces the out-of-sample RMSE the greatest. Once chosen, remove the variable from the candidate set and select the next variable from the remaining candidates, until the addition of a new variable no longer reduces the out-of-sample RMSE.

- The backward stepwise selection starts off with the whole set of candidate variables in the model. Remove one at a time, and select the one that reduces the out-of-sample RMSE the greatest and remove it. Screen the remaining variables in the model following the same fashion and stop when the deletion of an existing variable no longer reduces the out-of-sample RMSE.

Use the AMK as the power curve model and the WT6 data in the Offshore Wind Farm Dataset2. Test both the forward stepwise selection strategy and the backward stepwise selection strategy and see what subset of variables they select.

5.10 Take the BVK model, which is the same as AMK with two inputs (V, D), and the WT1 data in the `Inland Wind Farm Dataset1`. Build a BVK model using the original wind speed and then build another BVK model using the air-density-corrected wind speed, while all other things are kept the same. Denote the latter BVK model by BVK_a, with the subscript indicating air density correction. Conduct an out-of-sample test on BVK and BVK_a and observe what type of effect the air density correction has on the kernel regression. Also compare the RMSE of BVK_a with that of $AMK(V, D, \rho)$ in Table 5.6, i.e., the column under "BVK+air density," and see which one performs better. If $AMK(V, D, \rho)$ performs better, what does that tell us?

CHAPTER 6

Production Efficiency Analysis and Power Curve

DOI: 10.1201/9780429490972-6

The use of efficiency metrics for wind turbines is important for evaluating their productivity and quantifying the effectiveness of actions that are meant to improve their energy production. The IEC [102] recommends using (1) annual energy production (AEP), (2) the power curve, or (3) the power coefficient, for the purpose of performance evaluation of wind turbines. The drawback of using power output directly, as in the case of AEP, is obvious, because wind power output is affected by wind input conditions, which are variable and not controllable. While the total output does matter in an owner/operator's decisions, a wind turbine's efficiency should be evaluated while the input conditions are controlled for or set to comparable levels. Generally speaking, productive efficiency metrics used in the wind industry take the form of a ratio, which is often the observed wind power production normalized by a benchmark denominator. Different metrics apparently use different denominators. Power curves as we discuss in Chapter 5 are useful in producing some of the denominators.

6.1 THREE EFFICIENCY METRICS

We describe three efficiency metrics commonly used for wind power production—availability, power generation ratio (PGR), and power coefficient. Please be reminded that the wind speed used is adjusted through the air density correction in Eq. 5.3, unless otherwise noted.

The efficiency of a wind turbine is usually measured for a specific time duration, be it a week, a month, or a year, in which the turbine's efficiency is assumed constant. Consider the weekly resolution as an example. Analysts calculate a single value for the chosen efficiency metric for every single week and evaluate the time series of that metric. The same calculation can be easily extended to other time resolutions. Denote by (V_t, ρ_t, y_t), $t = 1, \ldots, n$, the data pairs of wind speed, air density, and wind power, observed during a unit period under the given resolution.

6.1.1 Availability

One of the efficiency metrics used broadly in the wind industry is *availability* [39, 209] described in the industry standard IEC Technical Specifications (TS) 61400-26-1 [103]. The availability tracks the amount of time in which power is produced by a turbine and then compares it to the amount of time when the turbine could have produced power. A wind turbine is supposed to produce power when the wind speed is between the cut-in and cut-out wind speeds, the two design characteristics of a turbine as described in connection with Fig. 1.2. Turbines are expected to produce power at all times when the recorded wind speed is within these two limits. If a turbine does not produce power when the wind conditions allow, the turbine is then deemed unavailable. The availability is thus defined as

$$\text{Availability} = \frac{\#\{(V_t, \rho_t, y_t) \quad \text{s.t.} \quad y_t > 0, V_{ci} \le V_t \le V_{co}\}}{\#\{(V_t, \rho_t, y_t) \quad \text{s. t.} \quad V_{ci} \le V_t \le V_{co}\}}, \tag{6.1}$$

where s.t. means *such that* and $\#\{\cdot\}$ counts the number of elements in the set defined by the brackets. The denominator in Eq. 6.1 approximates the total time, in terms of the number of 10-min intervals, that a turbine is expected to produce power, whereas the numerator approximates the total time that a turbine does produce power.

6.1.2 Power Generation Ratio

While the availability calculates a ratio in terms of the amount of up running time, the power generation ratio defines a ratio relevant to the amount of power output. The idea is similar to that of *production-based availability*, recently advocated by the industry standard IEC TS 61400-26-2 [105]. By contrast, the availability discussed in Section 6.1.1 is referred to as the *time-based availability*. The production-based availability calculates the ratio of actual energy production relative to the potential energy production, where the potential energy production is the sum of the actual energy production and the lost production caused by an abnormal operational status of a turbine (e.g., downtime, curtailment). The lost production needs to be estimated and its estimation requires detailed information about a turbine's operating status, not something easily accessible to anyone outside the immediate operator of a wind turbine or a wind farm.

Instead of estimating the lost production, let us make a revision so that the assessment is easier to carry out. The revision is to use a power curve to provide the value of potential energy production under a given wind or weather condition. The power curve used could be the turbine manufacturer's nominal power curve for its simplicity, or the advanced multi-dimensional power curves as described in Chapter 5 for better accuracy. The resulting ratio is referred to as PGR, which is in spirit similar to the production-based availability.

Let $\hat{y}(\boldsymbol{x})$ denote the potential energy production under input condition $\boldsymbol{x}$ and $y(\boldsymbol{x})$ denote the actual energy production under the same condition. In fact, $y(\boldsymbol{x}_t) = y_t$. The PGR of a given time duration (including n observations) can then be computed as

$$\text{PGR} = \frac{\sum_{t=1}^{n} y(\boldsymbol{x}_t)}{\sum_{t=1}^{n} \hat{y}(\boldsymbol{x}_t)} = \frac{\sum_{t=1}^{n} y_t}{\sum_{t=1}^{n} \hat{y}(\boldsymbol{x}_t)}. \tag{6.2}$$

If only the wind speed is considered, then $\boldsymbol{x} = (V)$ and the potential and actual energy production are, respectively, $\hat{y}(V_t)$ and $y(V_t)$.

6.1.3 Power Coefficient

Different from the availability and PGR, the power coefficient explicitly reflects the law of wind energy production, as described in Eq. 5.2, and measures the aerodynamic efficiency of a wind turbine. Power coefficient, C_p, refers to the ratio of actual energy production to the energy available in the ambient wind flowing into the turbine blades [229]. Based on Eq. 5.2, C_p can be expressed as

$$C_p(t) = \frac{2y_t}{\rho_t \cdot \pi R^2 \cdot V_t^3}, \tag{6.3}$$

for any given observation t. Note here that the C_p calculation uses the wind speed without the air density correction since the calculation itself involves air density.

Power coefficient, C_p, is typically modeled as a function of the tip speed ratio (i.e., the ratio of the tangential speed of the tip of a blade and the hub height wind speed), attack angle (related to wind direction), and air density. This dependency of C_p on weather-related inputs makes the power coefficient a *functional curve*, often plotted against the tip speed ratio. Like in the binning-based estimation of power curves, analysts bin individual C_p values by groups of one meter per second according to their respective wind speed and average the power coefficients in each bin to produce a C_p curve. In practice, the largest power coefficient on the curve, as the representative of the whole curve, is used for quantification of the aerodynamic efficiency [123, 126]. The peak power coefficient is a popular efficiency measure used to evaluate wind turbine designs and various control schemes including pitch and torque controls. The theoretical upper limit for the power coefficient is known as the Betz limit (=0.593) [18].

Fig. 6.1 presents a plot of two power curves and a power coefficient curve. In the left panel, the relative position of a power curve suggests relative pro-

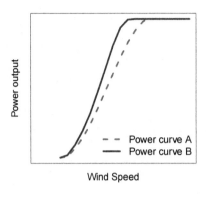 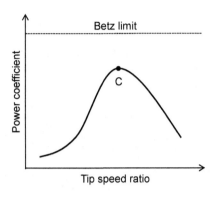

FIGURE 6.1 Left panel: two power curves indicating relative efficiencies of wind turbines, in which curve B suggests a higher productive efficiency; right panel: power coefficient curve and the Betz limit. (Reprinted with permission from Hwangbo. et al. [96].)

ductive efficiency, whereas in the right panel, point C corresponds to the peak power coefficient, used to represent a turbine's efficiency.

6.2 COMPARISON OF EFFICIENCY METRICS

When evaluating the efficiency based on multiple metrics, an immediate question to be addressed is whether or not the evaluation from each metric draws the same conclusion. If the metrics do not always agree with one another (they indeed do not), then subsequent questions are how consistent the results are based on the different metrics and which metric provides a better insight into turbine efficiency.

Niu et al. [155] compare the metrics described in the previous section by using the `Offshore Wind Farm Dataset2`. The layout of the offshore wind farm is sketched in Fig. 6.2.

The wind power data in all datasets associated with the book are normalized by a turbine's rated power. But to compute the power coefficient, the actual power output is needed. For the offshore wind turbines in the `Offshore Wind Farm Dataset2`, their characteristics follow what is presented in Table 5.1, meaning that the rated power of these offshore turbines is around 3 MW. So, we use 3 MW as the rated power to recover the actual power output in the subsequent calculation.

Temporal resolutions that are examined include weekly, monthly, quarterly, and yearly resolutions, with a primary focus on weekly and monthly as they provide greater numbers of data points and details.

For each temporal resolution, Niu et al. [155] calculate the three metrics of availability, PGR, and power coefficient as described in Section 6.1; hereafter

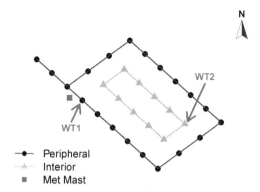

FIGURE 6.2 Basic layout of the offshore wind farm used in the study. Peripheral turbines are located along the black lines and interior turbines along the gray lines. A meteorological mast is indicated by a point along the left edge of the farm. (Reprinted with permission from Niu et al. [155].)

denoted as M1, M2, and M3, respectively. While the averages of M1 and those of M2 calculated for each turbine are within a similar range (0.75–1), the averages of M3 are noticeably lower, at the 0.35–0.5 range, about half the values of M1 and M2. This is understandable as the power coefficient (M3) is limited by the Betz limit to a theoretical maximum of 0.593, and commercial turbines realistically operate around 0.45 [19, page 16]. To make all three metrics comparable in magnitude, Niu et al. multiply M3 by two and use the rescaled metric (2×M3) for the subsequent analysis.

Fig. 6.3, left panel, presents the time-series plots of the three metrics at the monthly resolution over a four-year span. The figure demonstrates that the metrics follow similar overall trends, with peaks and troughs at similar periods of time. The level of variation associated with the three metrics looks similar. In fact, all the three metrics have similar coefficients of variation, though the one for M2 tends to be slightly higher—on average, 0.264 for M2 compared to 0.254 and 0.252 for M1 and 2×M3, respectively. These patterns and characteristics are consistently observed in the other turbines on the wind farm. Similar insights can be drawn for the weekly resolution.

Table 6.1 presents the correlation coefficients between the metrics for the peripheral turbine. As shown in the first two rows, the correlation coefficients are above 0.9, indicating strong correlations between the metrics. By considering the well-aligned time-series and the high correlation coefficients, one may impetuously conclude that the three metrics are consistent with each other and they can substitute for one another when evaluating turbine efficiency. However, if we eliminate some periods of nearly zero power production (for

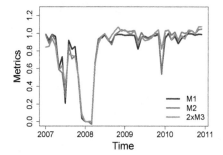 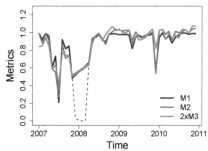

FIGURE 6.3 All three metrics plotted at monthly time resolution for one of the peripheral turbines closest to the meteorological mast. Left panel: for the full period; right panel: after eliminating the periods in which the turbine does not operate for most of the time (dashed line). (Reprinted with permission from Niu et al. [155].)

TABLE 6.1 Correlation between metrics for a peripheral turbine. Weekly and monthly temporal resolutions are shown below.

	M1 vs. M2	M1 vs. 2×M3	M2 vs. 2×M3
Weekly resolution (full)	0.975	0.946	0.959
Monthly resolution (full)	0.986	0.966	0.978
Weekly resolution (reduced)	0.843	0.661	0.785
Monthly resolution (reduced)	0.956	0.876	0.929

Source: Niu et al. [155]. With permission.

example, a period for which any metric is below 0.2; see Fig. 6.3, right panel), the metrics based on such a reduced period produce significantly lower correlation coefficients—for this particular turbine, as low as 0.661 between M1 and 2×M3 at the weekly time resolution. This implies that the original high correlation derived from the full period data could be contributed to substantially by the non-operating periods of the turbine. The lower correlation based on the reduced period further suggests possible disparity between the metrics under typical operating conditions. In the following subsections, the metric values presented are calculated for the reduced period only, for better differentiating the metrics' capability in quantifying turbine efficiency.

6.2.1 Distributions

Fig. 6.4 demonstrates the distributions of the calculated metrics for a single turbine, but it is representative of the other turbines as they all show similar distribution spreads. While M2 and 2×M3 both have relatively broad spreads

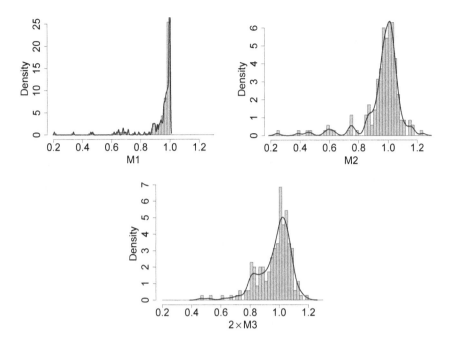

FIGURE 6.4 Histograms and probability density curves of the metric values at the weekly time resolution for the peripheral wind turbine. Top-left panel: availability (M1), top-right panel: power generation ratio (M2), bottom-middle panel: power coefficient (2×M3). (Reprinted with permission from Niu et al. [155].)

of data, M1 has a much narrower range. A significant portion of its density is concentrated near one at which the distribution is truncated with a steep taper to lower values. In contrast, M2 and 2×M3 both take a shape similar to the bell-shaped curve with smoother tapers in both directions. M1's concentration of values makes it difficult to differentiate between the turbine efficiency at different time periods, because as more values are within the same range, the variations in turbine performance are concealed. This can potentially mislead turbine operators into a conclusion that the turbines operate at a similar efficiency level, even though the underlying turbine efficiencies differ.

Such a unique distributional characteristic of M1 can be inferred by its calculation procedure. As expressed in Eq. 6.1, the numerator of M1 counts the number of elements in a set that is a subset of the one associated with the denominator, so it has a maximum value of one at all points in time. This is a desired property for an efficiency metric, which is not observed in M2 or 2×M3. M2 can exceed one because a power curve displays the expected power values as an averaged measure but particular instances of wind power production could exceed the expected productions. The power coefficient itself is smaller than one, but doubling the power coefficient value, namely $2 \times M3$, is bounded from above by twice the Betz limit at 1.186, which itself is greater than one. It is interesting to observe that M2 appears to be bounded by a value similar to 1.186.

The unique property of M1 when combined with its binary quantification of whether or not power was generated, however, adversely affects its quantification capability. As long as a turbine is generating power at a point in time, that point would be counted as a one. Even the cases when the power production is significantly lower than expected would still be counted as ones. Averaging over these counts produces the metric weighted heavily towards one. Using the availability metric, M1, periods with high actual efficiency, in terms of the amount of actual power production, look the same as low efficiency periods as long as the power produced exceeds a low threshold.

The methods calculating M2 and M3, on the other hand, allow for a sliding scale measure of power production so that they account for how much power is produced. Values of M2 and $2 \times M3$ thus have greater spread and do not concentrate around any particular value as narrowly as M1 does. This ability to better distinguish between time periods of differing performance as well as the distributional features render M2 and $2 \times M3$ stronger metrics than M1. They allow for a fuller portrayal of a turbine's efficiency over time as opposed to M1's more general overview of whether or not the turbine is in operation.

6.2.2 Pairwise Differences

Fig. 6.5 illustrates the absolute difference between the calculated metrics on a weekly basis. Darker bars indicate the periods of significantly large differences, whereas lighter bars are the periods of smaller differences.

Fig. 6.5, bottom panel, shows that the large differences between M2 and $2 \times M3$ are sparsely distributed through the four years. In contrast, as shown in Fig. 6.5, top and middle panels, there are significantly more instances of large value differences between M1 and either of the other metrics, especially between M1 and $2 \times M3$. This implies that both M1 and $2 \times M3$ are more similar to M2 than to each other. M1 and M2 calculate a ratio of the actual performance over the expected performance, although M1 focuses on the amount of time and M2 examines the amount of power. This sets $2 \times M3$ apart from M1 and M2. On the other hand, M2 and $2 \times M3$ quantify turbine efficiency with respect to the amount of power production, whereas M1 concerns the amount of operational time, which makes M1 distinct from the other two.

In Fig. 6.5, the large or medium differences tend to be heavily concentrated within some specific periods, notably in the second half of 2007 and the first half of 2010. In fact, these periods represent those in which turbines' true efficiencies are relatively low. There are two different aspects describing this phenomenon.

First, recall from Fig. 6.4 that M1 tends to be heavily weighted towards

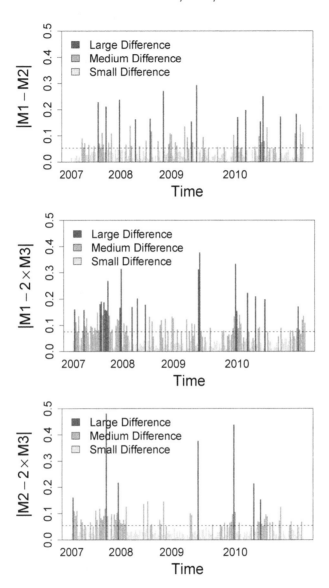

FIGURE 6.5 Magnitudes of absolute pairwise difference between metric values at weekly resolution for the peripheral wind turbine. Top panel: availability (M1) versus PGR (M2), middle panel: availability (M1) versus power coefficient (2×M3), bottom panel: PGR (M2) versus power coefficient (2×M3). The dashed line in each plot is the average of the absolute differences in that plot. An absolute difference is considered as a small difference, if its value is smaller than 0.05, as a large difference, if its value is greater than 0.15, and as a medium difference, if its value is in between. (Reprinted with permission from Niu et al. [155].)

its maximum, overestimating the turbine's efficiency in the relative scale. If a turbine produces some power for most time instances within a given period, its availability should be close to one. The large differences between M1 and the other two metrics then imply that the turbine is producing some power for most of the times but the amount of the power production is considerably lower relative to the expectation. If one refers to Fig. 6.3, one may notice that M1 is higher than the other two in the later part of 2007.

Secondly, recall that M3 represents a *maximum effect* (on the C_p curve), whereas M2 is an *integration effect*. For a functional response, the two effects can be understandably different. The large differences between M2 and $2 \times$M3 suggest that a turbine produces a sufficient amount of power only for a small portion of the given time period. In this case, the turbine's maximum efficiency measured by $2 \times$M3 is relatively high, but M2 is relatively low because the turbine does not produce much power on average during the same period (see the middle of 2007 and the beginning of 2010 in Fig. 6.3). M1 also measures an *integration effect*, but in terms of the operational time, so that the same argument is applicable to explaining the difference between M1 and $2 \times$M3. Most of the time when there is a large difference between M2 and $2 \times$M3, a large difference between M1 and $2 \times$M3 is also observed (see Fig. 6.5, middle and bottom panels).

All of these observations can be found in the cases of other turbines as well. Although the concentration periods of large and medium differences vary, all turbines display the clustering pattern, and such clusters are closely related to different characteristics of the metrics.

When comparing the mean of the absolute differences between the metrics, indicated by the dashed horizontal lines in Fig. 6.5, the disparity between the metrics becomes less pronounced. While a metric pair with the smallest mean difference varies by different turbines, the metric pair of the largest mean difference is consistently observed as between M1 and $2 \times$M3. This suggests that M2 has comparably closer values to M1 and $2 \times$M3. As such, M2 is more consistent in value with either of M1 and $2 \times$M3 and its values are a better reflection of all three metrics.

6.2.3 Correlations and Linear Relationships

Table 6.1 shows that the correlation calculated using the reduced period is the highest between M1 and M2 for most turbines. The correlations between M2 and $2 \times$M3 (or equivalently, between M2 and M3) are also relatively high. For most turbines, the correlation coefficients between M1 and M2 remain within the 0.8 range at weekly resolution, whereas those between M2 and M3 are generally in the 0.7 range.

The lowest correlations are found between M1 and M3 for all turbines and time resolutions, with the correlation coefficient values usually around 0.5–0.6 but dipping sometimes into the 0.4 range. The values displayed in Table 6.1 are among the higher values of M1–M3 correlation of turbines. Another turbine has an M1–M3 correlation of just 0.417 for the reduced weekly data. This indicates that the relationship between these two metrics is much weaker, highlighting the strength of M2 for its much stronger relationship with either of the other metrics.

Weekly time resolution is best for highlighting differences in correlation between metrics. Correlations rise as the time resolution becomes coarse; monthly, quarterly, and yearly resolutions in general return a correlation in the range of 0.9. Niu et al. [155] state that the averaging effect when using a coarse time resolution irons out a certain degree of details, making the metrics based on the coarse time resolutions less differentiating.

To analyze the consistency of the metrics, Niu et al. [155] also evaluate the linearity between any pair of the metrics around the $y = x$ line. Let us generate data points (x, y) paired by the values of two selected metrics. If the data points perfectly fit to the $y = x$ line, an increase in one metric implies the same amount of increase in the other metric. As such, their ability to capture change in efficiency is identical, or equivalently, they are consistent.

However, as noted earlier, the scales of the metrics are not the same, e.g., M1 and M2 are about twice the unscaled M3. To assess the extent of linearity around the $y = x$ line requires us to estimate the exact scale between the metrics.

To align the scales, Niu et al. [155] perform a linear regression upon the different metric pairs. For example, for the M1–M2 pair, fit a linear model of M1 $= \beta \cdot$ M2 $+ \varepsilon$ to estimate β, where ε is the random noise. Let $\hat{\beta}$ denote the coefficient estimate. Then, the estimate, $\hat{\beta}$, is used to rescale the values of M2, generating the scale-adjusted data points (M1, $\hat{\beta}\cdot$M2). With the scale adjustment, the data points should be centered about the $y = x$ line. If they show strong linearity around the $y = x$ line, one can conclude that the metrics for the corresponding pair are consistent with each other. To determine the extent of linearity, the average magnitude of the data points' vertical distance from the $y = x$ line (in an absolute value) is computed.

Fig. 6.6 presents the scatter plots of the scale-adjusted metrics and the $y = x$ line. For illustration purposes, two scatter plots are presented, one for the peripheral turbine used previously and the other is an interior turbine. For the metrics calculated for the peripheral turbine, the scale adjustment coefficients ($\hat{\beta}$) are 0.97, 1.93, and 1.99 for M1–M2, M1–M3, and M2–M3 pairs, respectively. The coefficient of 0.97 for the M1–M2 pair, for instance, implies that M2 will have the same scale with M1 after multiplying it by 0.97. For the interior turbine, the scale adjustment coefficients are 0.98, 2.01, and 2.06, for the three pairs of metrics in the same order, respectively.

In the figure, points are more concentrated near where x and y equal one. Whenever x refers to M1, there is a very apparent clustering of points at $x = 1$ due to the truncation of the distribution of M1 at one. On the other hand, the data points for the M2–M3 pair are well spread around the region, a characteristic reminiscent of the metrics distributions examined earlier.

After the scale-adjustment, whenever the y-axis represents a rescaled M3

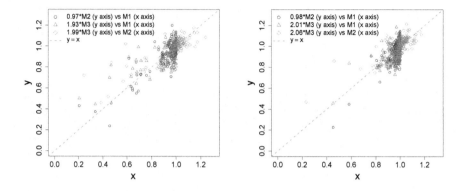

FIGURE 6.6 Linear relationships between metrics at the weekly time resolution. Left panel: for a peripheral turbine, right panel: for an interior turbine. Plots generated from scaling values by x-to-y ratio. The dashed line illustrates the $y = x$ line. (Reprinted with permission from Niu et al. [155].)

TABLE 6.2   Linearity between a pair of performance metrics measured by the average absolute vertical distances from the $y = x$ line.

	M1 vs $\hat{\beta}\cdot$M2	M1 vs $\hat{\beta}\cdot$M3	M2 vs $\hat{\beta}\cdot$M3
A peripheral turbine	0.050	0.068	0.055
An interior turbine	0.046	0.068	0.052

Source: Niu et al. [155]. With permission.

(triangles and diamonds), the data points tend to be placed above the $y = x$ line for relatively low x values, e.g., less than 0.8. This confirms the difference between the maximum effect (for M3) and the integration effect (for M1 and M2) discussed earlier.

As shown in Table 6.2, the average distances between the points and the $y = x$ line is the greatest for the M1–M3 pair for both turbines, suggesting that the M1–M3 pair has the weakest extent of linearity. This reinforces the understanding from the analysis of absolute differences that M1 and M3 are the least consistent metrics, while M2 has a stronger relationship with both other metrics.

6.2.4   Overall Insight

According to the previous analyses, while all metrics display some level of consistency, PGR (M2) is the most consistent with the other metrics. The absolute differences in metric values demonstrate that PGR produces values that are more representative of the three metrics. Correlations between the metrics also suggest that changes in turbine performance mapped by PGR are illustrative of such trends displayed by other metrics. Moreover, evaluation of the linearity between the metrics shows that availability (M1) or power coefficient (M3) has a stronger relation with PGR (M2) than with each other. It is not too far fetched to reach the conclusion that PGR better represents all three metrics. Various aspects of the analysis have shown availability's deficiency in discriminating changes in turbine performance. Practitioners are well aware of availability's deficiency, which becomes the chief reason for adopting the production-based availability measure recently.

6.3   A SHAPE-CONSTRAINED POWER CURVE MODEL

As said earlier in this chapter, efficiency metrics used in the wind industry take the form of a ratio, which is often the observed wind power production normalized by a benchmark denominator. Availability, albeit a ratio, does not use power output in either numerator or denominator. The metric that resembles availability and does use power output in both numerator and denominator is the capacity factor (CF). We did not include the capacity factor in the discussion of the previous section, because it is typically used for wind farms and much less often used for individual wind turbines. Its concept, however, is indeed applicable to wind turbines.

The capacity factor of a turbine is the ratio of the observed power output over the turbine's maximum production capacity at the rated power. When calculating the capacity factor, one assumes that the turbine is operating at its full capacity all the time. The use of the capacity factor entirely ignores the wind condition, so much so that its denominator calculates the absolute maximally possible wind power that can be produced for a given period for the specific design of the said wind turbine. In this sense, the capacity factor's denominator is even more aggressive than that used in availability, as availability only counts the time when the wind speed is between the cut-in and cut-out speeds.

Nevertheless, if analysts put all the metrics that use powers in their numerator and denominator side by side, as shown in Fig. 6.7, one can notice that they indeed have the same numerator, which is the observed power output. But the denominators are different, meaning that different benchmarks are used in computing the respective ratio. This in fact raises a question—what should be used as a performance benchmark? Hwangbo et al. [96] argue that to quantify a turbine's productive efficiency, one would need to estimate the best achievable performance as a benchmark, so that the ratio of the current performance to the best achievable performance quantifies the degree to which the turbine has performed relative to its full potential. In order to estimate the best achievable performance of a wind turbine, Hwangbo et al. look into the field of production economics [81], which refers to the "best achievable performance" as an *efficient frontier*. To facilitate this line of discussion, we start with some background on production economics.

Capacity Factor	Power Coefficient	Production-based Availability
Observed power output / *Maximum power output*	*Observed power output* / *Energy available in the wind*	*Observed power output* / *Expected power output*

FIGURE 6.7 Capacity factor, power coefficient, and the production-based availability or power production ratio.

6.3.1 Background of Production Economics

Efficiency analysis is a primary focus in production economics. Efficiency quantification is based on the estimation of a production function and the explicit modeling of systematic inefficiency, using input and output data for a set of production units, be it firms, factories, hospitals or power plants. Consider a set of production units (e.g., a wind farm) using x input (e.g., investment in a wind energy project) and producing y output (e.g., revenue from power generation). Analysts can create a scatter plot of many x-y data pairs coming from different production units or the same production unit but over different periods; see Fig. 6.8. Assuming no measurement errors associated with x and y, a common estimator in production economics—data envelopment analysis (DEA) [11]—estimates the efficient frontier enveloping all the observations.

The concept of an efficient frontier is understood as follows—a production unit whose input-output is on the frontier is more efficient than the production units whose input-output is being enveloped by the frontier. Consider observation D. Using the same input, the production unit associated with D produces less output than the production unit associated with point E; while to produce the same output, the production unit associated with D needs more input than the production unit associated with point F. For this reason, the production unit associated with D must be inefficient.

The efficient frontier is also called the *production function*, denoted by $Q(x)$. The production function characterizes producible output given input x in the absence of inefficiency. Using the production function, the output of the inefficient production unit D can be expressed as

$$y_D = Q(x_D) - u_D, \tag{6.4}$$

where $u_D \geq 0$ denotes the systematic inefficiency.

To estimate the production function $Q(x)$, certain assumptions are made restricting the shape of the frontier. The most common assumption is that the frontier forms a monotone increasing concave function consistent with basic stylized characteristics of production [222]. When the data are assumed noise free, the tightest boundary enveloping all observations and maintaining monotonicity and concavity is a piecewise linear function.

Let us consider the context of power production of a wind turbine, in which a wind turbine is a power production unit, wind speed is the dominating

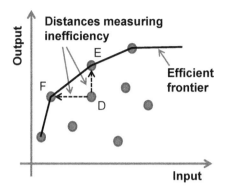

FIGURE 6.8 Production data and efficient frontier. (Reprinted with permission from Hwangbo et al. [96].)

input driving power production, and the generated power is the output. When applied to the wind turbine data, the use of convex or concave piecewise linear methods assuming noise-free data encounter some problems.

The first is the noise-free assumption. The frontier analysis approaches assuming noise-free observations are referred to as *deterministic*. But the wind turbine data, like all other physical measurements, are inevitably contaminated by noises. The problem with applying a deterministic approach to noisy wind production data is that it tends to overestimate the best performance benchmark because every observation is assumed to be achievable.

The second difference is that the shape of the wind-power scatter plot is not concave. When discussing Fig. 1.2, we show that the wind-power data appears to follow an S-shape, comprising a convex region, followed by a concave region, and the two segments of curves are connected at an inflection point. Fig. 6.9 makes this point clearer with its right panel illustrating the meaning of convexity and concavity.

In production economics, the need to model noise is established, promoting the *stochastic* frontier analysis (SFA) [5], which adds a random noise term ε to Eq. 6.4. When applying the SFA modeling idea to wind turbine data and replacing the generic input variable x with wind speed variable V, Hwangbo et al. [96] define their production function as

$$y = Q(V) - u(V) + \varepsilon, \qquad (6.5)$$

where ε is assumed having a zero mean, while the systematic inefficiency term $u(V)$ is a non-negative random variable with positive mean, i.e., $\mu(V) := \mathbb{E}[u(V)] > 0$. Note that $u(V)$ is a function of V, meaning that the amount of inefficiency varies as the input changes, known as a *heteroskedastic* inefficiency term.

While the SFA research considers the noise effect in observational data,

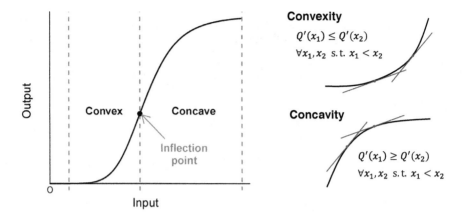

FIGURE 6.9 S-shaped curve, convexity and concavity. $Q'(x)$ is the first derivative of $Q(x)$ with respect to x. (Left panel reprinted with permission from Hwangbo et al. [96].)

analysts still need to address the second challenge mentioned above, namely, the S-shaped curve exhibited in the wind turbine data. In the vernacular of economists, the S-shape constraint is known as the regular ultra passum (RUP) law [81], which is motivated by production units having an increasing marginal rate of productivity followed by a decreasing rate of marginal productivity. Olesen and Ruggiero [156] develop a RUP law satisfying frontier analysis but their production function estimator is a deterministic DEA-type estimator, enveloping all observations from above, and consequently suffers from the overestimation problem that all other deterministic production function estimators suffer.

We want to note that methods from production economics have recently been used in wind energy applications [33, 96, 97, 162]. Two of them are deterministic, with one being a DEA type estimator [33] and the other called a free disposal hall estimator [162]. The production function estimator proposed by Hwangbo et al. [96, 97] is stochastic and attempting to be RUP law satisfying, subject to a rigorous proof in terms of consistency.

6.3.2 Average Performance Curve

The basic production function model in Eq. 6.5 can be re-written as,

$$
\begin{aligned}
y_i &= [Q(V_i) - \mu(V_i)] + [\mu(V_i) - u(V_i) + \varepsilon_i] \\
&= g(V_i) + e_i, \qquad i = 1, \ldots, n,
\end{aligned}
\tag{6.6}
$$

by letting $g(V_i) := Q(V_i) - \mu(V_i)$ and $e_i := \mu(V_i) - u(V_i) + \varepsilon_i$. The error term e is a redefinition of the error term with expectation zero. The above expression connects the power curve with the production function because $g(V)$ is effectively the power curve. As the power curve passes through the middle of the wind-power data, it is a curve representing the average performance of a turbine, also known as the average-practice function in the production economics literature. Understandably, the production frontier function, $Q(V)$, differs from the power curve, $g(V)$, by the mean of the inefficiency varying by V.

This connection helps lay out the intuition behind the procedure of estimating the production frontier function, $Q(V)$. One would start with a power curve from the wind turbine data. Then estimate the mean function of the inefficiency and use it to rotate the average performance curve to the new position to be the production frontier function.

Hwangbo et al. [96] note that because the final $Q(V)$ needs to satisfy the RUP law, i.e., the S-shape constraint, the average performance power curve, $g(V)$, that comes before the production function must, too, satisfy the same shape constraint. This requirement makes this specific power curve estimation procedure different from those currently used in practice, including all discussed in Chapter 5 because none of them imposes the S-shape constraint explicitly.

Estimating the shape-constrained power curve, $g(V)$, requires imposing convexity and concavity in the low and high wind speed regions, respectively. The convex segment should connect to the concave segment at the inflection point, which itself needs to be estimated from the data. Hwangbo et al. [97] formulate the estimation of the average performance curve as a constrained least squares estimation problem, which is to minimize the residual sum of squares, $\sum_{i=1}^{n}(y_i - g(x_i))^2$, subject to the constraints imposing monotonicity and S-shape on $g(\cdot)$.

In the absence of prior knowledge of the location of the inflection point, x^*, Hwangbo et al. [97] make use of a set of grid points and treat each of the grid points as a potential inflection point location. Provided a grid point, Hwangbo et al. [97] partition the functional domain left of the grid point as the convex region and the domain right of the grid point as the concave region and then construct a function g, minimizing the aforementioned residual sum of squares, while satisfying the sets of constraints applicable to the partitioned regions. The estimation of the convex segment or the concave segment can be done individually by using the method of convex nonparametric least squares (CNLS) [127]. Then, Hwangbo et al. [97] choose the grid point resulting in the smallest residual sum of squares as the estimate of the inflection point and the corresponding g as the estimator of the average performance power curve.

Suppose that m grid points, $t_1, \ldots, t_m$, are given. Also, assume that x_i's are distinct (no duplicated values) and arranged in a non-decreasing order for a given n, i.e., $x_i \leq x_j$ whenever $i < j$. For each t_k for $k = 1, \ldots, m$, let $\mathcal{V}^{(t_k)}$ and $\mathcal{C}^{(t_k)}$ be the sets of input points that belong to the (imposed) convex and concave regions, respectively, i.e.,

$$\mathcal{V}^{(t_k)} = \{x_i \mid x_i < t_k, \ i = 1, \ldots, n\},$$

and

$$\mathcal{C}^{(t_k)} = \{x_i \mid x_i \geq t_k, \ i = 1, \ldots, n\}.$$

Then, for each t_k, solve the following quadratic programming with respect to $\mathbf{g} = (g(x_1), \ldots, g(x_n))$ and $\boldsymbol{\beta} = (\beta_1, \ldots, \beta_n)$:

$$\min_{\mathbf{g}, \boldsymbol{\beta}} \ z^{(t_k)} = \sum_{i=1}^{n} (y_i - g(x_i))^2 \tag{6.7a}$$

$$\text{s.t.} \quad \beta_i = \frac{g(x_{i+1}) - g(x_i)}{x_{i+1} - x_i}, \ \forall i \text{ such that } x_i \in \mathcal{V}^{(t_k)}, \tag{6.7b}$$

$$\beta_i \leq \beta_{i+1}, \ \forall i \text{ such that } x_{i+1} \in \mathcal{V}^{(t_k)}, \tag{6.7c}$$

$$\beta_i = \frac{g(x_i) - g(x_{i-1})}{x_i - x_{i-1}}, \ \forall i \text{ such that } x_i \in \mathcal{C}^{(t_k)}, \tag{6.7d}$$

$$\beta_{i-1} \geq \beta_i, \ \forall i \text{ such that } x_{i-1} \in \mathcal{C}^{(t_k)}, \tag{6.7e}$$

$$\beta_i \geq 0, \ \forall i = 1, \ldots, n. \tag{6.7f}$$

The constraints in Eq. 6.7d–6.7e, together with the objective function in Eq. 6.7a, are equivalent to the Hildreth type estimator [92] of a function that is concave over $[\max_{x_i \in \mathcal{V}^{(t_k)}} x_i, x_n]$, the constraints in Eq. 6.7b–6.7c, together with the objective function in Eq. 6.7a, describe the estimator for a convex function defined over $[x_1, \min_{x_i \in \mathcal{C}^{(t_k)}} x_i]$, and the inequalities in Eq. 6.7f ensure the monotonicity of g.

Let the minimizer of Eq. 6.7a for a given t_k be $\mathbf{g}^{(t_k)} = (g^{(t_k)}(x_1), \ldots, g^{(t_k)}(x_n))$ and $\boldsymbol{\beta}^{(t_k)} = (\beta_1^{(t_k)}, \ldots, \beta_n^{(t_k)})$, and let the corresponding optimal objective function value be $z^{(t_k)}$. The vector, $\mathbf{g}^{(t_k)}$, provides estimates only at the given locations, i.e., the g-values at x_i's. For the functional estimator over the interval between two observational data points, x_i and x_j, Hwangbo et al. [97] use a hyperplane to interpolate between the two locations, i.e., they define $\hat{g}^{(t_k)}(x)$ as

$$\hat{g}^{(t_k)}(x) = \begin{cases} \max\{\alpha_i^{(t_k)} + \beta_i^{(t_k)} x \mid \forall i \text{ such that } x_i \in \mathcal{V}^{(t_k)}\}, & \text{if } x < t_k \\ \min\{\alpha_i^{(t_k)} + \beta_i^{(t_k)} x \mid \forall i \text{ such that } x_i \in \mathcal{C}^{(t_k)}\}, & \text{if } x \geq t_k, \end{cases} \tag{6.8}$$

where $\alpha_i^{(t_k)} := g^{(t_k)}(x_i) - \beta_i^{(t_k)} x_i$ for $i = 1, \ldots, n$. Apparently, $\hat{g}^{(t_k)}$ is a piecewise linear function, connecting two adjacent points in the set of $\{(x_i, g^{(t_k)}(x_i)), \forall i = 1, \ldots, n\}$ and extending the hyperplanes, $\alpha_1^{(t_k)} + \beta_1^{(t_k)} x$ and $\alpha_n^{(t_k)} + \beta_n^{(t_k)} x$, each toward the adjacent boundary of the input domain. As such, $\hat{g}^{(t_k)}$ is convex on $[\min x, \max_{x_i \in \mathcal{V}^{(t_k)}} x_i]$, concave on

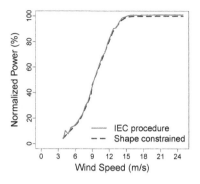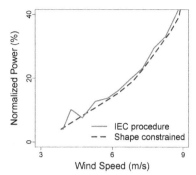

FIGURE 6.10 Illustration of shape-constrained and unconstrained power curves. Left panel: shaped-constrained power curve compared to the IEC binning power curve, right panel: a zoomed view at the wind speed from 3 m/s to 9 m/s. (Reprinted with permission from Hwangbo et al. [96].)

$[\min_{x_i \in \mathcal{C}^{(t_k)}} x_i, \max x]$, and linear within $[\max_{x_i \in \mathcal{V}^{(t_k)}} x_i, \min_{x_i \in \mathcal{C}^{(t_k)}} x_i]$. By letting $t^* = \mathrm{argmin}_{t_k} z^{(t_k)}$, the final average performance curve is simply $\hat{g}^{(t^*)}(x)$. Fig. 6.10 presents a comparison between the shape-constrained power curve versus its unconstrained counterpart using the IEC binning method.

6.3.3 Production Frontier Function and Efficiency Metric

After the average performance power curve, $g(V)$, is estimated, one can take differences between the fitted power curve and the output y. According to the relationship in Eq. 6.6, the resulting residuals are the summation of two random components: $\mu - u$ and ε. The modeling assumption used here is that u is non-negative and ε is symmetrically distributed with respect to a zero mean. As such, one can expect to see a significant decrease at the value of μ in the density curve of the residuals. This understanding is used to estimate μ—if one can locate where the greatest decrease in the residual distribution occurs, it provides an estimate of μ. Hwangbo et al. [96] use the technique in [84] for this estimation. An illustration is given in Fig. 6.11, but we skip the procedure and refer interested readers to [97] for technical details.

The following summarizes the steps in estimating the shape-constrained stochastic production function, $Q(V)$:

1. Use the wind turbine data (wind speed and power) to estimate $g(V)$ while imposing the shape constraints and the continuity requirement at the inflection point. Denote the estimated power curve by $\hat{g}(V)$.

2. Estimate $\mu(V)$, the mean function of the inefficiency term.

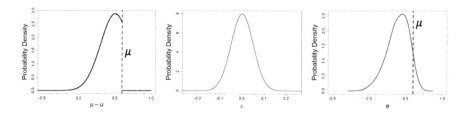

FIGURE 6.11 Estimation of the mean function of the inefficiency term μ. Left panel: density curve of $\mu - u$, middle panel: density curve of ε, right panel: density curve of the residual e.

3. Estimate the production function, $Q(V)$, based on the relationship of $Q(V) = g(V) + \mu(V)$. Denote the estimated frontier curve by $\hat{Q}(V)$.

The estimated production frontier function, $\hat{Q}(V)$, provides a performance benchmark for wind turbines. Hwangbo et al. [96] propose to quantify the productive efficiency of a wind turbine by using the estimated performance benchmark curve and the average performance power curve. Specifically, they propose the following production efficiency measure, PE, which is the ratio of the energy produced under the average performance curve over that under the performance benchmark curve, integrated over the whole wind speed spectrum,

$$ \text{PE} = \frac{\int_{V_{ci}}^{V_{co}} \hat{g}(V)dV}{\int_{V_{ci}}^{V_{co}} \hat{Q}(V)dV}. \tag{6.9} $$

Apparently, PE takes a value between zero and one; the closer a PE is to one, the closer the wind turbine performs to its full potential.

One may have noticed that the discussion above treats the production functions, both the frontier function and the average performance function, as univariate. This is because a multivariate production function satisfying the RUP law is still not fully developed. On the other hand, however, besides wind speed, air density and several other environmental variables, including wind direction, humidity, turbulence intensity and wind shear, all potentially affect wind power production, as seen in the analysis in Chapter 5. These environmental influences are not controllable but their existence does play a role affecting the inefficiency estimated from the power output data. Consequently, when comparing the productive efficiency of different turbines or the same turbine over different operational periods, analysts may need to control for the influence of these environmental factors; otherwise, one may wonder what part of inefficiency is due to a turbine's intrinsic differences and what part of inefficiency comes from differences in environmental characteristics such as air density.

This sort of ambiguity can be alleviated if the comparison periods have comparable environmental profiles. The environmental variables are referred to as *covariates* in the statistical literature. Hwangbo et al. [96] use a co-variate matching procedure to select a subset of the data, in order to make the environmental profiles across different time periods as similar as possible, thus removing the effect of environmental influences from the efficiency analysis. The detail of the matching process is described in Section 7.2. One thing worth noting is that the covariate matching does not produce an exact match but a good match instead, subject to the degree of dissimilarity allowed by a prescribed threshold, ϖ. To confirm the quality of the matches, Hwangbo et al. suggest plotting the pdfs of each environmental variable, empirically estimated from the data and visually inspected for assessing how well the pdfs match across the comparison periods.

6.4 CASE STUDY

In this case study, data from two wind turbines in the `Inland Wind Farm Dataset2` and two turbines from the `Offshore Wind Farm Dataset2` are used. The two inland turbines are referred to as WT1 and WT2, respectively, and the two offshore ones as WT3 and WT4, respectively. These turbine names do not imply relationship with the turbines of the same names in Chapter 5. But they do come from the same wind farms, so that the characteristics in Table 5.1 can be referenced for respective turbines. Table 5.1 states that the rated power for the offshore turbines is around 3 MW and that for the inland turbines is between 1.5 MW and 2 MW. In the following numerical analysis, in order to compute the power coefficient, we use 3 MW as the rated power for the offshore turbines and 1.65 MW as the rated power for the inland turbines.

Hwangbo et al. [96] analyze the wind turbine data on an annual basis, which means that they divide the four-year data into four consecutive annual periods. The number of periods is denoted by $T = 4$ and the period index is $t = 1, 2, 3, 4$. Hwangbo et al. evaluate turbine efficiency for each year.

The first step of data processing is to control for the influence of environmental factors, which is to select the subset of data with comparable environmental profiles through the covariate matching method described in Section 7.2. For inland wind turbines, the covariates to be matched include $\boldsymbol{x} = (V, D, \rho, I, S)^T$, whereas for offshore wind turbines, $\boldsymbol{x} = (V, D, \rho, H, I)^T$. The wind shear, S, is left out in the offshore cases because the study in Chapter 5 suggest that conditioned on the inclusion of (V, D, ρ, H, I), the effect of the two-height vertical wind shear on the offshore turbine's power output appears weak. For all turbine cases, Hwangbo et al. [96] set the similarity threshold as $\varpi = 0.25$. Before the covariate matching, the number of observations in each annual dataset ranges from 14,000 to 37,000, and these numbers reduce to 1,400–2,300 after the matching. The matched dataset includes thousands of observations which is still large enough for estimating the performance benchmark function curve as well as the average performance curve.

Figs. 6.12 and 6.13 present, for inland turbine WT1 and offshore turbine

FIGURE 6.12 Probability density function plots of the matched covariates over the four comparison periods for inland turbine WT1. (Reprinted with permission from Hwangbo et al. [96].)

FIGURE 6.13 Probability density function plots of the matched covariates over the four comparison periods for offshore turbine WT3. (Reprinted with permission from Hwangbo et al. [96].)

WT3, respectively, the pdfs of each environmental variable across the four comparison periods after the covariate matching. The same plots for WT2 and WT4 are omitted because they convey similar messages. One can notice that the choice of $\varpi = 0.25$ leads to sufficiently good matching as demonstrated in the pdf plots.

Subsequently, Hwangbo et al. [96] use the matched subset of data to estimate the productive efficiency measure for each comparison period, as defined in Eq. 6.9. To add a confidence interval, 100 bootstrapping samples are randomly drawn from a respective original dataset, and for each resampled dataset, the efficiency metric is computed once. Doing this 100 times allows the calculation of the 90% confidence intervals for the productive efficiency metric. The bootstrap procedure can be performed on any other performance metrics as well; for more details about the bootstrap technique, please refer to [55].

Fig. 6.14 shows the PEs and its confidence intervals for the four comparison periods, which happen to be the first four years of a turbine's operation. Interestingly, one can notice that for all four turbines, their productive efficiency appears to have increased slightly, rather than deteriorated, during the early stage of operation. This pattern is more obvious for offshore turbines. This initial increase in efficiency was also recognized by Staffell and Green (2014) [203, Figure 9b]. Staffell and Green plot the fleet-level performance degradation of wind turbines over a twenty-year period using the fleet's load

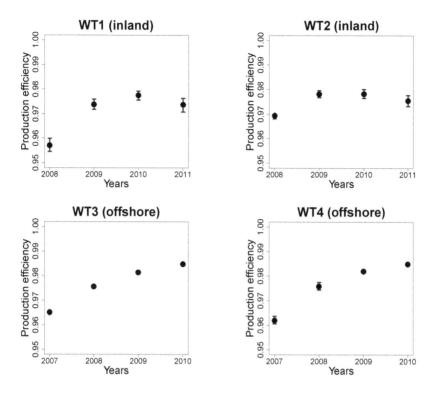

FIGURE 6.14 Productive efficiency, PE_t, $t = 1, 2, 3, 4$. The bars represent 90% confidence intervals and the dots denote the mean values of the efficiency. For offshore wind turbines, some of the confidence intervals are very narrow and are thus not visible. (Adapted with permission from Hwangbo et al. [96].)

factor as the performance measure. Staffell and Green's study appears to suggest an initial period of four to five years before any noticeable degradation is witnessed, as well as an increase in turbine performance for the first one and one-half years, which is rather consistent with what is observed here.

This increasing efficiency phenomenon, however slight, is perhaps counterintuitive. Hwangbo et al. [96] theorize that this could be due to the rational behavior of the operator when faced with initial start-up risk. Recall the typical bathtub curve used in reliability engineering [90], in which there is a short "infant mortality" period at the beginning of a system's operation. In this period, the failure rate of a system appears to be higher than that in the subsequent stable operation period but the failure rate declines rapidly as the components in the system break in with each other. Flipping the bathtub curve upside down shows the reverse effect of failures, or rather, the effective

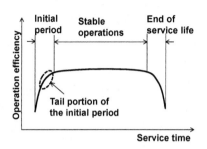 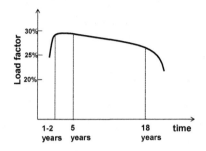

FIGURE 6.15 Flipped bathtub curve and its relevance to turbine reliability. Left panel: a flipped bathtub curve, right panel: load factor average curve implied by Figure 9b in [203].

production of a system. On the flipped curve, one expects to see an increase in production efficiency in the initial period; see Fig. 6.15.

Aware of the existence of this period of increased risk of failure, engineers and operators take proactive actions to reduce risk in their operational policies. One typical action is called "burn-in," in which the manufacturers break in key components before the final system is put together and shipped off to the end users, so that a system installed at a user's site can skip the rapid ascending phase of the initial period. But a system after burn-in can still experience the tail portion of the initial period, seeing a slight increase in efficiency.

Another common recommendation is for operators to ramp up the production of a system slowly in its initial operation period. Wind turbine operators may have operated the turbines following a similar ramping-up strategy. Consequently, the initial increase in production efficiency could be the combined effect of components breaking in and a strategic ramping up.

An interesting observation of Staffell and Green's Figure 9b is that for a wind turbine fleet, there is a noticeable decline even during its stable operation period (until around 18 years in service). This differs apparently from the typical bathtub curve which has a flat stable operation period. Analysts indeed have been arguing [27] that wind turbines, subject to non-steady loads as a result of the uncontrollable nature of wind, degrade faster than a turbine machinery operating under relatively long periods of steady loads (such as the gas turbines in fossil fuel power plants). This tilted stable operation line may be a testimonial to this faster degradation argument.

Hwangbo et al. [96] compare their productive efficiency metric, PE, and power coefficient, C_p. Table 6.3 presents the power coefficient values and the PE values for the four turbines in the four periods. The values in the parentheses are the respective confidence intervals.

As mentioned before, the theoretical maximum of power coefficient is the Betz limit. One could divide a power coefficient by the Betz limit to get an

TABLE 6.3 Comparison between the productive efficiency and the (peak) power coefficient.

	Year 1	Year 2	Year 3	Year 4
	Power coefficient C_p			
WT1	0.374	0.390	0.394	0.393
	(0.369, 0.380)	(0.387, 0.394)	(0.390, 0.399)	(0.390, 0.398)
WT2	0.442	0.463	0.463	0.461
	(0.436, 0.449)	(0.459, 0.468)	(0.458, 0.473)	(0.456, 0.468)
WT3	0.419	0.464	0.482	0.506
	(0.417, 0.422)	(0.460, 0.470)	(0.479, 0.489)	(0.498, 0.515)
WT4	0.418	0.467	0.483	0.504
	(0.415, 0.422)	(0.461, 0.474)	(0.477, 0.491)	(0.497, 0.511)
	Productive efficiency PE			
WT1	0.957	0.974	0.978	0.974
	(0.955, 0.960)	(0.972, 0.976)	(0.976, 0.979)	(0.971, 0.976)
WT2	0.969	0.978	0.978	0.976
	(0.968, 0.971)	(0.977, 0.980)	(0.976, 0.980)	(0.973, 0.978)
WT3	0.965	0.976	0.981	0.985
	(0.964, 0.966)	(0.975, 0.976)	(0.981, 0.982)	(0.984, 0.986)
WT4	0.962	0.976	0.982	0.985
	(0.961, 0.964)	(0.974, 0.977)	(0.981, 0.983)	(0.984, 0.986)

interpretation of how much the turbine performs relative to its potential, similar to what the productive efficiency attempts to do. But the Betz limit is not practically achievable, so that power coefficients are generally below 0.45. When normalizing by the Betz limit, the resulting power coefficient-based efficiency measure never approaches one. Considering the yearly power coefficient in Table 6.3, ranging from 0.371 to 0.506, the relative power coefficient efficiency is therefore between 63% and 85%. These low percentages should not be interpreted as saying that power production of the wind turbines is inefficient. Looking at the PE values, the wind turbine operations are actually reasonably efficient, relative to their full potentials.

Using the power coefficient values, one also notices a general upward trend and a leveling off. This message appears to reinforce what is observed using the PE measure. In fact, there appears a fairly obvious positive correlation between the two metrics. Using all the average values in Table 6.3 yields a correlation of 0.75 between C_p and PE values. This positive correlation suggests that the productive efficiency metric measures a turbine's performance on a broad common ground with the power coefficient.

GLOSSARY

AEP: Annual energy production

CF: Capacity factor

CNLS: Convex nonparametric least squares

DEA: Data envelope analysis

IEC: International Electrotechnical Commission

pdf: Probability density distribution

PE: Productive efficiency

PGR: Power generation ratio

RUP: Regular ultra passum

SFA: Stochastic frontier analysis

s.t.: Such that

TS: Technical specifications

EXERCISES

6.1 Use the `Offshore Wind Farm Dataset2`, select one of the turbines, and conduct the following exercise.

   a. Calculate the capacity factor for this wind turbine on the weekly time resolution, and let us call this capacity factor M4.

   b. Plot the histogram and the empirically estimated density curve of M4. Compare M4 with the other three metrics in Fig. 6.4.

   c. Replicate the results in Table 6.1 but now it is M4 versus M1, M2, and M3, respectively.

6.2 Use the `Inland Wind Farm Dataset2`, select one of the turbines, and replicate the analysis in Section 6.2 but for this inland turbine. Are the overall insights concerning the three performance metrics still valid?

6.3 Generate a plot like Fig. 6.14 but using the data of the (peak) power coefficient in Table 6.3.

6.4 The power coefficient computed in Section 6.2 is the peak power coefficient, i.e., that the largest value on a power coefficient curve is used to represent the whole curve. Let us compute the average power coefficient instead, i.e., the average of all values on the power coefficient curve. Then, use the average power coefficient to re-do the analysis in Tables 6.1 and 6.2 and Figs. 6.4–6.6. Is the average power coefficient closer to the other two performance metrics than the peak power coefficient?

6.5 Khalfallah and Koliub [121] study the effect of dust accumulation on turbine blades on power production performance of the affected turbine. They reckon that the power production, in the presence of dust accumulation, deteriorates more significantly for wind speed higher than 9 m/s than the lower wind speeds. They estimate the annual loss is around 3%.

    a. Take the WT1's 2008 data and modify it by decreasing the power output value by 3% for those power outputs corresponding to wind speed of 9 m/s or higher. Treat this as the 2009 data. Then, reduce the 2008 power data by 6% and 9% and use them as the substitute of 2010 and 2011 power data, respectively. This gives us a set of simulated wind-power data, mimicking the dust accumulation effect over four years. This data simulation is first suggested by Hwangbo et al. [96].

    b. Compute the (peak) power coefficient for the four years on an annual basis.

    c. Use 100 bootstrap samples to compute the 90% confidence interval for each of the point estimates of power coefficient obtained in (b).

    d. Plot the point estimates and the associated 90% confidence intervals in the fashion similar to Fig. 6.14. What do you observe? Does it tell you certain limitation of using the power coefficient as a performance metric?

6.6 Find some other real-life examples, if possible, supporting or illustrating the flipped bathtub curve. Do any of your examples have an accelerated deterioration even during its supposedly stable operation period, like the curve in the right panel of Fig. 6.15?

6.7 The details of the covariate matching procedure can be found in [96, Section 3] or in Chapter 7 of this book. Please read the material and understand how it works. Choose different ϖ values and see how it affects the matching quality and the resulting data amount. Choices of ϖ can be $0.1, 0.15, 0.2, 0.3$ or 0.5.

6.8 Use the 10-min resolution data in the Wind Time Series Dataset and split the data into 80% for training and 20% for testing. Conduct the following exercise.

    a. Use the training data to construct a V-versus-y power curve using the IEC binning method.

    b. Use the same training data to construct a shape-constrained power curve using the method outlined in Section 6.3.2.

c. Use the test data to perform an out-of-sample test on the two power
    curves. Please compare the two power curve estimates in terms of
    both RMSE and MAE.

CHAPTER 7

Quantification of Turbine Upgrade

DOI: 10.1201/9780429490972-7

Turbine performance assessment, as discussed in Chapter 6, plays an important role in wind turbine maintenance, equipment procurement and wind energy planning. Over time, a wind turbine naturally degrades, losing efficiency in power generation. To maintain the production efficiency of a wind turbine, the owners or operators sometimes perform a retrofit to an existing wind turbine, also known as an upgrade, in the hope to restore or enhance the production efficiency of the existing asset. But upgrading can be costly. Owners or operators of wind farms understandably wonder whether the performance of a wind turbine is improved enough to justify the cost of upgrading. This chapter presents several data science methods aiming at addressing this issue.

7.1   PASSIVE DEVICE INSTALLATION UPGRADE

Power output from a wind turbine is driven by wind input. It therefore makes little sense to compare, without controlling for the input conditions, the difference in wind power production before and after an upgrade. The output-only difference, even if present, would not reveal whether the difference comes from upgrading the turbine or from the occurrence of a strong wind after the upgrade.

The output-only comparison could be effective for some of the upgrades that change the control logic without necessarily installing or adjusting physical devices on a turbine. For these cases, analysts suggest switching back and forth between two operational options in 30-second intervals and recording the power production under each option, respectively. Conducting such test for a long enough duration and under a broad variety of environmental conditions, and comparing the power outputs under respective operational options sheds light on which option leads to better power production. The assumption here is that the environmental conditions as well as the turbine's own conditions, besides the operational option under test, are unlikely different in a duration as short as 30 seconds apart. The difference between the power output, if existing, must thus be attributed to the difference in the operational options.

Not all the upgrades can be tested in the aforementioned manner. Many upgrades involve installing a passive device to the existing turbine or adjusting existing turbine components physically. One such upgrade is the vortex generator (VG) installation. The wind industry has long been aware of the VG technology and the potential benefit that VG installation may bring to wind power production, as past studies [151, 157, 224] claim that having VGs could improve the lift characteristics of the blades. Installing vortex generators requires retrofitting turbine blades, incurs material and labor costs, and halts energy production during installation. Once installed, owners and operators would rather not take them off, as doing so incurs even more costs. Turbine upgrades like VG installation are definitely not candidates for conducting the aforementioned 30-second operational switching test.

Although the precise magnitude of the benefit from VG installation is unknown, the general feeling in the industry is that it would be moderate, likely resulting in 1–5% extra wind energy production under the same wind and environmental conditions. Detecting this moderate improvement in the turbine operational data, with the presence of large amounts of noise, is not a trivial task. The IEC binning method for power curve modeling, as explained in Section 5.1, is probably the most widely used approach in the wind industry for estimating and quantifying a turbine's performance before and after VG installation. The IEC standard method is, however, ineffective in this endeavor, which has been noticed by industrial practitioners and documented in previous studies [50, 133]. The IEC admits that *"Depending on site conditions and climate, the uncertainty may amount to several percent"* [102].

A second difficulty in quantifying the benefit of a turbine upgrade lies in the lack of a good method to validate the estimated effect. In order to validate the estimated VG effect, one ought to know the ground truth of the actual effect. For that purpose, one would ideally conduct a controlled experiment, in which all environmental conditions are set the same before and after VGs are installed, so that the difference in power outputs before and after the installation signifies the VG effect. The problem is that such a controlled experiment is impractical and will probably never be feasible, considering the sheer physical size of commercial wind turbine generators. Analysts could conduct small-scale experiments in a wind tunnel, but the amount of uncertainty encountered in the extrapolation of the small-scale wind tunnel test to commercial operations makes such results much less credible to use.

We want to caution readers that our purpose here is not to advocate a specific type of turbine upgrade or retrofit option but to present some options that may better serve the purpose of upgrade quantification. When a quantification method, for example, the IEC binning method, indicates that there is no difference in a turbine's power output before and after an upgrade, a question remains: Is there really no benefit to have that type of upgrade for this particular turbine, or is it possible that the method used is incapable of detecting small to moderate changes due to the method's inherent limitations?

7.2   COVARIATE MATCHING-BASED APPROACH

The IEC method's ineffectiveness is rooted in its lack of control of the influence of multiple environmental factors other than wind speed. Shin et al. [198] present a covariate matching approach to select a subset of data from datasets before and after an action of upgrade and to ensure the environmental covariates of the selected subset to have comparable distribution profiles. Recall from Chapter 5 that x denotes the vector of environmental variables, including wind speed, and the variables in x are called *covariates*. Once the covariates are matched, Shin et al. then quantify the benefit of the upgrade by taking the difference of power outputs under the matched environmental condition and apply a paired t-test for testing the significance of the upgrade effect.

Covariate matching methods are rooted in the statistical literature [206]. In stabilizing the non-experimental discrepancy between non-treated and treated subjects of observational data, Rubin [185] adjusts covariate distributions by selecting non-treated subjects that have a similar covariate condition as that of treated ones. Through the process of matching, *non-treated* and *treated* groups become only randomly different on all background covariates, as if these covariates were designed by experimenters. As a result, the outcomes of the matched non-treated and treated groups, which keep the originally observed values, are comparable under the matched covariate conditions.

Fig. 7.1 demonstrates the discrepancy of the covariate distributions of the un-matched or non-treated data in the `Turbine Upgrade Dataset`. It presents for each covariate the difference between the pre-upgrade and post-upgrade periods using the empirically estimated density functions. The last subplot in both the upper and lower panels is the density function of the power output of the respective control turbine. For the control turbine, as it is not modified, the distribution of its power output is supposed to be comparable, should the environmental conditions be maintained the same. But the data show otherwise, signifying the impact of the confounding environmental influence.

7.2.1   Hierarchical Subgrouping

In the context of wind turbine upgrade, the data records collected before the upgrade form the non-treated data group, whereas those collected after the upgrade form the treated group. Let Q_{bef} and Q_{aft} be the index set of the data records in the non-treated and treated group, respectively. Let x_Q denote the values of a covariate x for data indices in Q. For example, $V_{Q_{\text{bef}}}$ is the vector of all wind speed values that are observed before the upgrade.

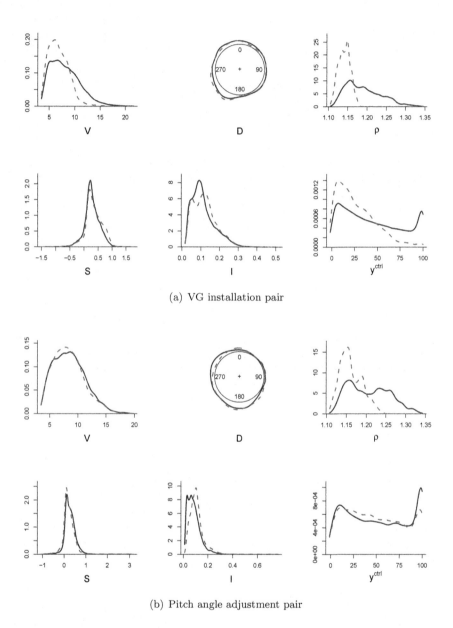

(a) VG installation pair

(b) Pitch angle adjustment pair

FIGURE 7.1 Overlaid density functions of the unmatched covariates and the density functions of control turbine power output. Solid line = before upgrade (non-treated) and dashed line = after upgrade (treated). (Reprinted with permission from Shin et al. [198].)

The first action of preprocessing is to narrow down the dataset and create a subset on which one subsequently performs the data records matching. The reason for this preprocessing is to alleviate a computational demand arising from too many pairwise combinations when comparing two large size datasets. This goal is fulfilled through a procedure labeled as *hierarchical subgrouping*. The idea is summarized in Algorithm 7.1.

Algorithm 7.1 Hierarchical subgrouping procedure to match covariates.

---

1. Locate a data record in the treated group, Q_{aft}, and label it by the index j.

2. Select one of the covariates, for instance, wind speed, V, and designate it as the variable of which the similarity between two data records is computed.

3. Go through the data records in the non-treated group, Q_{bef}, by selecting the subset of data records such that the difference, in terms of the designated covariate, between the data record j in Q_{aft} and any one of the records in Q_{bef} is smaller than a pre-specified threshold, ϖ. When V is in fact the one designated in Step 2, the resulting subset is then labeled by placing V as the subscript to Q, namely Q_V.

4. Next, designate another covariate and use it to prune Q_V in the same way as one prunes Q_{bef} into Q_V in Step 3. Doing so produces a smaller subset nested within Q_V. Then continue with another covariate until all covariates are used.

The order of the covariates in the above hierarchical subgrouping procedure is based on the importance of them in affecting wind power outputs. According to the analysis in Chapter 5, V, D, and ρ are more important, whereas S and I are less so. Shin et al. [198] use the order, V, D, ρ, S, and I. While it is generally a good idea to follow the physically meaningful order when conducting the hierarchical subgrouping, Shin et al. present a numerical analysis based on the reserve order, i.e., I, S, ρ, D, and V, and find that using the reserve order hardly affects the downstream analysis outcome, suggesting a certain degree of robustness in the overall analysis procedure. The robustness appears to be a result of the checks and diagnostics included in the later stage of procedure and to be explained in the sequel. If a priority order of the covariates is unknown, it is recommended to perform some statistical analysis using, for example, random forests [21], which can measure the importance of covariates before applying the matching method.

Note that wind direction, D, is a circular variable, so that an absolute difference between two angular degrees is between 0 and $180°$. The circular variable formula, $180° - |180° - (|D_i - D_j|)|$, is used to calculate the difference between two D values.

The above process can also be written in set representation. For a data record j in Q_{aft}, define subsets of data records in Q_{bef}, hierarchically chosen, as

$$
\begin{aligned}
Q_V &:= \{i \in Q_{\text{bef}} : |V_i - V_j| < \varpi \cdot \sigma(V_{Q_{\text{bef}}})\}, \\
Q_D &:= \{i \in Q_V : 180° - |180° - (|D_i - D_j|)| < \varpi \cdot \sigma(D_{Q_V})\}, \\
Q_\rho &:= \{i \in Q_D : |\rho_i - \rho_j| < \varpi \cdot \sigma(\rho_{Q_D})\}, \\
Q_S &:= \{i \in Q_\rho : |S_i - S_j| < \varpi \cdot \sigma(S_{Q_\rho})\}, \\
Q_I &:= \{i \in Q_S : |I_i - I_j| < \varpi \cdot \sigma(I_{Q_S})\},
\end{aligned}
\tag{7.1}
$$

where $\sigma(x)$ is the standard deviation of x in the specified dataset. The thresholding coefficient, ϖ, can be different at each layer but to make thing simple, analysts usually select a single constant threshold for the whole procedure. This hierarchical subgrouping establishes the subsets nested as such: $Q_I \subset Q_S \subset Q_\rho \subset Q_D \subset Q_V \subset Q_{\text{bef}}$. Consequently, the data records in the last hierarchical set, Q_I, have the closest environmental conditions as compared with the data record j in Q_{aft}.

There could be other conditions, in addition to the five variables mentioned above, which may affect wind power production while not measured. The possible existence of unmeasured factors presents the risk of causing a distortion in comparison, even when the aforementioned measured environmental factors are matched between the treated and non-treated groups. In order to alleviate this risk, Shin et al. [198] make use of the power output, y^{ctrl}, of the control turbine in each turbine pair. The idea is to further narrow down from the most nested subset, produced above, by taking the following action—select records from Q_I whose y^{ctrl} values are comparable to the y^{ctrl} value of a data record j in Q_{aft}. Doing this actually amounts to continuing the hierarchical subgrouping action to produce a Q_y, a subset of Q_I, based on y^{ctrl}, such that

$$
Q_y := \{i \in Q_I : |y_i^{\text{ctrl}} - y_j^{\text{ctrl}}| < \varpi \sigma(y_{Q_I}^{\text{ctrl}})\}.
\tag{7.2}
$$

Shin et al. [198] perform this procedure for all data records in the treated group so that each record j in Q_{aft} has its matched set, which is denoted by $Q_{y,j}$. The set of $Q_{y,j}$ is in fact the set of data records in the non-treated group matched to the data record j in the treated group. In the case that $Q_{y,j}$ is an empty set, one should then discard the respective index j from Q_{aft}. Because of this, Q_{aft} may shrink after the subgrouping steps.

7.2.2 One-to-One Matching

The next action is to choose a data record in $Q_{y,j}$ that is the closest to the data record j. For this purpose, analysts need to define a dissimilarity measure to quantify the closeness between two data records. Shin et al. [198] decided to use the Mahalanobis distance [140] as the dissimilarity measure, which is popularly used in the context of multivariate analysis. A Mahalanobis distance re-weighs the Euclidean distance between two covariate vectors with the reciprocal of a variance-covariance matrix. Before presenting the definition of the Mahalanobis distance between two wind turbine data records, Shin et al. first transform $\boldsymbol{x}$ into $\boldsymbol{x}^*$, such that

$$\boldsymbol{x}^* := (V \cos D, V \sin D, \rho, S, I)^T.$$

Using $\boldsymbol{x}^*$ makes it easier to deal with the circular wind direction variable, D. The Mahalanobis distance, MD_{ij}, between a data record j in Q_{aft} and a data record i in $Q_{y,j}$, is defined as

$$\mathrm{MD}_{ij} := \sqrt{(\boldsymbol{x}_i^* - \boldsymbol{x}_j^*)^T \boldsymbol{\Sigma}^{-1} (\boldsymbol{x}_i^* - \boldsymbol{x}_j^*)}, \tag{7.3}$$

where $\boldsymbol{\Sigma}$ is the covariance matrix of $\boldsymbol{x}^*$. Obviously, the larger an MD value, the less similar the two data records.

With the Mahalanobis distance defined, one can simply select the data record i_j in $Q_{y,j}$ that has the smallest Mahalanobis distance as the best match to data record j in Q_{aft}. In other words, i_j is found such that

$$i_j = \arg \min_{i \in Q_{y,j}} \mathrm{MD}_{ij},$$

for each j in Q_{aft}. In case two or more are tied for the smallest value, Shin et al. [198] choose one of them randomly.

After this step, each data record j in the treated group is matched to a non-treated counterpart i_j, with the exception of those already discarded during the subgrouping step. Let us define the index set of the matched data records from the non-treated group as

$$Q_{\mathrm{bef}}^* := \{i_j \in Q_{\mathrm{bef}} \,|\, j \in Q_{\mathrm{aft}}\}.$$

It should be noted that Shin et al. [198] allow replacement in the matching procedure. In other words, i_j is not eliminated from the candidate set Q_{bef}, even though it has matched to j once. When the next data record $j + 1$ is selected from Q_{aft}, the same non-treated data i is possible to be matched again.

7.2.3 Diagnostics

After performing the matching procedure, it is important to diagnose how much of the discrepancy in the covariate distributions has been removed. Only after the diagnostics signifies a sufficient improvement, can an outcome analysis be performed in the next step.

Shin et al. [198] measure the discrepancy of distributions in two ways, numerically and graphically. For the numerical diagnostics, the standardized

TABLE 7.1 SDM values serve as the numerical diagnostics for covariate matching.

VG installation pair					
	V	D	ρ	S	I	y^{ctrl}
Unmatched	0.6685	0.0803	3.2715	0.2312	0.1382	0.8132
Matched	0.0142	0.0026	0.0589	0.0721	0.0003	0.0083

Pitch angle adjustment pair					
	V	D	ρ	S	I	y^{ctrl}
Unmatched	0.0605	0.1647	1.6060	0.2759	0.4141	0.0798
Matched	0.0077	0.0029	0.0263	0.0158	0.0111	0.0036

Source: Shin et al. [198]. With permission.

difference of means (SDM) is used as a measure of dissimilarity of a covariate between the treated and non-treated groups, i.e.,

$$\text{SDM} := \frac{\overline{x}_{Q_{\text{aft}}} - \overline{x}_{Q_{\text{bef}}}}{\sigma(x_{Q_{\text{aft}}})}, \tag{7.4}$$

where x is one of the covariates and $\overline{x}_Q$ denotes the average of x in the set of Q. The SDM decreases if the matching procedure indeed reduces the discrepancy between the two groups. As shown in Table 7.1, SDM decreases significantly for all covariates after matching. A previous study [186] suggests that SDM should be less than 0.25 to render the two distributions in question comparable.

The graphical diagnostics uses the pdf plots just like in Figs. 6.12 and 6.13, in which the empirical density functions before and after the upgrade are overlaid on top of each other. Then, a visual inspection is conducted to check and verify that the two respective density functions are similar enough. Fig. 7.2 presents the well-matched distributions of covariates after the matching process. The improvements in terms of distribution similarity are apparent when compared to Fig. 7.1.

It should also be noted that, if the size of Q_{aft} after the matching loses too many data records, and this can happen when too small ϖ's are applied, Shin et al. [198] suggest to enlarge the size of the original Q_{aft} prior to the matching process, in order to secure a sufficient amount of representative weather conditions in the matched Q_{aft}. Enlarging Q_{aft} can be done by extending the post-upgrade data collection period, for instance.

7.2.4 Paired t-tests and Upgrade Quantification

The matching procedure produces a set of paired data records of the two groups, each pair denoted by (i_j, j), where $i_j \in Q^*_{\text{bef}}$ and $j \in Q_{\text{aft}}$. Using these paired indices, Shin et al. [198] retrieve the paired power outputs for the test turbine, i.e., $(y^{\text{test}}_{i_j}, y^{\text{test}}_j)$. The power output pair can be interpreted

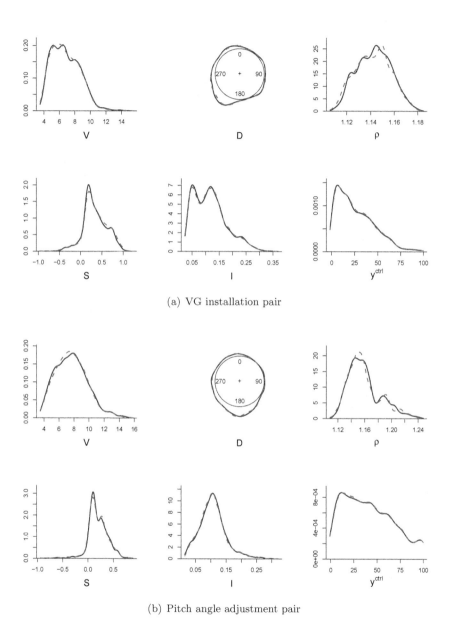

(a) VG installation pair

(b) Pitch angle adjustment pair

FIGURE 7.2   Overlaid density functions of the matched covariates and the density functions of the matched control turbine power output. Solid line = before upgrade (non-treated) and dashed line = after upgrade (treated). (Reprinted with permission from Shin et al. [198].)

TABLE 7.2   The results of paired t-tests and upgrade quantification.

VG installation pair			Pitch angle adjustment pair		
t-stat	p-value	UPG	t-stat	p-value	UPG
3.015	0.003	1.13%	7.447	< 0.0001	3.16%

Source: Shin et al. [198]. With permission.

as repeated measurements under comparable environmental conditions, thus making the power outputs also comparable. One can apply a t-test to analyze the difference of the paired power outputs, computed as $\delta_j = y_j^{\text{test}} - y_{i_j}^{\text{test}}$. The null hypothesis is that the expected mean of the difference is zero, that is, $H_0 : \mathbb{E}(\bar{\delta}) = 0$, where $\bar{\delta}$ is the sample mean of $\{\delta_j : j \in Q_{\text{aft}}\}$. Accordingly, the test statistic is

$$t\text{-stat} := \frac{\bar{\delta}}{s_\delta/\sqrt{n_\delta}}, \tag{7.5}$$

where s_δ and n_δ are the sample standard deviation and the sample size of $\{\delta_j : j \in Q_{\text{aft}}\}$, respectively. If the test concludes a significant positive mean difference, the upgrade on the test turbine is deemed effective. Table 7.2 presents the results from the paired t-test of both datasets, which show a significant upgrade effect at the 0.05 level.

Shin et al. [198] quantify the upgrade effect, denoted by UPG, by computing

$$\text{UPG} := \frac{\sum_{j \in Q_{\text{aft}}} (y_j^{\text{test}} - y_{i_j}^{\text{test}})}{\sum_{j \in Q_{\text{aft}}} y_{i_j}^{\text{test}}} \times 100. \tag{7.6}$$

The quantification results are shown in Table 7.2 as well. Recall that in the pitch angel adjustment pair, the test turbine's power is increased by 5% for wind speeds of 9 m/s and above, which translates to a 3.11% increase for the whole wind spectrum. The quantification outcome shows an improvement of 3.16% overall, which appears to present a fair agreement with the simulated amount.

7.2.5   Sensitivity Analysis

The pitch angle adjustment pair is analyzed for the purpose of getting a sense of how well a proposed method can estimate a power production change. Recall that the upgrade in the pitch angle adjustment pair is simulated, so we know the true upgrade amount and can use that as a reference for comparison. In Section 7.2.4, however, only a single simulated improvement value (5%) is used. To have a fuller sense, this section conducts the matching-based quantification on various degrees of the simulated improvement.

Denote by r the nominal power increase rate. Because the nominal power increase rate is applied only to the partial range of wind power corresponding to wind speed higher than 9 m/s, the effective power increase rate for the

TABLE 7.3 Sensitivity analysis of covariate matching-based turbine upgrade quantification.

r	2%	3%	4%	5%	6%	7%	8%	9%
r'	1.25%	1.87%	2.49%	3.11%	3.74%	4.36%	4.98%	5.60%
UPG	1.74%	2.21%	2.68%	3.16%	3.63%	4.11%	4.58%	5.05%
UPG/r'	1.39	1.18	1.08	1.02	0.97	0.94	0.92	0.90

Source: Shin et al. [198]. With permission.

whole wind spectrum, denoted by r', is different. When it comes to verifying the upgrade effect for the simulated case, the effective power increase rate r' is computed through

$$r' := \frac{\sum_{j \in Q_{\text{aft}}} y_j^{\text{test}}\{1 + r \cdot \mathbb{1}(V_j^{\text{test}} > 9)\} - \sum_{j \in Q_{\text{aft}}} y_j^{\text{test}}}{\sum_{j \in Q_{\text{aft}}} y_j^{\text{test}}}. \quad (7.7)$$

Shown in Table 7.3, as r changes from 2% to 9%, r' changes from 1.25% to 5.6%. This range of power increases is practical for the detection purpose. If an improvement is smaller than 1%, it is going to be considerably hard to detect. On the other hand, when an improvement is greater than 6%, it is possible that even the IEC binning method can detect that level of change.

Table 7.3 presents the UPGs corresponding to the respective r'. One can observe that UPG noticeably overestimates r' when r' is small (smaller than 2%)—the overestimation is as much as 40% for the smallest change in the table. But the quantification quality using UPG gets stabilized as r' increases. For the last six cases in Table 7.3, the differences between UPG and r' are within 10%.

7.3 POWER CURVE-BASED APPROACH

The multi-dimensional power curve methods, explained in Chapter 5, can account for the influence of environmental variables on power output. It is thus not surprising that upgrade quantification approaches are developed based on power curve models.

The basic idea is as follows. Once a power curve model is established using the pre-upgrade data, it captures the power production characteristics of the old turbine before the upgrade. Feeding the post-upgrade wind and environmental data to the power curve model is analogous to running the old, unmodified turbine under the new conditions. Comparing the model outputs with the actual physical outputs under the same input conditions is supposed to reveal the difference that an upgrade makes.

Using the IEC binning method to quantify the benefit of an upgrade is in fact a power curve-based approach. The drawback of that specific approach lies in the fact that IEC binning controls for practically only the wind speed effect, which accounts for roughly 85% of the variation in the power data.

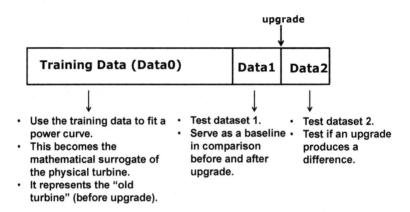

FIGURE 7.3 Dataset partition used in Kernel Plus for quantifying a turbine upgrade. (Reprinted with permission from Lee et al. [133].)

The remaining unaccounted variation is still too much relative to the typical upgrade effect, and without accounting for that, the resulting method is rendered ineffective, which is what happens to the IEC binning-based quantification method [50].

Lee et al. [133] present a turbine upgrade quantification method based on the additive-multiplicative kernel power curve model, introduced in Section 5.2.3. The resulting method is nicknamed *Kernel Plus*.

7.3.1 The Kernel Plus Method

The central element in Kernel Plus is the AMK model in Eq. 5.16. In the context of upgrade quantification, only the mean prediction equation, $\hat{y}(x)$, is used. To make it explicit that the power curve model is from AMK, let us denote it with an AMK superscript, namely $\hat{y}^{\text{AMK}}(x)$.

In its procedure to quantify a turbine upgrade, Kernel Plus involves three datasets, as illustrated in Fig. 7.3. The three datasets include a training dataset of historical observations of (x, y) pairs and two test datasets before and after an upgrade. The training dataset is referred to as "Data0" and is used to fit the power curve model. Data0 should be collected from a reasonable duration of a turbine's operation, for instance, one year, such that the seasonal weather effects are well represented in the data. The two test sets, referred to as "Data1" and "Data2," respectively, are collected for the same length of duration before and after the upgrade. They are used to detect and quantify the upgrade. Their corresponding data duration can be much shorter than that of Data0; a few weeks to a few months may be sufficient.

In the Kernel Plus method, Lee et al. [133] introduce a self-calibration procedure to alleviate the bias associated with a Nadaraya-Watson kernel estimator when it is used on a new dataset. The existence of bias based on finite samples is a common problem in statistical prediction. Hastie et al. present a full illustration of biases and variances involved in statistical prediction [86, Figure 7.2].

The self-calibration procedure is done by using subsets of the training data in Data0. To select a calibration set of data that has similar weather conditions to those in Data1 and Data2, Lee et al. [133] define a distance measure, which is in spirit similar to the Mahalanobis distance—recall that the Mahalanobis distance is used in the covariate matching procedure in Section 7.2.2. Like a Mahalanobis distance, the distance measure in the self-calibration procedure is a weighted distance but unlike a Mahalanobis distance, it is not weighted by the reciprocal of the corresponding variance-covariance matrix. Instead, the weighting matrix is a diagonal matrix whose diagonal elements are from the bandwidth vector $\boldsymbol{\lambda}$. Let us denote this diagonal matrix by $\boldsymbol{\Lambda}$, such that $\boldsymbol{\Lambda}_{i,i} = \lambda_i$ and $\boldsymbol{\Lambda}_{i,j} = 0$ $\forall i \neq j$. The resulting distance measure between a training data point, $\boldsymbol{x}_i \in$ Data0, and a test data point, $\boldsymbol{x}_j$, in either Data1 or Data2, is

$$D(\boldsymbol{x}_i, \boldsymbol{x}_j) = \sqrt{(\boldsymbol{x}_i - \boldsymbol{x}_j)^T \boldsymbol{\Lambda}^{-1} (\boldsymbol{x}_i - \boldsymbol{x}_j)}. \tag{7.8}$$

Lee et al. [133] elaborate that the reason to choose this distance measure is because a simple Euclidean distance does not reflect well the similarity between the $\boldsymbol{x}$'s, as different elements in $\boldsymbol{x}$ have different physical units, leading to different value ranges. To define a sensible similarity measure, a key issue is to weigh different elements in $\boldsymbol{x}$ consistently with their relative importance pertinent to the power output. The original Mahalanobis distance does not serve this purpose because the squared distance associated with an input variable is weighted by the inverse of its variance. In a power curve model, wind speed is arguably the most important variable, yet it has a large variance. Because of this large variance, using the Mahalanobis distance will in fact diminish the importance of wind speed relative to other variables that have a smaller variance. The choice in Eq. 7.8 that uses the kernel bandwidth parameters as the weighting coefficients in $\boldsymbol{\Lambda}$ is consistent with the goal of weighting each element according to its relative importance, because the bandwidth parameters are selected based on how sensitive the power output is to a unit change in the corresponding input variable. If an input variable has a small bandwidth, it means that the power output could produce an appreciable difference with a small change in the corresponding input, suggesting that this variable is relatively important. On the other hand, a large bandwidth indicates a less important input variable.

For any test data point $\boldsymbol{x}_j$, one can choose a calibration data point, $\boldsymbol{x}_i^{\text{cal}}$, from Data0, which has the minimum $D(\boldsymbol{x}_i^{\text{cal}}, \boldsymbol{x}_j)$. The calibration procedure proceeds as described in Algorithm 7.2.

Algorithm 7.2 Self-calibration procedure in Kernel Plus.

1. For $\boldsymbol{x}_i^{\text{cal}} \in \text{Data0}$, compute $\hat{y}^{\text{AMK}}(\boldsymbol{x}_i^{\text{cal}})$.

2. Compute the calibration value $\mathfrak{R}^{\text{cal}}(\boldsymbol{x}_j) = y(\boldsymbol{x}_i^{\text{cal}}) - \hat{y}^{\text{AMK}}(\boldsymbol{x}_i^{\text{cal}})$, where $\boldsymbol{x}_j$ is paired to the calibration data point $\boldsymbol{x}_i^{\text{cal}}$.

3. For any test data point $\boldsymbol{x}_j$, the final, calibrated power estimate from the Kernel Plus method is $\hat{y}^{\text{KP}}(\boldsymbol{x}_j) = \hat{y}^{\text{AMK}}(\boldsymbol{x}_j) + \mathfrak{R}^{\text{cal}}(\boldsymbol{x}_j)$.

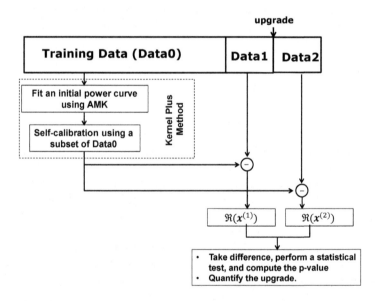

FIGURE 7.4 Diagram of quantifying a turbine upgrade using the Kernel Plus method. (Reprinted with permission from Lee et al. [133].)

7.3.2 Kernel Plus Quantification Procedure

Fig. 7.4 outlines the procedure for detecting and quantifying an upgrade using the Kernel Plus method and the three sets of data. To start, one establishes the Kernel Plus model, which includes both AMK and self-calibration. Then, this Kernel Plus model, representing the "old" turbine, is used to make a prediction/estimation of power output under a new weather profile x in either Data1 or Data2—the result is denoted as $\hat{y}(x^{(1)})$ and $\hat{y}(x^{(2)})$, respectively. Here $\hat{y}(x^{(1)})$ and $\hat{y}(x^{(2)})$ are $\hat{y}^{KP}(\cdot)$ but for notational simplicity and without ambiguity, the superscript, KP, is dropped. Consequently, the corresponding power output residuals can be computed. Had a turbine undergone an upgrade, one would expect the residuals before and after the upgrade to be different. A t-test is used to detect a potential difference in the residuals. Suppose that n_1 and n_2 are the number of data points in Data1 and Data2, respectively. The statistical test procedure is presented in Algorithm 7.3.

Algorithm 7.3 Statistical test procedure for upgrade detection.

---

1. Compute the residuals before and after an upgrade. For Data1, $\Re(x^{(1)}) := y(x^{(1)}) - \hat{y}(x^{(1)})$, and for Data2, $\Re(x^{(2)}) := y(x^{(2)}) - \hat{y}(x^{(2)})$;

2. Compute the two sample means and the corresponding standard deviations by using the following formula,

$$\bar{\Re}_k = \frac{\sum_{j=1}^{n_k} \Re(x_j^{(k)})}{n_k}, \quad k = 1, 2, \quad \text{and,}$$

$$s_k = \sqrt{\frac{\sum_{j=1}^{n_k} (\Re(x_j^{(k)}) - \bar{\Re}_k)^2}{n_k - 1}}, \quad k = 1, 2. \tag{7.9}$$

3. Then, calculate the pooled estimate of standard deviation, $\hat{\sigma}_r$, by

$$\hat{\sigma}_r = \sqrt{\frac{(n_1 - 1)s_1^2 + (n_2 - 1)s_2^2}{n_1 + n_2 - 2}}. \tag{7.10}$$

4. The t statistic is calculated by

$$t = \frac{\bar{\Re}_2 - \bar{\Re}_1}{\hat{\sigma}_r \cdot \sqrt{\frac{1}{n_1} + \frac{1}{n_2}}}. \tag{7.11}$$

5. Finally, calculate the p-value of the t statistic. The smaller the p-value, the more significant the difference.

The procedure in Algorithm 7.3 is devised to confirm any detectable difference resulting from an upgrade. The output is binary: either the upgrade produces a statistically significant difference in a turbine's performance or it does not.

If the t-test above does indicate a significant difference, how much difference in terms of power generation does the upgrade produce? To answer this question, Lee et al. [133] define a quantifier as follows,

$$
\text{DIFF}(\boldsymbol{x}) = \frac{\sum_{\boldsymbol{x} \in \mathbb{D}_{test}} \left(y(\boldsymbol{x}) - \hat{y}(\boldsymbol{x}) \right)}{\sum_{\boldsymbol{x} \in \mathbb{D}_{test}} y(\boldsymbol{x})} \times 100\%, \tag{7.12}
$$

where $\mathbb{D}_{test}$ is a test dataset and can be either Data1 or Data2, so that $\boldsymbol{x}$ can accordingly be either $\boldsymbol{x}^{(1)}$ or $\boldsymbol{x}^{(2)}$. Similar to the residual analysis described above, comparing $\text{DIFF}(\boldsymbol{x}^{(2)})$ with $\text{DIFF}(\boldsymbol{x}^{(1)})$ for the test turbine, i.e., $\text{DIFF}^{test} = \text{DIFF}(\boldsymbol{x}^{(2)}) - \text{DIFF}(\boldsymbol{x}^{(1)})$, produces the difference demonstrated in the test turbine before and after the upgrade. If one conducts the same analysis to the control turbine, it produces a DIFF^{crtl}. The final quantification is the difference between the two turbines, i.e., $\text{DIFF} = \text{DIFF}^{test} - \text{DIFF}^{crtl}$.

7.3.3 Upgrade Detection

Using the `Turbine Upgrade Dataset` but just two weeks' worth of data for Data1 and Data2, Lee et al. [133] apply both the Kernel Plus method and the binning method to the two pairs of turbines and conduct a residual analysis. When using the binning method for upgrade quantification, one simply replaces the dashed-line rectangle in Fig. 7.4 with the binning method (the version with air density adjustment).

For an upgraded turbine, a method is supposed to produce a large t-statistic (in its absolute value), which further leads to a small p-value that signifies the difference between the residuals, whereas for a turbine without upgrade, a small t statistic, or equivalently, a large p-value is expected. The commonly used threshold of a p-value to indicate significance is 0.05, which is what Lee et al. [133] use in their analysis.

Table 7.4 presents the outcomes from the residual analysis of both pairs of turbines. The Kernel Plus method has significant outcomes consistent with the upgrade action while the binning method does not.

The outcome of the binning method is attributable to the still large amount of uncertainty unaccounted for in its residuals. To intuitively understand the outcomes of the statistical tests, Lee et al. [133] present the residual plots using data from the test turbine in the VG installation pair when applying the binning method and the Kernel Plus method, respectively. The residual plots are presented in Fig. 7.5. The residuals of the binning method exhibit an obvious pattern (leading to bias) and have a large dispersion, suggesting a poor model fit and large uncertainty, whereas the residuals of the Kernel Plus method have a considerably smaller dispersion and exhibit a random pattern, indicating an adequate model fit and reduced uncertainty.

TABLE 7.4 Comparing the Kernel Plus and binning methods on their ability to detect turbine upgrade.

Turbine		Binning		Kernel Plus	
		t statistic	p-value	t statistic	p-value
VG	Test	−0.46	0.65	2.24	0.025
installation pair	Control	−2.54	0.01	−0.14	0.89
Pitch angle	Test	5.09	3.89×10^{-7}	3.18	0.002
adjustment pair	Control	4.51	6.84×10^{-6}	−1.71	0.09

Source: Lee et al. [133]. With permission.

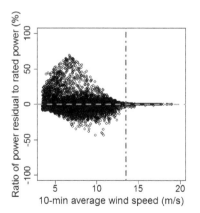 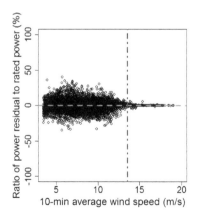

FIGURE 7.5 Residual plots. Left panel: after IEC binning is applied; right panel: after Kernel Plus is applied. The vertical dashed line indicates the rated wind speed. (Reprinted with permission from Lee et al. [133].)

TABLE 7.5 Sensitivity analysis of Kernel Plus-based turbine upgrade quantification.

r	2%	3%	4%	5%	6%	7%	8%	9%
r'	1.25%	1.87%	2.49%	3.11%	3.74%	4.36%	4.98%	5.60%
DIFF	1.97%	2.56%	3.15%	3.73%	4.30%	4.86%	5.42%	5.97%
DIFF/r'	1.58	1.37	1.27	1.20	1.15	1.11	1.09	1.07

Source: Shin et al. [198]. With permission.

7.3.4 Upgrade Quantification

As it is done for the covariate matching in Table 7.3, let us conduct a sensitivity analysis for the Kernel Plus-based method, again using the data from the pitch angle adjustment pair. The outcome is presented in Table 7.5. One can see that the Kernel Plus method does an adequate job, but performs slightly worse than the covariate matching on this simulated case.

When applying to the VG installation pair, the DIFF value is 1.48%. Recall that the quantification from the covariate matching is UPG = 1.13%. It seems that the Kernel Plus method tends to over-estimate the upgrade effect. Please note that the DIFF values reported here are different from those reported in [133], because of the difference in data. Lee et al. [133] use two weeks' worth of data in the post-upgrade period, whereas the results above are obtained using eight weeks of data. Should the post-upgrade period be shortened to two weeks, the DIFF value is 1.81%.

To visualize a multi-dimensional response surface, like the response from the Kernel Plus method, analysts can condition some of the covariates on a constant value and average the others over all possible values. Lee et al. [133] produce a series of one-dimensional power curves under different combinations of ρ and I. Three settings each are chosen for ρ and I, respectively, which are $\rho = (1.15, 1.18, 1.21)$ and $I = (0.08, 0.12, 0.16)$. Altogether, there are nine combinations. The power curves presented in Fig. 7.6 are produced based on the data from the test turbine in the VG installation pair and include those produced by using both the Kernel Plus method and the binning method.

When using the Kernel Plus method, there are observable differences in several subplots between the power curves before and after the upgrade. The difference is pronounced around the rated wind speed. By comparison, the binning method produces power curves with no visually detectable difference. This result is consistent with the message of the previous subsections.

7.4 AN ACADEMIA-INDUSTRY CASE STUDY

In Section 7.1, we mention that one difficulty in quantifying the benefit of a turbine upgrade is due to the lack of a good method to validate the estimated effect. Currently, it appears that a viable way to address this issue, yet still

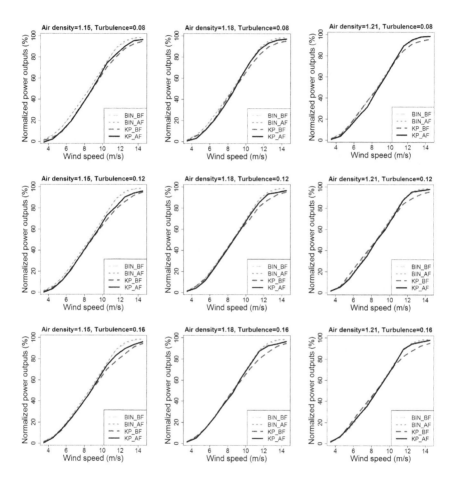

FIGURE 7.6 Power curves conditioned on air density and turbulence intensity and averaged across wind directions. BIN_BF: binning method before upgrade; BIN_AF: binning method after upgrade; KP_BF: Kernel Plus method before upgrade; KP_AF: Kernel Plus method after upgrade. (Reprinted with permission from Lee et al. [133].)

indirect, is to use two different methods to cross validate each other. Comparison between the covariate matching method and the Kernel Plus method serves that purpose.

Hwangbo et al. [95] present an academia-industry joint case study, in which an academic institution and a wind technology company use their respective method to analyze eight pairs of turbines from two wind farms, four pairs per farm. The academic institution's method is the Kernel Plus method described in Section 7.3, whereas the company's method uses high-frequency data via primarily a direct power comparison approach that relies less on the environmental data. The company method is referred to as the power-vs-power method.

In the next subsections, we briefly explain, based on the materials in [95], the power-vs-power method and then present the joint case study. The data used in this joint study is proprietary, and is therefore not included as one of the datasets associated with the book.

7.4.1 The Power-vs-Power Method

The basic idea behind the power-vs-power approach is similar in spirit to that of the covariate matching. The difference is that the company's specific method used in the joint study relies on the high-frequency historian data, usually a data point per a few seconds to a few data points per second. By contrast, the covariate matching in Section 7.2 and Kernel Plus in Section 7.3 both use the 10-min data. The amount of the high-frequency data could be as much as 600 times more than the 10-min data for the same time duration. The power-vs-power method uses additional mechanisms to control for the environmental influence—the controlling mechanisms are often called *filters* in industry practice.

The power-vs-power approach entails five main steps, outlined in Algorithm 7.4.

Step 1 is to ensure the validity of the assumption that when two turbines are close enough in space, it is likely that the wind and environmental conditions they are subject to are comparable. This assumption may not be reasonable for the situation when one turbine is in the wake of the other one. Step 1 is to identify the free wind sectors in the turbine operational data and then use only the free sector data in the subsequent analysis.

Step 2 performs an air density normalization. The thought behind this is similar to that of using the density-normalized wind speed, as recommended by the IEC [102]. In the case of the power-vs-power approach, no wind speed measurements are involved in the latter power comparison steps. For this reason, density normalization must be accomplished by direct normalization of the power values for the below rated region.

Step 3 is another step designed to verify and uphold the assumption that both turbines must "see" the same conditions and must operate similarly.

Algorithm 7.4 Five main steps in the power-vs-power approach.

---

1. Determine the valid wind sectors and eliminate the wind and power measurements taken under wake conditions. Also apply all other data filters (Status.Flag, Yaw.Error, etc.).

2. Apply a power density normalization, namely, normalize the wind power output through $y \times \rho/\rho_0$, where ρ_0 is the sea-level dry air density. Use the density-normalized power in the subsequent analysis.

3. If necessary, verify whether there is any other source of variation significantly affecting the power difference between the pre-upgrade and post-upgrade periods. If such a source of variation is identified, further reduce the dataset so that its effect is controlled for.

4. Compute the bin-wise power difference, namely, calculate the power production difference of the test turbine, relative to that of the control turbine, for each of the power output bins, for both the pre-upgrade and post-upgrade periods.

5. Compute the power difference produced by the VG installation over the whole power output spectrum.

When there are obvious sources of variation, additional filters may be needed to split the data into sets of equal conditions.

After completion of the pre-processing steps that filter, clean, and normalize the data, Step 4 of the power-vs-power approach is to compute the bin-wise power difference between the two turbines. Specifically,

- Take the high-frequency power output data of the control turbine and partition the data into B bins by using a bin width of, say, 100 kW. The bin width can be adjusted for other applications, but, for megawatts capacity turbines, 100 kW appears to be a reasonable default number.

- For each bin, calculate the median of the power difference between the test turbine and the control turbine.

- Conduct the above two steps for the pre-upgrade and post-upgrade periods individually. Denote the resulting power differences by $\Delta \bar{y}_b^{\mathrm{PRE}}$ and $\Delta \bar{y}_b^{\mathrm{POST}}$, respectively, for $b = 1, \ldots, B$.

- Conduct a bin-wise comparison between the control and test turbine for the pre-upgrade period to verify the performance similarities between the pair of turbines.

- Calculate the bin-wise power difference as $\Delta \bar{y}_b = \Delta \bar{y}_b^{\mathrm{POST}} - \Delta \bar{y}_b^{\mathrm{PRE}}$, for $b = 1, \ldots, B$.

Finally, Step 5 of the power-vs-power approach combines all the bin-wise power differences by using the weights derived from the power distribution over a given year; the resulting metric serves as the estimate of the upgrade effect. The detailed procedure is:

- Compute a power curve as a function of wind speed using the measurements taken from the control turbine. Alternatively, one can use the turbine manufacturer's certified reference power curve.

- Using the power curve, find the specific wind speeds, $V_{b,left}$ and $V_{b,right}$, that correspond, respectively, to the lower and upper bound of the b-th power bin. Convert the wind speed distribution into a power distribution through

$$P(y_b) = F_V(V_{b,right}) - F_V(V_{b,left}),$$

where y_b is the midpoint of the b-th power bin, $F_V(\cdot)$ is the cumulative distribution function of wind speed, and $P(y_b)$ is the probability of the b-th power bin or, intuitively, the relative occurrence frequency of that particular power bin in the period of evaluation (i.e., a given year).

- Estimate the overall upgrade effect as

$$\Delta_{\text{upgrade}} = \frac{\sum_{b=1}^{B} \Delta \bar{y}_b \cdot P(y_b)}{\sum_{b=1}^{B} y_b \cdot P(y_b)} \times 100\%. \tag{7.13}$$

7.4.2 Joint Case Study

In this case study, the upgrade action is VG installation. Four turbine pairs are taken from each of two wind farms, making a total of eight pairs. Both wind farms are inland but of different terrain complexity. The historian data is collected in high temporal resolution (about one Hertz) with no averaging applied; this is the high-frequency data referred to earlier. The 10-min data is produced from the historian data. Periods that are known to be under curtailment are manually excluded prior to the analysis.

Wind Farm #1
The layout of the four turbine pairs on the first wind farm is illustrated in Fig. 7.7. The wind farm is on a terrain of medium complexity. The turbines on the farm belong to the general 2 MW turbine class. The VG installation took place in a summer month of 2014, but it was conducted on different days for each of the four VG turbines. There are six months of turbine data, including wind speed and wind power, in the pre-upgrade period and 13 months of the data in the post-upgrade period. Several of the environmental measurements, such as air density and humidity, are taken from the mast. Missing data is common in all datasets and in both periods. Other details of the datasets and turbines are withheld due to the confidentiality agreement in place.

The estimated VG effect on the four pairs of turbines is presented in

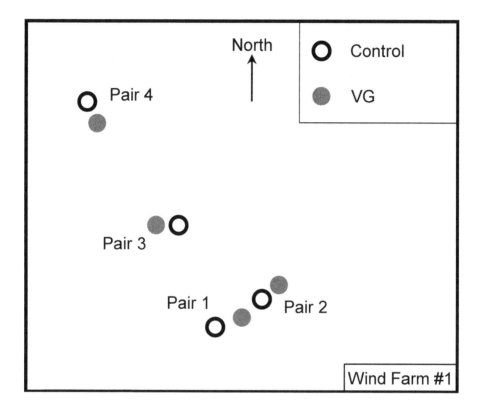

FIGURE 7.7 Layout of the four turbine pairs on wind farm #1. The distance among the turbines are not scaled precisely, but their relative positions, as well as their locations on the farm, reflect the reality. The between-turbine distances are expressed as multiples of the rotor radius, R, as follows: Pair 1, $14R$; Pair 2, $11R$; Pair 3, $6R$; and Pair 4, $9R$. The met mast is directly north of all turbine pairs. Its distance to the turbine pairs are: Pair 1 & Pair 2, 11 km; Pair 3, 8.8 km; and Pair 4, 6 km. (Reprinted with permission from Hwangbo et al. [95].)

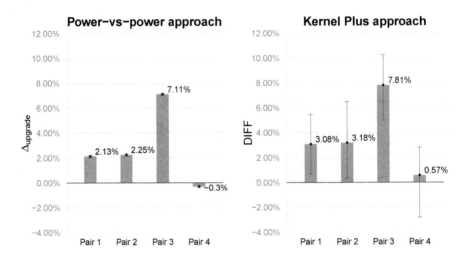

FIGURE 7.8 Estimates of the VG effect, together with the respective 90% confidence intervals, on the four pairs of turbines on wind farm #1. (Reprinted with permission from Hwangbo et al. [95].)

Fig. 7.8. Uncertainty quantification is conducted via the bootstrap resampling method, so that the 90% confidence intervals are added in the plot on top of the respective mean estimates. Understandably, the two sets of estimates are not exactly the same, but they are reasonably consistent, especially in terms of the relative significance of the VG effect on a specific turbine. The difference between the two sets of estimates are within the margin of error, and the overall difference between the two methods, averaged over the four pairs of turbines, is about 0.86%, with the Kernel Plus slightly overestimating relative to the power-vs-power approach.

Wind Farm #2
The layout of the four turbine pairs on the second wind farm is illustrated in Fig. 7.9. The wind farm is in a coastal area and on relatively flat terrain. The turbines on the second farm belong also to the general 2 MW turbine class. The VG installation took place in December of 2015 and was also conducted on different days for each of the four VG turbines. The duration of the common period where both the turbine data and mast data are available is 3.5 months in the pre-upgrade period and one month in the post-upgrade period. In this analysis, because the mast is close to the turbines, the wind speed measurements, together with the rest of the environmental measurements, are taken from the mast. Humidity is not measured on site. The average of the humidity measurements from two nearby weather stations is thus used, one located at 10 km north of the wind farm and the other at 10 km east of the farm. Missing data is also common in all datasets and in both periods. Other

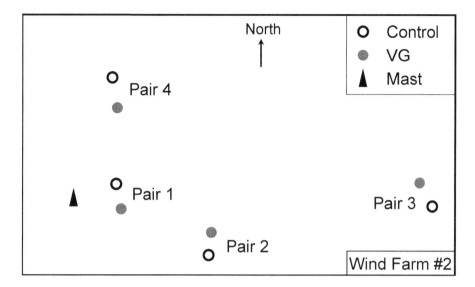

FIGURE 7.9 Layout of the four turbine pairs on wind farm #2. The between-turbine distances are: Pair 1, $6R$; Pair 2, $6.6R$; Pair 3, $6.6R$; and Pair 4, $7.4R$. The met mast's distance to the turbine pairs are: Pair 1, 0.2 km, Pair 2, 1.3 km; Pair 3, 3.6 km; and Pair 4, 1.3 km. (Reprinted with permission from Hwangbo et al. [95].)

details of the datasets and turbines are withheld due to the confidentiality agreement in place.

The estimated VG effect on the four pairs of turbines is presented in Fig. 7.10. Again, one can observe consistent outcomes from the two methods: the overall difference between the two methods, averaged over the four pairs of turbines, is about 0.15%, with the Kernel Plus still slightly overestimating relative to the power-vs-power approach.

7.4.3 Discussion

This academia-industry joint exercise presents a pair of upgrade quantification methods that are profoundly different in their respective underlying design and data usage. The profound difference in these two methods in fact lends more credibility in cross validation when they are employed to evaluate the same upgrade cases. The upgrade effects estimated by the two respective methods differ, on average, 0.86% and 0.15%, respectively, suggesting a good degree of consistence between them.

The power-vs-power method is simple to understand. But the data filtering procedure appears to be *ad hoc* and relies heavily on domain expertise and field judgment. By using the high-frequency data and having a larger sample

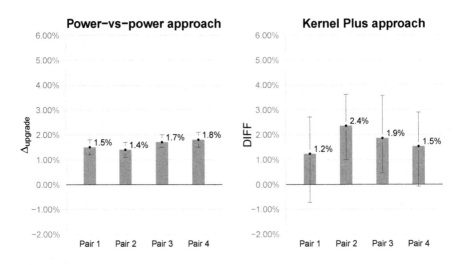

FIGURE 7.10 Estimates of the VG effects, together with the respective 90% confidence intervals, on the four pairs of turbines on wind farm #2. (Reprinted with permission from Hwangbo et al. [95].)

size, the power-vs-power approach enjoys the benefit of producing upgrade effect estimates with tighter confidence intervals.

The Kernel Plus method can possibly be applied to a single turbine when a control turbine does not exist. This explains why the Kernel Plus passed a blind test [50] in which no prior knowledge of control and VG turbines was given and no turbine pairs were provided. Yet, it is always beneficial to have a control turbine, whenever possible, as an additional reference. The premise of Kernel Plus is that it controls for the influence of the environmental factors through the learning of a multi-dimensional power curve model, but the inputs currently included in the power curve model may not be comprehensive enough. It is possible, of course, that measurements of certain important environmental factors are not available on a wind farm or analysts may not have realized yet the importance of certain other environmental factors. Improving the capability and accuracy of the underlying power curve method is always desirable.

While the general understanding of the VG effect is an extra 1–5% wind power production, it is a bit surprising to see that the quantification of Pair #3 on Farm #1 yields a greater than 7% improvement. As noted before, Wind Farm #1 is a medium complexity site that makes the wind inflow conditions complicated. Hwangbo et al. [95] believe that the VG effect tends to be greater when the wind inflow condition is more turbulent on a complex terrain. Whether this is accurate needs future studies.

It is generally a good idea to test on more than one pair of turbines and get a site average to represent the turbine upgrade effect for a specific farm.

The averaged upgrade effect from a few turbine pairs on the same farm is more stable, as the farm-level averaging irons out potential biases and reduces variability. The decision for wind farm owners/operators is not whether to install VGs on a particular turbine, but rather, whether to install VGs on the tens or even hundreds of turbines on their wind farm. For that purpose, the site-specific average is a more indicative metric. On Farm #1, the site average of VG effect is about 2.80% based on the power-vs-power approach and 3.66% based on the Kernel Plus. As for the performance on Farm #2, which is flat and at which wind inflow conditions are simpler and measured with higher confidence, the VG effects fall into a narrower range, with the site average at 1.60% based on the power-vs-power approach and at 1.75% based on the Kernel Plus method. The range of the site-averaged VG effects is consistent with what is anticipated in industrial practice.

7.5   COMPLEXITIES IN UPGRADE QUANTIFICATION

To conclude this chapter, we would like to discuss a few general issues encountered in the upgrade quantification effort. Most of the issues do not have a perfect solution yet, making the continuous effort in solving the upgrade quantification problem still much appreciated.

Bias Correction

One reason that the Kernel Plus uses a self-calibration procedure is to correct potential biases in upgrade effect estimation. It turns out that the bias issue is not only limited to the model-based approach like Kernel Plus. It happens also to the covariate matching approach and the power-vs-power approach. After years of research, it becomes clearer to us that whichever method can correct the bias in upgrade effect estimation outperforms the peer methods that do not do so very well.

When applying a quantification method to the control turbine data, one would ideally expect a zero upgrade effect, since the turbine undergoes no change. Apparently, this is not achieved by most of the existing methods, and because of this, a control turbine is needed as the datum to adjust the estimation of upgrade effect for the test turbine. Consider a simple case in which the data is taken from a control turbine and then duplicated and treated as the data for the test turbine. In this fashion, one in fact has two identical datasets. When an upgrade quantification method is applied to these two sets of data, it should presumably yield a zero upgrade effect. To much a surprise, many of the methods still do not. While it is easy to see the existence of the bias, how best to correct the bias eludes the analysts. The self-calibration procedure plus the control/test difference used in the Kernel Plus method provides certain degrees of safeguard.

Wind Speed Adjustment

When building the Kernel Plus model or aligning the covariates in the matching approach, wind speed measurements are used. The wind speed measurements may be from either a nearby mast or the nacelle. The wind measurement, if from the nacelle, is in the wake of the rotor. Installation of vortex generators alters wind flow separation behind the rotor, so much so that for the same free inflow wind in front of the rotor, the wind speed measurements taken by the nacelle anemometer before and after the VG installation are most likely different. This difference could introduce a degree of inaccuracy if left unaddressed.

IEC 61400-12-2 [104] deals with nacelle measurements through a nacelle transfer function (NTF), which is the relation between the free inflow wind speed and that measured at the nacelle anemometer. Typically an NTF can be obtained by comparing the nacelle measurements with that on a nearby mast or with a nacelle mounted LIDAR (light detection and ranging) sensor. Some operators establish an NTF for a VG turbine, so that the wind speed after the VG installation can be adjusted using the NTF. In practice, however, an NTF is rarely available, because obtaining it and continuously calibrating it are costly.

In the absence of a nacelle transfer function, Hwangbo et al. [95] introduce a wind speed adjustment procedure, acting similarly as an NTF. The shortcoming of the procedure in [95] is that the adjustment quality and accuracy is not yet verified with actual physical measurements.

Another idea to address the wind speed measurement issue is to avoid using the wind speed measured on the test turbine when building a power curve model. If wind speed measurements are available on a met mast in its physical vicinity, that would be the best. Even without a met mast nearby, analysts can consider using the wind speed measured on a neutral turbine in its vicinity, which is not affected by the VG installation as much as the wind speed measured on the test turbine.

Annualization

When an upgrade quantification is conducted, wind farm owners or operators would like to know the benefit in terms of annual energy production. As one has seen in the examples presented earlier, there are not always a full year's worth of data available in the post-upgrade period for conducting such comparison. For the Kernel Plus method, having a full year of data in the post-upgrade period poses another problem. The post-upgrade period is known as Data2 in the Kernel Plus method, which is supposedly of the same length as that of Data1. If Data2 is one year's worth, then the total amount of data needed, combining Data0 through Data2, would have been three years' worth. This data amount requirement is too demanding. On top of that, as we see in Chapter 6, a turbine's own production efficiency characteristics may change in the span of three years even in the absence of any upgrade action, adding additional confounding effect to be shielded off in the already difficult task of estimating the upgrade benefit.

When using data from a shorter period, or a subset of data from one year span, one may extrapolate the estimated upgrade effect to the whole year. This process is called *annualization*. The process of annualization is actually explained in Steps 4 and 5 of the power-vs-power approach. The idea is that when estimating the upgrade effect, conduct the estimation for a set of power bins. Assuming that the bin-wise upgrade effect stay more or less the same for the whole year, analysts extrapolate the estimated upgrade effect to AEP by re-weighting the bin-wise upgrade effect with the empirical distribution of wind power. For the re-weighting formula, please see Eq. 7.13.

GLOSSARY

AEP: Annual energy production

AMK: Additive-multiplicative kernel method

BIN: Binning method

BIN_AF: Binning method, after upgrade

BIN_BF: Binning method, before upgrade

DIFF: Upgrade effect quantification when using Kernel Plus

IEC: International Electrotechnical Commission

KP: Kernel Plus

KP_AF: Kernel Plus, after upgrade

KP_BF: Kernel Plus, before upgrade

LIDAR: Light detection and ranging

MD: Mahalanobis distance

NTF: Nacelle transfer function

pdf: Probability density function

SDM: Standardized difference of means

UPG: Upgrade effect

VG: Vortex generator

EXERCISES

7.1 Using the `Turbine Upgrade Dataset`, please present the boxplots of y^{test} for the pre-upgrade and post-upgrade periods, respectively. Please do this for the unmatched data and matched data and for the respective test turbine in both turbine pairs. What do you observe?

7.2 The current quantification outcome of the covariate matching is not annualized. To get an AEP, we need to go through the following procedure.

    a. Conduct the covariate matching analysis using the same parameters as used in this chapter, but compute the bin-wise upgrade effect. Use the bin width of 100 kW. Do this only for the VG installation pair.

    b. Take the `Inland Wind Farm Dataset1`, which has more than one year's worth of data of four turbines from the same wind farm, and estimate the distribution of wind power output by pooling the data from the four turbines. Still use the bin width of 100 kW.

    c. Use the re-weighting formula to compute the AEP effect due to the VG installation. How much is it different from the eight-week outcome?

7.3 Conduct the covariate matching analysis using the pitch angle adjustment pair.

    a. Estimate the upgrade effect by using, respectively, two weeks', five weeks', or eight weeks' worth of post-upgrade data. Observe how sensitive the method is to the length of the post-upgrade period.

    b. Conduct annualization using the power distribution estimated in Exercise 7.2(b). Apply the annualization to the above three post-upgrade period choices. Are the differences in the AEPs greater than that in the upgrade effect estimation in (a)?

7.4 Conduct the covariate matching analysis using the pitch angle adjustment pair, but use the reverse priority order among the covariates, i.e., y^{ctrl}, I, S, ρ, D, and V. Compute the SDMs and present them in a table similar to the lower half of Table 7.1. Does the matching procedure still significantly reduce the SDM? Go ahead and estimate the UPG again. How much is the new UPG different from what was estimated in this chapter (which is 3.16%)?

7.5 The current quantification outcome of the Kernel Plus is not annualized. To get an AEP, we need to go through the following procedure.

    a. Conduct the Kernel Plus, but compute the bin-wise upgrade effect. Again, use the bin width of 100 kW and do this only for the VG installation pair.

    b. Take the power distribution estimated in Exercise 7.2(b) and use the re-weighting formula to compute the AEP of the VG installation. How much is it different from the eight-week outcome?

7.6 Conduct the Kernel Plus-based analysis using the pitch angle adjustment pair.

    a. Estimate the upgrade effect by using, respectively, two weeks', five weeks', or eight weeks' worth of post-upgrade data. Observe how sensitive the method is to the length of the post-upgrade period.

    b. Conduct annualization using the power distribution estimated in Exercise 7.2(b). Apply the annualization to the above three post-upgrade period choices. Are the differences in the AEPs greater than that in the upgrade effect estimation in (a)?

7.7 Replace Eq. 7.8 with the Mahalanobis distance and conduct the Kernel Plus-based analysis using the pitch angle adjustment pair. What difference does it make when this distance measure is changed? What if you use a simple Euclidean distance (unweighted)?

7.8 Use the binning method to replace the dashed-line rectangular box in Fig. 7.4 and treat that as a binning-based quantification method. Apply the binning-based quantification method to the two pairs of turbines and estimate the respective upgrade effect. How much are they different from the covariate matching and Kernel Plus?

7.9 Conduct the sensitivity analysis for the binning-based quantification, as it is done in Tables 7.3 and 7.5. Compare your results with those in Tables 7.3 and 7.5.

7.10 Take the control turbine data from the pitch angle adjustment pair, duplicate the data and treat it as if it were the test turbine data. Now, you have two identical datasets.

    a. Multiply the test turbine power by r (for all power values), for $r = 0, 1, 2, 3, 4,$ and 5%.

    b. For each r, employ, respectively, the covariate matching, Kernel Plus, and binning methods to estimate the upgrade effect. Tabulate the outcomes similar to the presentation in Table 7.3.

    c. For each original power value in the test turbine set (before being multiplied by an r in (a)), multiply it by a random number, drawn uniformly from the range $[0, 5\%]$. Compute the effective power increase rate for the test turbine power data. Employ, respectively, the covariate matching, Kernel Plus, and binning methods to estimate the upgrade effect. Compare the estimated upgrade effect and the effective power increase rate.

CHAPTER 8

Wake Effect Analysis

DOI: 10.1201/9780429490972-8

While a wind turbine is operating, the rotating blades disturb the natural flow of wind and create turbulence for the downstream turbines. During this process, the turbine absorbs kinetic energy in wind and converts the energy into electricity. As a result, the wind loses some of its original kinetic energy after the rotor, exhibiting reduction in its speed. Such a phenomenon differentiating the after-rotor wind flow from the free-stream one (before the rotor) is referred to as wake effect.

Understanding and quantifying the wake effect plays an important role in improving wind turbine designs and operations as well as wind farm layout planning. Being a physical phenomenon, the majority of the wake effect models are understandably physics based. Modelers resort in particular to sophisticated, computational fluid dynamics (CFD) models that can achieve a higher accuracy [129]. However, using the CFD models entails significant computational challenges. For example, running a large eddy simulation, one of the popular CFD methods, requires days or even weeks of computation on supercomputers for analyzing a single-wake situation [192]. The abundance of wind farm operational data motivates the development of data science methods for analyzing and estimating wake effect, which is the focus of this chapter.

8.1 CHARACTERISTICS OF WAKE EFFECT

The wake of a turbine propagates with a certain range of angles, and its impact remains effective up to a certain distance from the turbine. Fig. 8.1 illustrates a snapshot of a single-wake situation. A single wake refers to the circumstance in which two operating turbines are involved and one is in the wake of the other for a given wind direction. In the figure, θ denotes an acute angle between the wind direction and the line connecting the two turbines. For the wind direction shown in Fig. 8.1, left panel, the wind passes through Turbine 1 along the center line. The wake caused by Turbine 1 affects the downstream region with a range of angles (the shaded area). The wind speed loss due to the wake is greater for locations closer to the upstream turbine

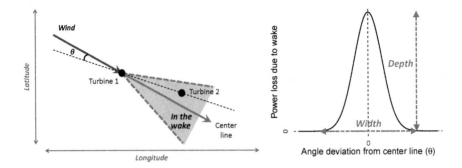

FIGURE 8.1 Characteristics of wind turbine wake effect. Left panel: wake region and θ; right panel: wake depth and wake width. (Reprinted with permission from Hwangbo et al. [98].)

(Turbine 1) and closer to the center line. Turbine 2, its position fixed, is subject to the greatest power loss when $\theta = 0$. The power loss amount decreases as θ deviates from zero. After θ exceeds a certain value, Turbine 2 is no longer in the wake of Turbine 1. The maximum power loss when $\theta = 0$ is referred to as the *wake depth*, whereas the range of θ for which a turbine is in the wake of another turbine (with positive power losses) is referred to as the *wake width*. Wake depth and width are expected to remain constant when the relative positions between two turbines are fixed.

Fig. 8.2 illustrates power output of a wind turbine when it is wake free versus when it is in the wake of another turbine.

Knowledge of wake characteristics is crucial for improving power generation performance on wind farms. As wake width and depth strongly depend on the relative positions of turbines, characterizing the turbine specific wake effect facilitates the layout planning [56, 128], particularly when using the same turbine model in future wind projects. Understanding the wake characteristics also supports effective operational control of wind turbines through pitch and yaw controls [70, 146]. The pitch control can regulate the magnitude of wind speed loss in a downstream region by adjusting the energy absorption level of an upstream turbine. The yaw control can change the amount of the wind speed loss by tilting the downstream wake region. By carefully controlling the yaw of Turbine 1, Turbine 2 may be as nearly wake free as possible for a given wind direction.

8.2 JENSEN'S MODEL

As mentioned earlier in this chapter, sophisticated CFD wake models take long computational time to run and their use is less practical for commercial wind farm operation. A widely used, physics-based model is Jensen's model [108], due to its simplicity and easiness to compute. Jensen's model is derived by

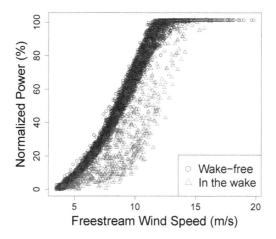

FIGURE 8.2 Power output in the wake versus that under a free-stream condition. (Reprinted with permission from Hwangbo et al. [98].)

solving an equation based on the balance of momentum. The resulting wake effect, in terms of wind speed, is expressed as

$$V_{\text{wake}} = \left\{ 1 - \frac{2}{3} \left(\frac{R}{R + \kappa \cdot d} \right)^2 \right\} \cdot V, \tag{8.1}$$

where R is the rotor radius of a wind turbine, d is the down-wind distance from the rotor, and κ is known as the entrainment constant. For wake effect, the entrainment constant, κ, is approximately 0.1. Based on Eq. 8.1, the wind speed immediately after the rotor where $d = 0$, is one-third of the free-stream wind speed. On the other hand, the wind speed at the down-wind distance of ten rotor diameters, i.e., $d = 20R$, is about $0.926V$. This is part of the reason that analysts deem $20R$ a safe boundary beyond which the wake effect by and large weans off.

Eq. 8.1 can be simplified to $V_{\text{wake}} = (1 - \kappa_{\text{deficit}}) \cdot V$ [232, Eq. 23], where κ_{deficit} depends generally on the down-wind distance. Some analysts further simplify κ_{deficit} to be a constant of 0.075, which, according to Eq. 8.1, corresponds roughly to $d = 20R$.

Apparently, Jensen's model does not directly estimate the power loss. In fact, nearly all other physics-based wake effect models do not do so, either. Instead, they primarily focus on estimating the reduced wind speed due to wake. To quantify wake power loss, these models then require an additional layer of converting the wind speed estimates into a corresponding power output— a conversion can be done by using a simple power curve model such as the IEC binning method or more complicated power curve models as presented in Chapter 5. Data science methods, on the other hand, can connect the wind

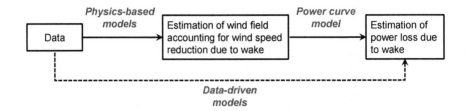

FIGURE 8.3 Wake power loss estimation procedures. (Reprinted with permission from Hwangbo et al. [98].)

speed data directly to the power output in a single step; see Fig. 8.3. In the case study section, the data-driven models are compared with the two-step approach that has Jensen's model as its first step.

8.3 A DATA BINNING APPROACH

The binning approach is rather popular in industrial practice, acting as a robust, nonparametric method, easy to understand and easy to use. Not surprisingly, it is used for estimating the wake characteristics as well. A common data binning approach for wake effect estimation is presented in Algorithm 8.1.

Algorithm 8.1 Data binning approach.

---

1. Gather the power output data from two turbines.

2. Choose a specific range of wind speed where the maximum power loss is expected, e.g., 8.0 ± 0.5 m/s [14], or extend the coverage of wind speed to a wider range, e.g., 5.0–11.0 m/s [146] or even to the whole wind spectrum, which is the choice used in Section 8.6.1.

3. Plot the power difference between the two turbines under the above-specified wind speeds against the wind direction (0 degree means due north)

4. To smooth out the noise, apply the action of binning, namely, partition the wind direction by a bin width, say 5°, and then average all the power difference data in a specific bin. Use the bin-wise averages as the representative of the original data.

In Fig. 8.4, the dark dots constitute a scatter plot of the power differences against wind direction. Once applying the data binning approach to the raw power differences, it produces the solid line passing through the data cloud. The solid line is the estimated wake effect. The wake depth can be read from

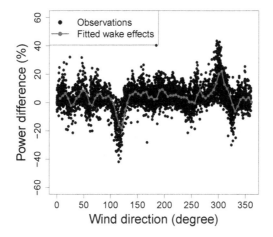

FIGURE 8.4 Data binning approach for estimating the wake effect between a pair of turbines. The between-turbine distance is four times the rotor diameter. The distance from this pair to other turbines is more than $20R$. (Reprinted with permission from Hwangbo et al. [98].)

the plot by observing the two peaks around 120° and 300°, respectively. As one moves along the wind direction from 0° to 360°, the roles of the two turbines, i.e., which one is wake free and which one is in the wake, are reversed. That is why one observes that one of the peaks is downward. In Fig. 8.4, the wake width is not immediately obvious. Analysts usually impose a large enough angle coverage, say, $\theta \in (-25°, 25°)$, and then verify with the estimated curve if the angle range is broad enough to represent the wake width [146]. To refine the estimation, analysts sometimes find a wind direction value on each side of the center line at which the power loss estimate is within a certain level, for example, $\pm 5\%$ of the free-stream power, and use the angle coverage formed by these wind direction values as the estimate of wake width [14]. When using this data binning approach, the power difference from the wake-free turbine to the in-the-wake turbine, which estimates wake power loss, is not guaranteed to be positive. As a matter of fact, previous studies [169, 215] often show that some of the bin-wise estimates of the power difference is negative even after θ moves beyond the obvious wake width region—this phenomenon is indeed evident in Fig. 8.4.

8.4 SPLINE-BASED SINGLE-WAKE MODEL

Hwangbo et al. [98] develop a wake effect model based on splines. Their model is intended to estimate wake effect characteristics, such as wake width and wake depth, under single-wake situations arising between two turbines of which modeling assumptions are easier to justify. To facilitate a successful transition from physics-based models to data-driven modeling, Hwangbo et al. incorporate certain physical understandings and considerations as constraints in the model fitting process. Because of this, the resulting single-wake model is a physics-constrained, rather than a purely, data-driven model.

8.4.1 Baseline Power Production Model

Hwangbo et al. [98] start with the production economics model in Eq. 6.5 but make some changes to it. Recall that Eq. 6.5 reads as,

$$y(V) = Q(V) - u(V) + \varepsilon, \tag{8.2}$$

where $Q(\cdot)$ is the production frontier function and $u(\cdot)$ is the systematic in-efficiency term. Eq. 6.5 is expressed as a univariate function of wind speed, V.

The baseline power production model used for wake effect modeling reads as,

$$y_t(\boldsymbol{x}) = Q_t(\boldsymbol{x}) - \eta_t(\boldsymbol{x}) - \omega_t(\boldsymbol{x}) + \varepsilon_t, \quad t = 1, \dots, N, \tag{8.3}$$

where t is the turbine index and N is the number of turbines. In the above model, the inefficiency term, $u(\cdot)$, is split into two terms—$\eta_t(\cdot)$ and $\omega_t(\cdot)$—such that $\eta_t(\cdot)$ represents a turbine's inherent inefficiency independent of wake, whereas $\omega_t(\cdot)$ represents the turbine's power loss due to wake. Also, the input variable is now a vector rather than wind speed only. Furthermore, Hwangbo et al. [98] postulate that both power loss terms in the above model are non-negative, i.e., $\eta_t(\cdot) \geq 0$ and $\omega_t(\cdot) \geq 0$, $\forall t = 1, \dots, N$, to be consistent with the physical understanding of the phenomenon.

We said in Chapter 6 that estimating $Q_t(\cdot)$ under a multivariate setting while satisfying the S-shape constraint is not trivial. Luckily, for the single-wake situation, $Q_t(\cdot)$ does not have to be estimated explicitly. For a pair of turbines, one can pool together the two turbines' power production data and estimate a common production frontier. As it will become clear in the next section, Hwangbo et al. [98] establish a power difference model, which takes the *power difference* between a pair of turbines, and in doing so, the common frontier function cancels each other in the resulting model. This is to say, the production frontier function, $Q_t(\cdot)$, does not appear in the final wake effect model.

8.4.2 Power Difference Model for Two Turbines

For a single-wake situation with two turbines, two angle variables, θ_1 and θ_2, are used and associated, respectively, with the two turbines. Specifically, θ_1 is related to the wind direction causing power loss on Turbine 1 and θ_2 is with the wind direction under which Turbine 2 endures power loss. As illustrated in Fig. 8.5, the wind directions associated with θ_1 and θ_2 can take any value

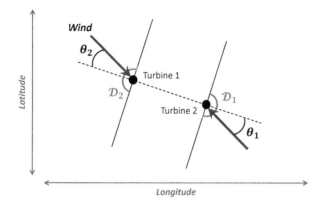

FIGURE 8.5 Notations of wind direction in wake analysis. The union of $\mathcal{D}_1$ and $\mathcal{D}_2$ covers the entire $360°$ wind direction. (Reprinted with permission from Hwangbo et al. [98].)

in the sets, $\mathcal{D}_1$ and $\mathcal{D}_2$, respectively, given the definition of these sets stated below. For the purpose of analyzing the wake effect, θ_1 and θ_2 only need to vary in the $180°$ outer hemisphere surrounding their respective turbine. Note that θ_1 is actually on the side of Turbine 2, whereas θ_2 is on the side of Turbine 1. If one positions the zero degree of θ_1 and θ_2 at the line connecting the two turbines, then $\theta_1, \theta_2 \in (-90°, 90°)$. Denote by $\mathcal{D}_1$ the set of wind directions corresponding to the support of θ_1, and likewise, by $\mathcal{D}_2$ the set of directions in which θ_2 is defined.

With this notation, applying the baseline power production model in Eq. 8.3 individually to the two turbines yields

$$y_1(\boldsymbol{x}) = Q(\boldsymbol{x}) - \eta_1(\boldsymbol{x}) - \omega_1(\boldsymbol{x}) \cdot \mathbb{1}_{\mathcal{D}_1}(\boldsymbol{x}) + \varepsilon_1,$$
$$y_2(\boldsymbol{x}) = Q(\boldsymbol{x}) - \eta_2(\boldsymbol{x}) - \omega_2(\boldsymbol{x}) \cdot \mathbb{1}_{\mathcal{D}_2}(\boldsymbol{x}) + \varepsilon_2,$$
(8.4)

where $\mathbb{1}_{\mathcal{D}_t}(\boldsymbol{x})$ is an indicator function taking the values of one, if the wind direction belongs to $\mathcal{D}_t$, or zero otherwise. Here, the production frontier function, $Q(\boldsymbol{x})$, is assumed to be common to the same type of turbines, and for this reason, it does not use a turbine-differentiating subscript. Taking the difference between the two equations in Eq. 8.4 leads to

$$\tilde{y}_{1\text{-}2}(\boldsymbol{x}) = \tilde{\eta}_{2\text{-}1}(\boldsymbol{x}) - \omega_1(\boldsymbol{x}) \cdot \mathbb{1}_{\mathcal{D}_1}(\boldsymbol{x}) + \omega_2(\boldsymbol{x}) \cdot \mathbb{1}_{\mathcal{D}_2}(\boldsymbol{x}) + \tilde{\varepsilon},$$
(8.5)

where the tilde indicates a turbine difference term and the subscripts 1-2 and 2-1 signify the specific order of the difference. The above model is interpreted as follows: the power difference of Turbine 1 over Turbine 2 is due to the inherent production difference between the two turbines, $\tilde{\eta}_{2\text{-}1}(\cdot)$, and the power loss caused by the wake effect, characterized by either $\omega_1(\cdot)$ or $\omega_2(\cdot)$, both depending on specific wind conditions. Because the sets, $\mathcal{D}_1$ and $\mathcal{D}_2$, are mutually exclusive, $\omega_1(\cdot)$ and $\omega_2(\cdot)$ do not appear at the same time.

It is well known that the dominating input factors for wind power production are wind speed, V, and wind direction, D. The analysis in Chapter 5 shows that environmental factors other than wind speed and direction, such as air density, turbulence, and humidity, may also have an impact on wind power output. One advantage of using the power difference model in Eq. 8.5 is that one no longer needs to consider other environmental factors because once the power difference is taken of the two turbines, the impact of the environmental factors other than that of the wind is neutralized. Still, to be consistent with the IEC standard procedure and to further neutralize the effect of air density, Hwangbo et al. [98] decided to use the normalized wind speed, following the air density correction formula in Eq. 5.3.

For the turbine difference term and the indicator function terms in Eq. 8.5, either wind speed or wind direction, but not both, is needed as an input. The input to the indicator function $\mathbb{1}_{\mathcal{D}_1}(\cdot)$ or $\mathbb{1}_{\mathcal{D}_2}(\cdot)$ is wind direction. The turbine difference term, $\tilde{\eta}_{2\text{-}1}(\cdot)$, represents the between-turbine production difference independent of wake. Hwangbo et al. [98] thereby assume that it is only a function of wind speed, not of wind direction, as the portion of the power difference, $\tilde{y}_{1\text{-}2}$ or $\tilde{y}_{2\text{-}1}$, related to wind direction should be included in the wake related term ω.

As such, the power difference model can be expressed as:

$$\tilde{y}_{1\text{-}2}(V, D) = \tilde{\eta}_{2\text{-}1}(V) - \omega_1(V, D) \cdot \mathbb{1}_{\mathcal{D}_1}(D) + \omega_2(V, D) \cdot \mathbb{1}_{\mathcal{D}_2}(D) + \tilde{\varepsilon}, \quad (8.6)$$

where $\tilde{\varepsilon}$ is still an i.i.d. noise, assumed to follow a normal distribution.

8.4.3 Spline Model with Non-negativity Constraint

In order to empirically estimate the power difference model in Eq. 8.6, Hwangbo et al. [98] assume the following model structure for the three functional terms: (a) $\tilde{\eta}$ is modeled by a univariate cubic smoothing spline and (b) the two wake power loss terms, ω_1 and ω_2, are modeled by bivariate thin plate splines [228], a multidimensional generalization of the smoothing splines. Recall that when Eq. 8.3 is presented, non-negativity constraints are imposed on the power loss terms, which state that $\eta(\cdot) \geq 0$ and $\omega(\cdot) \geq 0$. After taking the power difference, the turbine difference term, $\tilde{\eta}_{2\text{-}1}(\cdot)$, no longer needs to be non-negative; in fact, it can be positive, zero, or negative. But the wake power loss terms, ω_1 and ω_2, are still supposed to be non-negative. As such, the power difference model in Eq. 8.6 becomes a spline model with the non-negativity constraint imposed on ω_1 and ω_2. The resulting model is referred to as the thin plate regression spline model with non-negativity (TPRS-N).

To estimate the spline-based power difference model, Hwangbo et al. [98] follow the generalized additive model (GAM) scheme [87]. GAMs represent a univariate response as an additive sum of multiple smooth functions, each having its own predictor variables. Estimation of GAMs can be performed by implementing the backfitting algorithm for which each smooth function is fitted for the residuals of all the others, iteratively one at a time until the fitted functions converge.

The learning formulas for smoothing spline and thin plate splines follow the regularized learning formulation in Eq. 5.20. Because of the use of the iterative backfitting algorithm, y is replaced by the residual variable r in that formulation. When the algorithm starts, the residual r is the same as y (differing by the average of y). When the algorithm proceeds, the residual from the preceding steps is used in the learning formulation, instead of the original y.

Consider n data pairs for which a residual r is paired with $\boldsymbol{x}$, i.e., $(\boldsymbol{x}_i, r_i)$ for $i = 1, \ldots, n$. The formula for the smoothing spline can be found in Eq. 5.22 in which y_i is replaced by r_i. For thin plate splines with two predictors, x_1 and x_2, the first part of the loss function is the same as in the smoothing spline, but the penalty function reads as

$$\text{Penalty}(g) = \int \int_{\mathbb{R}^2} \left[\left(\frac{\partial^2 g(\boldsymbol{x})}{\partial x_1^2} \right)^2 + 2 \left(\frac{\partial^2 g(\boldsymbol{x})}{\partial x_1 \partial x_2} \right)^2 + \left(\frac{\partial^2 g(\boldsymbol{x})}{\partial x_2^2} \right)^2 \right] dx_1 \, dx_2. \quad (8.7)$$

For the smoothing splines, let us express the corresponding $g(x)$ as (recall Exercise 5.4)

$$g(x) = \sum_{j=1}^{n} h_j(x) \beta_j, \quad (8.8)$$

where $h_j(x)$ is the jth basis function of a natural cubic spline and β_j is the corresponding coefficient. Then, Eq. 5.22 can be expressed as

$$\underset{\boldsymbol{\beta}}{\text{argmin}} \, (\mathbf{r} - \mathbf{H}\boldsymbol{\beta})^T (\mathbf{r} - \mathbf{H}\boldsymbol{\beta}) + \gamma \boldsymbol{\beta}^T \boldsymbol{\Omega}_h \boldsymbol{\beta}, \quad (8.9)$$

where $\mathbf{H}$ is a matrix whose $(i, j)^{th}$ element is $h_j(x_i)$ and $\boldsymbol{\Omega}_h$ is a matrix derived from $h(\cdot)$ (therefore the subscript) whose $(j, k)^{th}$ element is $\int h_j''(t) h_k''(t) dt$. The solution is

$$\hat{\boldsymbol{\beta}} = (\mathbf{H}^T \mathbf{H} + \gamma \boldsymbol{\Omega}_h)^{-1} \mathbf{H}^T \mathbf{r}. \quad (8.10)$$

Different from the smoothing splines, the $g(\boldsymbol{x})$ of the thin plate splines is expressed as

$$g(\boldsymbol{x}) = \mathbf{X}\boldsymbol{\beta}^{(\text{tp})} + \sum_{i=1}^{n} \delta_i \phi(\|\boldsymbol{x} - \boldsymbol{x}_i\|),$$

where the $n \times 3$ matrix $\mathbf{X} = [\mathbf{1}_n; \boldsymbol{x}_1; \boldsymbol{x}_2]$ includes the unit vector of size n as its first column and the n observations for the two covariates as its second and third columns, and $\phi(\|\boldsymbol{x} - \boldsymbol{x}_i\|)$ is a radial basis function. Here the radial basis function is involved because analysts find that thin plate splines have a natural representation in terms of radial basis functions [228]. The three-dimensional vector $\boldsymbol{\beta}^{(\text{tp})}$ and the n-dimensional vector $\boldsymbol{\delta} = (\delta_1, \ldots, \delta_n)^T$ are, respectively, the coefficients associated with $\mathbf{X}$ and those associated with the radial basis functions, and both sets of coefficients need to be estimated. The superscript "tp" is added to $\boldsymbol{\beta}$ to differentiate this vector in the thin plate splines from that in the smoothing splines.

Using the thin plate spline's $g(\boldsymbol{x})$ and its penalty function (as in Eq. 8.7) in a regularized learning formulation, like in Eq. 5.22, leads to the estimation of the model coefficients in the thin plate spline. It turns out (details skipped) that the solution is equivalent to solving

$$\min \|\mathbf{r} - \mathbf{X}\boldsymbol{\beta}^{(\text{tp})} - \boldsymbol{\Phi}\boldsymbol{\delta}\|^2 + \gamma\boldsymbol{\delta}^T\boldsymbol{\Phi}\boldsymbol{\delta}, \quad \text{subject to } \mathbf{X}^T\boldsymbol{\delta} = \mathbf{0}, \qquad (8.11)$$

where $\boldsymbol{\Phi}$ is the radial basis matrix, defined by $\boldsymbol{\Phi}_{ji} = \phi(\|\boldsymbol{x}_j - \boldsymbol{x}_i\|) = \|\boldsymbol{x}_j - \boldsymbol{x}_i\|^2 \log\|\boldsymbol{x}_j - \boldsymbol{x}_i\|$ for $i, j = 1, \ldots, n$.

Different from the univariate spline problem that can be solved by $O(n)$ operations, the computations for the thin plate splines require $O(n^3)$ operations [87]. To overcome the computational problem, Wood [228] proposes the thin plate regression splines (TPRS), which uses only k eigenbasis functions ($k \ll n$) corresponding to the largest k eigenvalues of the basis matrix $\boldsymbol{\Phi}$. Doing so reduces the rank of the basis matrix significantly.

TPRS can be fitted as follows. First, applying the eigen decomposition of $\boldsymbol{\Phi}$ leads to $\boldsymbol{\Phi} = \mathbf{U}\boldsymbol{\Lambda}\mathbf{U}^T$ where $\boldsymbol{\Lambda}$ is a diagonal matrix whose diagonal elements are the eigenvalues of $\boldsymbol{\Phi}$ and arranged in a non-increasing order, i.e., $\boldsymbol{\Lambda}_{i,i} \geq \boldsymbol{\Lambda}_{i+1,i+1}$ for $i = 1, \ldots, n-1$. Matrix $\mathbf{U}$ is an orthogonal matrix whose columns are the eigenvectors ordered accordingly. Then, TPRS considers the first k columns of $\mathbf{U}$, denoted by $\mathbf{U}_k$, and uses them to construct a rank k eigenbasis matrix $\boldsymbol{\Phi}_k = \mathbf{U}_k\boldsymbol{\Lambda}_k\mathbf{U}_k^T$, where $\boldsymbol{\Lambda}_k$ is a $k \times k$ diagonal matrix taking the first k rows and columns of $\boldsymbol{\Lambda}$.

By restricting $\boldsymbol{\delta}$ in the column space of $\mathbf{U}_k$, i.e., let $\boldsymbol{\delta} = \mathbf{U}_k\boldsymbol{\delta}_k$, Eq. 8.11 becomes

$$\min \|\mathbf{r} - \mathbf{X}\boldsymbol{\beta}^{(\text{tp})} - \mathbf{U}_k\boldsymbol{\Lambda}_k\boldsymbol{\delta}_k\|^2 + \gamma\boldsymbol{\delta}_k^T\boldsymbol{\Lambda}_k\boldsymbol{\delta}_k \quad \text{subject to } \mathbf{X}^T\mathbf{U}_k\boldsymbol{\delta}_k = \mathbf{0}.$$

In expressing the above equation, one needs $\mathbf{U}_k^T\mathbf{U}_k = \mathbf{I}$, which is true, due to the fact that columns in $\mathbf{U}_k$ are orthogonal by construction.

The constrained problem can be replaced by an unconstrained problem through the QR decomposition of $\mathbf{U}_k^T\mathbf{X}$. Specifically, form a $\mathbf{Z}_k$ that takes the last $k - 3$ columns of the orthogonal factor of the decomposition. Restricting $\boldsymbol{\delta}_k$ to the column space of $\mathbf{Z}_k$ by letting $\boldsymbol{\delta}_k = \mathbf{Z}_k\tilde{\boldsymbol{\delta}}$ renders the constraint satisfied. Then, the rank-k approximation can be used to fit TPRS by solving

$$\min \|\mathbf{r} - \mathbf{X}\boldsymbol{\beta}^{(\text{tp})} - \mathbf{U}_k\boldsymbol{\Lambda}_k\mathbf{Z}_k\tilde{\boldsymbol{\delta}}\|^2 + \gamma\tilde{\boldsymbol{\delta}}^T\mathbf{Z}_k^T\boldsymbol{\Lambda}_k\mathbf{Z}_k\tilde{\boldsymbol{\delta}}, \qquad (8.12)$$

for the unknown $\boldsymbol{\beta}^{(\text{tp})}$ and $\tilde{\boldsymbol{\delta}}$. The prediction for any given $\boldsymbol{x}$ can be achieved by calculating $\hat{\boldsymbol{\delta}} = \mathbf{U}_k\mathbf{Z}_k\tilde{\boldsymbol{\delta}}$ and plugging $\hat{\boldsymbol{\delta}}$ and $\hat{\boldsymbol{\beta}}^{(\text{tp})}$ into

$$\hat{g}(\boldsymbol{x}) = \mathbf{X}\hat{\boldsymbol{\beta}}^{(\text{tp})} + \sum_{i=1}^{n} \hat{\delta}_i\phi(\|\boldsymbol{x} - \boldsymbol{x}_i\|). \qquad (8.13)$$

Recall that the wake power loss term, ω_t, is assumed non-negative to be consistent with the physical understanding of the wake effect, but the modeling procedure of TPRS does not guarantee non-negativity. In order to make sure the wake power loss is indeed non-negative, Hwangbo et al. [98] apply an exponential transformation on top of the conventional TPRS estimation in Eq. 8.13, i.e., let

$$\hat{\omega}(\boldsymbol{x}) = \exp\left\{\mathbf{X}\hat{\boldsymbol{\beta}}^{(\text{tp})} + \sum_{i=1}^{n}\hat{\delta}_i\phi(\|\boldsymbol{x} - \boldsymbol{x}_i\|)\right\}. \tag{8.14}$$

Because of this change, instead of solving Eq. 8.12, one now aims at solving

$$\min\left\|\mathbf{r} - \exp\left\{\mathbf{X}\boldsymbol{\beta}^{(\text{tp})} + \mathbf{U}_k\boldsymbol{\Lambda}_k\mathbf{Z}_k\tilde{\boldsymbol{\delta}}_k\right\}\right\|^2 + \gamma\tilde{\boldsymbol{\delta}}_k^T\mathbf{Z}_k^T\boldsymbol{\Lambda}_k\mathbf{Z}_k\tilde{\boldsymbol{\delta}}_k, \tag{8.15}$$

with respect to $\boldsymbol{\beta}^{(\text{tp})}$ and $\tilde{\boldsymbol{\delta}}_k$.

When estimating a GAM, a constant term generally precedes the functional terms, and is estimated by using the global mean. In other words, the global mean is calculated and subtracted from the response in advance, before implementing the backfitting algorithm that estimates the rest of the functional terms. In the power difference model in Eq. 8.6, this constant term should be part of the turbine-difference term, $\tilde{\eta}(\cdot)$, meaning that a portion of the turbine difference is constant regardless of the wind conditions, while the other portion may change with the wind speed. For the implementation of the backfitting algorithm, Eq. 8.6 is re-expressed as

$$\tilde{y} = \alpha + [\tilde{\eta}(V) - \alpha] - \omega_1(V, D) \cdot \mathbb{1}_{\mathcal{D}_1}(D) + \omega_2(V, D) \cdot \mathbb{1}_{\mathcal{D}_2}(D) + \tilde{\varepsilon}. \tag{8.16}$$

One proceeds to estimate α using the global mean and estimate $[\tilde{\eta}(V) - \alpha]$ using a cubic smoothing spline (and the wake loss terms using TPRS-N). Once all the functional terms are estimated, $\tilde{\eta}(V)$ is restored by $\hat{\alpha} + \hat{\eta}(V)$ where $\hat{\eta}(V)$ is the estimate of $[\tilde{\eta}(V) - \alpha]$.

Before implementing the backfitting algorithm, some tuning parameters need to be set, including the smoothing parameter γ and the value of the reduced rank k used for improving the computational efficiency of TPRS-N. There are in fact three γ parameters, one each for the three smooth function estimations, associated, respectively, with $\tilde{\eta}(\cdot)$ and $\omega(\cdot)$'s. They are chosen based on a 10-fold cross validation while applying grid search. For the reduced rank k, Wood [228] states that the choice of k is not so critical as long as it is larger than the degrees of freedom required for the estimation. Hwangbo et al. [98] set $k = 30$ which turns out to be large enough for the wake effect analysis application. Finally, Hwangbo et al. set a threshold, $\epsilon_0 = 0.1$, which determines the convergence of the model fitting. The choice of 0.1 is believed to be sufficiently small considering the magnitude of the functional estimates changing exponentially due to the imposition of non-negativity.

The backfitting algorithm for the power difference model is summarized in Algorithm 8.2.

Algorithm 8.2 Backfitting algorithm for wake power loss estimation.

1: **Initialize:**
$$m \leftarrow 0; \; \hat{\alpha} \leftarrow \sum_{i=1}^{n} y_i/n; \; \hat{\boldsymbol{\eta}}^m \leftarrow \mathbf{0}; \; \hat{\omega}_1^m \leftarrow \mathbf{0}; \; \hat{\omega}_2^m \leftarrow \mathbf{0}$$

2: **repeat**

3:     Set $m \leftarrow m + 1$.

4:     **Estimation of $\hat{\eta}$**

5:         Calculate partial residuals: $\mathbf{r}_\eta \leftarrow \boldsymbol{y} - \hat{\alpha} + \hat{\omega}_1^{m-1} - \hat{\omega}_2^{m-1}$.

6:         Set $\hat{\boldsymbol{\eta}}^m$ by fitting smoothing spline to $\mathbf{r}_\eta$ with respect to $\mathbf{V}$.

7:     **Estimation of $\hat{\omega}_1$**

8:         Calculate partial residuals: $\mathbf{r}_{\omega_1} \leftarrow -(\boldsymbol{y} - \hat{\alpha} - \hat{\boldsymbol{\eta}}^m - \hat{\omega}_2^{m-1})$.

9:         Set $\hat{\omega}_1^m$ by fitting thin plate regression spline with non-negativity to $\mathbf{r}_{\omega_1}$ with respect to $\mathbf{V}$ and $\mathbf{D}$ for the data whose $D \in \mathcal{D}_1$.

10:     **Estimation of $\hat{\omega}_2$**

11:         Calculate partial residuals: $\mathbf{r}_{\omega_2} \leftarrow \boldsymbol{y} - \hat{\alpha} - \hat{\boldsymbol{\eta}}^m + \hat{\omega}_1^m$.

12:         Set $\hat{\omega}_2^m$ by fitting thin plate regression spline with non-negativity to $\mathbf{r}_{\omega_2}$ with respect to $\mathbf{V}$ and $\mathbf{D}$ for the data whose $D \in \mathcal{D}_2$.

13:     **Computation of convergence criterion**

14:     $\Delta \leftarrow \dfrac{||\hat{\boldsymbol{\eta}}^m - \hat{\boldsymbol{\eta}}^{m-1}|| + ||\hat{\omega}_1^m - \hat{\omega}_1^{m-1}|| + ||\hat{\omega}_2^m - \hat{\omega}_2^{m-1}||}{||\hat{\boldsymbol{\eta}}^{m-1}|| + ||\hat{\omega}_1^{m-1}|| + ||\hat{\omega}_2^{m-1}||}.$

15: **until** $\Delta \leq \epsilon_0$ where ϵ_0 is a prescribed threshold.

8.5 GAUSSIAN MARKOV RANDOM FIELD MODEL

The spline-based wake model in the preceding section is a single-wake model. You et al. [232] present a Gaussian Markov random field (GMRF) model that makes use of the spatial correlations among multiple turbines located close to one another and estimates simultaneously the heterogeneous power outputs from multiple turbines and the wake interactions. Apparently, this GMRF model is capable of modeling circumstances involving more than single wakes.

You et al. [232] do not directly model the wake loss, but model the power output with both a global term and a local term. The global term characterizes the average power production behavior of the turbines on a farm, as a function of environmental covariates, $\boldsymbol{x}$, similar to the power curve models presented in Chapter 5, whereas the local term characterizes the turbine-to-turbine variability unique to that specific turbine location and its neighboring turbines. The global term is not exactly the same as the power curve model. The difference is that the global term uses the same coefficient vector, $\boldsymbol{\beta}$, for all turbines on a farm, whereas the power curve model in Chapter 5 is supposed to be turbine specific; see also Exercise 8.2.

The GMRF model could be used to estimate the wake power loss *indirectly* by taking the difference of the maximum fitted value among all turbines and the power output fitted to a specific turbine. Please note that the GMRF model does not impose the constraint that the wake power loss should be non-negative.

The GMRF model takes the model structure as

$$y_t = G(\boldsymbol{x}, \boldsymbol{\beta}) + L(\boldsymbol{x}, \boldsymbol{\zeta}_t) + \varepsilon_t, \quad t = 1, \dots, N, \tag{8.17}$$

where $G(\cdot, \cdot)$ and $L(\cdot, \cdot)$ are the global and local terms, referred to, respectively, in the preceding paragraph. Consider q input variables in $\boldsymbol{x} = (x_1, \dots, x_q)^T$. You et al. [232] model $G(\cdot, \cdot)$ as an additive model of q terms, each of which is a set of univariate B-spline functions taking one of the input variables in $\boldsymbol{x}$ as its input. Specifically, You et al. express

$$G(\boldsymbol{x}, \boldsymbol{\beta}) = G_1(x_1, \boldsymbol{\beta}^{(1)}) + \cdots G_j(x_j, \boldsymbol{\beta}^{(j)}) + \cdots G_q(x_q, \boldsymbol{\beta}^{(q)}), \tag{8.18}$$

where $G_j(x_j, \boldsymbol{\beta}^{(j)})$ is further written as

$$G_j(x_j, \boldsymbol{\beta}^{(j)}) = \sum_{k=1}^{K^{(j)}} \beta_k^{(j)} g_k^{(j)}(x_j),$$

and the superscript, (j), is used to indicate that the B-spline basis functions are for the j-th input variable x_j, $K^{(j)}$ is the number of the basis functions, $g_k^{(j)}(\cdot)$ is the k-th univariate, global-term B-spline basis, as a function of x_j, and $\beta_k^{(j)}$ is the k-th spline regression coefficient.

The local term is likewise modeled as

$$L(\boldsymbol{x}, \boldsymbol{\zeta}_t) = L_1(x_1, \boldsymbol{\zeta}_t^{(1)}) + \cdots L_j(x_j, \boldsymbol{\zeta}_t^{(j)}) + \cdots L_q(x_q, \boldsymbol{\zeta}_t^{(q)}), \tag{8.19}$$

where $L_j(x_j, \boldsymbol{\zeta}_t^{(j)})$ is further expressed as

$$L_j(x_j, \boldsymbol{\zeta}_t^{(j)}) = \sum_{k=1}^{K^{(j)}} \zeta_{t,k}^{(j)} l_k^{(j)}(x_j),$$

and $l_k^{(j)}(\cdot)$ is the k-th univariate, local-term B-spline basis, and $\zeta_{t,k}^{(j)}$ is the k-th spline regression coefficient but specific to turbine t.

You et al. [232] treat $\boldsymbol{\zeta}_t^{(j)}$ as a random effect term and model it using GMRF. To reduce the modeling complexity, they further decompose $\boldsymbol{\zeta}_t^{(j)}$ into

$$\boldsymbol{\zeta}_t^{(j)} = \eta_t^{(j)} \boldsymbol{\zeta}^{(j)},$$

where the scalar term, $\eta_t^{(j)}$, captures the variations among individual turbines, while the vector term, $\boldsymbol{\zeta}^{(j)}$, becomes turbine-independent. With this decomposition, $\zeta_{t,k}^{(j)}$ in Eq. 8.19 can be expressed as $\zeta_{t,k}^{(j)} = \eta_t^{(j)} \cdot \zeta_k^{(j)}$. The scalar random effect term, $\eta_t^{(j)}$, is modeled by

$$\eta_t^{(j)} | \{\eta_{t'}^{(j)} : t' \in \mathfrak{N}_t\} \sim \mathcal{N}\left(\sum_{t' \in \mathfrak{N}_t} c_{t,t'} \eta_{t'}^{(j)}, \tau_j^2\right), \tag{8.20}$$

where $\mathfrak{N}_t$ denotes the neighborhood of turbine t, $c_{t,t'}$ captures inter-dependence between turbines t and t', and τ_j^2 is the variance for this conditional normal distribution, associated with the j-th input variable. Following the approach proposed in [115], You et al. [232] use the directional spatial dependence intensity to model $c_{t,t'}$ as

$$c_{t,t'} = \alpha_1 \sin^2(\theta_{t,t'}) \left(\frac{1}{d_{t,t'}}\right)^h + \alpha_2 \cos^2(\theta_{t,t'}) \left(\frac{1}{d_{t,t'}}\right)^h, \qquad (8.21)$$

where $d_{t,t'}$ is the distance between the two turbines in question, $\theta_{t,t'}$ is the angle between the wind direction and the line connecting turbine t and turbine t', α_1 and α_2 are the coefficients to be estimated by data, and h is a shape parameter, set to 0.5 by You et al. in their applications.

While implementing the method for wind applications, You et al. [232] include two primary inputs, wind speed and turbulence intensity. The global term in their GMRF model uses a B-spline function with degree 2 or higher for the wind speed input and a B-spline function of degree 1 or 2 for the turbulence intensity input. The local term uses B-spline functions of degree equaling to or smaller than their counterparts in the global term. In defining the knots for wind speeds, You et al. set five equal-distanced knots between 5 m/s and 17.5 m/s, resulting in four internal knots, respectively, at 7.5 m/s, 10 m/s, 12.5 m/s, and 15 m/s. In defining the knots for turbulence intensity, You et al. set two internal knots, with equal distance, in the turbulence intensity data range observed in their dataset. The turbulence intensity data range is $[0.2, 1.5]$, which yields two knots at 0.63 and 1.07, respectively.

The GMRF also needs to define a neighborhood, $\mathfrak{N}_t$, for each turbine t. You et al. [232] primarily use the first-order neighborhood turbines, which are defined as the eight nearest turbines surrounding turbine t.

You et al. [232] estimate the model parameters through a Bayesian hierarchical inference framework that is numerically solved by a Markov chain Monte Carlo (MCMC) sampling procedure. We will discuss MCMC more in Chapter 10.

8.6 CASE STUDY

8.6.1 Performance Comparison of Wake Models

In this section, a few wake models are compared in terms of their prediction performance of the power difference. Because directly measuring the actual wake power loss is difficult, the prediction or estimation of the power difference becomes an important proxy alluding to a model's capability of accounting for the wake effect in wind power production. Furthermore, power difference prediction could be in and by itself useful in a number of wind energy applications—for instance, the turbine upgrade quantification discussed in Chapter 7.

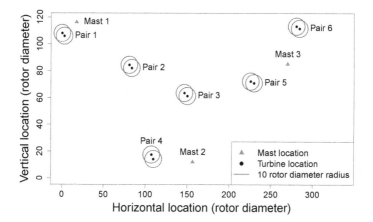

FIGURE 8.6 Locations of the six pairs of wind turbines and three met masts. The distances along both axes are expressed as a multiple of the rotor diameter of the turbines. All turbines have the same rotor diameter. (Reprinted with permission from Hwangbo et al. [98].)

TABLE 8.1 Between-turbine distances and relative positions of the six pairs of turbines. Bearing 1 to 2 indicates a relative direction of Turbine 1 to the location of Turbine 2, and Bearing 2 to 1 is similarly defined.

	Pair 1	Pair 2	Pair 3	Pair 4	Pair 5	Pair 6
Between-turbine distance	$6.8R$	$7.6R$	$8.4R$	$8.2R$	$8.2R$	$7.8R$
Bearing 1 to 2 ($^\circ$)	307.1	308.7	302.6	325.0	288.3	294.2
Bearing 2 to 1 ($^\circ$)	127.1	128.7	122.6	145.0	108.3	114.2

Source: Hwangbo et al. [98]. With permission.

This study uses the `Wake Effect Dataset`. Fig. 8.6 shows the relative locations of the six pairs of turbines and three met masts. The circle around each turbine is the $20R$ radius from, or the ten times rotor diameter centering at, the turbine. All turbine pairs happen to have the northwestern-to-southeastern orientation. Hwangbo et al. [98] designate, for all turbine pairs, the turbine on the northwestern side as Turbine 1 and the one on the southeastern side as Turbine 2.

Table 8.1 provides the between-turbine distances, in terms of a multiple of the rotor radius, and the relative positional angles between a pair of turbines. Based on the specific relative positions between a pair of turbines and the notations illustrated in Fig. 8.5, one can divide wind direction into two distinct sectors of $\mathcal{D}_1$ and $\mathcal{D}_2$ for each turbine pair. For a wind direction $D \in \mathcal{D}_2$, Turbine 1 is wake free and Turbine 2 is in the wake, whereas for $D \in \mathcal{D}_1$, Turbine 2 is wake free and Turbine 1 is in the wake.

TABLE 8.2   Comparison of prediction error in terms of RMSE. The value in the table is the power difference relative to the maximum power of the turbine. The boldface values are the smallest in each column.

	RMSE					
	Pair 1	Pair 2	Pair 3	Pair 4	Pair 5	Pair 6
Jensen's model	0.1103	0.0887	0.1109	0.0971	0.0956	0.1020
GMRF	0.0846	0.0752	0.0888	0.0797	0.0798	0.0877
Binning	0.0778	0.0667	0.0818	0.0800	0.0706	0.0751
TPRS-N	**0.0668**	**0.0627**	**0.0802**	**0.0758**	**0.0683**	**0.0699**

Source: Hwangbo et al. [98]. With permission.

TABLE 8.3   Comparison of prediction error in terms of MAE. The value in the table is the power difference relative to the maximum power of the turbine. The boldface values are the smallest in each column.

	MAE					
	Pair 1	Pair 2	Pair 3	Pair 4	Pair 5	Pair 6
Jensen's model	0.0544	0.0530	0.0673	0.0570	0.0631	0.0565
GMRF	0.0484	0.0469	0.0568	**0.0470**	0.0497	0.0544
Binning	0.0434	0.0435	0.0532	0.0504	0.0489	0.0457
TPRS-N	**0.0375**	**0.0408**	**0.0523**	0.0477	**0.0447**	**0.0434**

Source: Hwangbo et al. [98]. With permission.

Hwangbo et al. [98] evaluate the performance of a model with respect to its out-of-sample prediction errors. For this, each turbine pair's annual data are split into training and test subsets by a ratio of 80:20. In other words, 80% of a given dataset are randomly selected to train the model and the remaining 20% are used to calculate the prediction error. To measure the prediction error, Hwangbo et al. use both RMSE and MAE.

This section presents a performance comparison of four methods: Jensen's model, the data binning approach, the GMRF model, and the TPRS-N wake model, all under the single-wake situations.

Tables 8.2 and 8.3, respectively, present the RMSE and MAE values for the four methods and six turbine pairs. Relative to the Jensen's model, three data-driven methods significantly reduce the level of uncertainty by accounting for the variation observed in the data.

Recall that GMRF is not specifically developed for the single-wake situation. By construction, GMRF is designed to perform well with more turbines since it benefits from the spatial modeling of multiple turbines at different locations. Understandably, the method loses some of the benefits when being applied to a single pair of turbines. Still, the method shows significant improvement with, on average, an 18% reduction in RMSE and 14% in MAE as compared to Jensen's model.

The data binning approach, while fitting the trend of data without any restriction, in fact attains competitive prediction errors. This should not come as a surprise, as the binning approach is nonparametric and can adapt to local data features, as long as one uses a small enough binning resolution and there are dense enough data points to fit the binning model. The data binning approach is competitive in terms of out-of-sample prediction when compared with GMRF. The fact that its RMSE and MAE are larger than those of the TPRS-N model suggests, however, that the data binning approach overfits the (training) data. Another shortcoming of the data binning approach is that it is less insightful at providing wake characteristics.

The TPRS-N wake model, having incorporated physical constraints on wake power losses, demonstrates its superiority over other alternatives in terms of the prediction error of the power difference. It yields the smallest RMSE values across all six turbine pairs and the smallest MAE values for five among the six pairs. Its RMSE (MAE) is, on average, 30% (24%) smaller than that of Jensen's model, 15% (12%) smaller than GMRF, and 6% (7%) smaller than the data binning approach.

8.6.2 Analysis of Turbine Wake Effect

This section presents a study that quantifies annual wake power loss in actual wind turbine operations. Quantification of the wake power loss based on an annual period supports economic assessment of wake effect in terms of AEP. Doing so also provides practical insights into the economic impact of decisions and actions attempting to alleviate the wake power loss.

Fig. 8.7 illustrates the estimated wake characteristics using the `Wake Effect Dataset`. The wake loss is supposed to be strictly positive. What is shown in the plot is actually $-\hat{\omega}_1(V, D) \cdot \mathbb{1}_{\mathcal{D}_1}(D) + \hat{\omega}_2(V, D) \cdot \mathbb{1}_{\mathcal{D}_2}(D)$, so that one sees both positive and negative portions. The raw power differences of some pairs of turbines, when plotted against wind direction, exhibit large variation with several peaks and troughs. Even under such a noisy circumstance, the TPRS-N wake model captures the wake power loss signals well, by focusing on where the wake power loss is expected. In the figure, the vertical dashed lines indicate the bearings, i.e., $\theta_1 = 0$ or $\theta_2 = 0$.

Comparing Fig 8.7, bottom-left panel, to Fig. 8.4 (both generated from Pair 5), it is obvious that the TPRS-N wake loss estimation method captures the signals much better than the data binning approach could, making the subsequent derivation of the wake characteristics more convincing. One may also observe from Fig. 8.7 that the wind direction associated with the highest power loss is not exactly aligned with the bearings of the turbine pairs. This implies that there are measurement errors in wind direction. When applying the data binning approach, analysts typically generate angle bins starting from a bearing by making it the midpoint of an angle bin (and propagate with a resolution of 5 degrees, for example) and then regard the wake loss estimate of this specific bin as the wake depth. It turns out that, in the presence of measurements errors in wind direction, such a practice has an obvious disadvantage and will likely underestimate the wake depth due to the

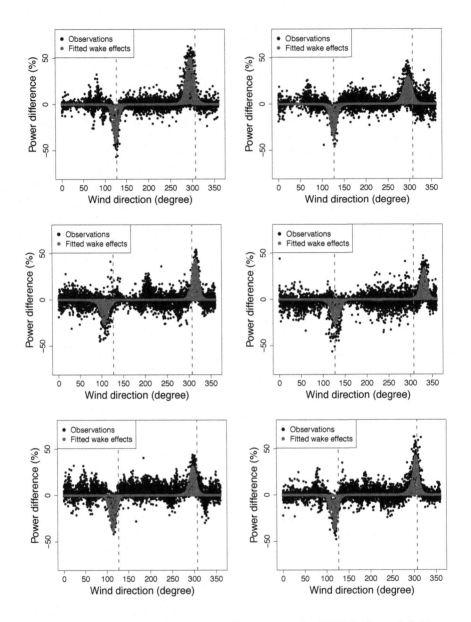

FIGURE 8.7 Estimates of the wake effect using the TPRS-N model. From top-left to bottom-right are, respectively, the estimation for Pair 1 through Pair 6. The shaded areas represent the fitted wake loss in terms of $-\hat{\omega}_1(V, D) \cdot \mathbb{1}_{\mathcal{D}_1}(D) + \hat{\omega}_2(V, D) \cdot \mathbb{1}_{\mathcal{D}_2}(D)$. Two dashed vertical lines indicate wind direction that is parallel to the line connecting the pair of turbines. (Reprinted with permission from Hwangbo et al. [98].)

TABLE 8.4   Wake depth and width for the six pairs of turbines.

	Pair 1	Pair 2	Pair 3	Pair 4	Pair 5	Pair 6
Depth: Turbine 1	41.6% (60.2%)	35.4% (41.8%)	26.6% (36.8%)	26.3% (41.1%)	40.0% (43.8%)	39.1% (44.3%)
Depth: Turbine 2	50.9% (56.2%)	29.2% (33.6%)	43.5% (44.8%)	33.5% (44.8%)	31.1% (44.5%)	42.1% (47.9%)
Width: Turbine 1	40.1°	42.7°	57.8°	51.6°	41.6°	57.5°
Width: Turbine 2	57.0°	52.8°	44.7°	45.6°	47.0°	49.9°

discrepancy between a bearing and the actual wind direction with the highest wake loss—see Fig. 8.7, middle-right panel, for an extreme example.

Table 8.4 shows the wake characteristics for the six turbine pairs. The first two rows are the estimates of wake depth, namely the magnitude of the wake power losses. The last two rows are the estimates of wake width. The wake depth is identified as the peak of the wake loss estimate representing the maximum power loss. The wake width is supposed to be determined by the angles around the bearings at which the power loss eventually becomes zero. However, given noisy signals spreading over a large range of wind directions, the fitted wake power loss is not completely zero even in the regions where it is unquestionably wake free. To estimate the wake width, Hwangbo et al. [98] use the range of wind direction for which loss is greater than 1% of the rated power of the turbine. For the wake depth, Table 8.4 presents two percentage values for each turbine. The one outside the parenthesis is the wake power loss relative to the rated power of that turbine, whereas the one inside the parenthesis is the loss relative to the free-stream equivalent power output.

In the literature, the wake power loss is often expressed as the ratio of the loss over the free-stream equivalent power output [2, 13, 85], which can be computed by

$$\frac{\hat{\omega}_t(V_i, D_i)}{\hat{y}_t(V_i, D_i) + \hat{\omega}_t(V_i, D_i)}, \quad t = 1, 2, \quad i = 1, \ldots, n, \tag{8.22}$$

where $\hat{y}_t(V_i, D_i)$ denotes the expected power generation given (V_i, D_i). Depending on (V_i, D_i), $\hat{y}_t(V_i, D_i)$ could be the expected power in the wake of another turbine, so that the free-stream equivalent power output is to be recovered by adding $\hat{y}_t(V_i, D_i)$ and $\hat{\omega}_t(V_i, D_i)$. To calculate $\hat{y}_t(V_i, D_i)$, Hwangbo et al. [98] define a neighborhood of (V_i, D_i), i.e., $\mathfrak{N}_i = \{(V, D) : V \in (V_i - \epsilon_V, V_i + \epsilon_V], D \in (D_i - \epsilon_D, D_i + \epsilon_D]\}$, and compute $\hat{y}_t(V_i, D_i)$ by taking the average of the power outputs whose corresponding wind speed and direction is a member of $\mathfrak{N}_i$. This is a two-dimensional binning with $2\epsilon_V$ and $2\epsilon_D$ as the respective bin width, where ϵ_V and ϵ_D are predetermined constants. In this application, Hwangbo et al. set $\epsilon_V = 0.25$ m/s and $\epsilon_D = 2.5°$. The second

TABLE 8.5 Annual power loss for the six turbine pairs.

	Percentage measure (%)					
	Pair 1	Pair 2	Pair 3	Pair 4	Pair 5	Pair 6
Turbine difference ($\tilde{\eta}_{2\text{-}1}$)	0.26	1.10	0.21	-1.40	3.41	0.94
Wake loss: Turbine 1 (ω_1)	0.85 (1.67)	0.78 (1.62)	0.51 (1.13)	0.64 (1.42)	0.59 (1.27)	0.66 (1.33)
Wake loss: Turbine 2 (ω_2)	2.00 (4.04)	1.24 (2.69)	1.39 (3.15)	1.11 (2.40)	1.01 (2.36)	1.75 (3.68)
Average wake loss for the pair	1.43 (2.84)	1.01 (2.14)	0.95 (2.13)	0.88 (1.92)	0.80 (1.19)	1.20 (2.48)

percentage values in Table 8.4, i.e., the ones inside the parentheses, are the wake power loss expressed in this conventional fashion.

The peak power loss relative to the free-stream equivalent (the value inside the parenthesis) ranges from 34% to 60%. The wake width for the 12 turbines ranges from 40° to 58° with concentration around 40°–53°. The wake depth commonly stated in the literature is in the range of 30%–40% [12, 14, 192], which appears to be at the lower side of the spline wake model-based estimates. In addition, the new wake width estimates are noticeably larger than the 25° to 40° range stated previously [14, 146, 215]. The difference can be attributed to two major factors. The first one is that the new estimation can identify the wake region more accurately, producing better estimates of the two main characteristics, whereas the methods in the literature rely on *ad hoc* data segmentation and partition and often use a partial set of data based on a pre-selected range of wind direction. Consequently, the previous wake power loss estimates do not capture the characteristics as well as the new estimator does. The second factor is that the historical estimates are usually the averages over multiple turbines, understandably leading to a narrower range.

Table 8.5 shows how each term in the power difference model of Eq. 8.6 affects the power generation of a turbine pair in an annual period, namely the AEP power difference or AEP loss. The first row is the between-turbine power production difference independent of wake effect, expressed relative to the rated power. The second and third rows present the wake loss. Similar to Table 8.4, the values outside the parentheses is the loss relative to the rated power, whereas the values inside the parentheses is the loss relative to the free-stream equivalent. Both percentages represent the AEP wake loss but use different baselines.

The wake loss relative to the rated power is in fact related to the capacity factor of a wind turbine. Recall that the capacity factor is the ratio of the actual power production of a turbine for a selected period of time, say, one year, over the supposed power production the turbine could have produced, had it operated at its maximum capacity (i.e., at the rated power) all the time; the typical range of the capacity factor is 25%–35%. The wake loss relative to the rated power, therefore, can be seen as the direct reduction to a turbine's capacity factor. Hwangbo et al. [98] refer to the corresponding AEP loss as the capacity factor AEP loss and refer to the AEP loss relative to the free-stream equivalent as the traditional AEP loss. The traditional AEP loss is computed, if using Turbine 1 group as an example, by

$$\frac{\sum_{i=1}^{n} \hat{\omega}_1(V_i, D_i)}{\sum_{i=1}^{n} \{\hat{y}_1(V_i, D_i) + \hat{\omega}_1(V_i, D_i)\}}. \tag{8.23}$$

The fourth row is the average AEP wake loss for a pair of turbines. The average is weighted by the number of data points in the respective wake regions to account for the annual distribution of the AEP loss for the turbine pairs. For this reason, the values in the fourth row may be slightly different from the simple average of the two individual losses. The average *traditional AEP loss* for a pair is computed by

$$\frac{\sum_{i=1}^{n} \{\hat{\omega}_1(V_i, D_i) + \hat{\omega}_2(V_i, D_i)\}}{\sum_{i=1}^{n} \{\hat{y}_1(V_i, D_i) + \hat{\omega}_1(V_i, D_i) + \hat{y}_2(V_i, D_i) + \hat{\omega}_2(V_i, D_i)\}}. \tag{8.24}$$

The average *capacity factor AEP loss* is computed by setting the denominator in the above equation to be $\sum_{i=1}^{n} \{(\text{rated power}) + (\text{rated power})\} = 2n \cdot (\text{rated power})$.

From Table 8.5, one may notice that the magnitude of the between-turbine difference is sizeable, sometimes even larger than that of the wake effect. This result suggests that modeling of the between-turbine difference as a separate term in the power difference model is important to the mission of estimating the wake effect; otherwise, the estimate of the wake effect can be biased considerably.

One can immediately observe that the AEP losses are much smaller than the peak power loss (wake depth). This is expected because the annual loss is the average over all kinds of wind speed and direction conditions in an entire year. Under many circumstances, the wake loss is much smaller than the peak loss. The capacity factor AEP loss is between 0.5–2.0%, meaning that if the turbine's actual capacity factor is 25%, then its ideal capacity factor, if the turbine always operated wake free, could have been 25.5% to 27%. This difference, while appearing as a small percentage, should not be taken lightly. Consider a wind farm housing 200 turbines all in the 2 MW turbine class. A 1% capacity factor AEP loss for the whole farm translates to $1.3 million annual loss in revenue at the wholesale price of $37 per MWh.

One may also notice that the wake loss endured by Turbine 2 in a pair is always greater than that of Turbine 1. This can be explained by the relative positions of the turbines and the prevailing wind direction over this farm during that particular year. Fig. 8.8 presents the wind rose plots for three pairs of

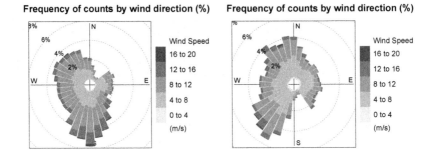

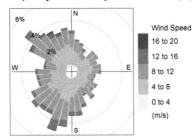

FIGURE 8.8 Wind rose plots illustrating the relative frequency of incoming wind for different direction sectors and for different speed ranges. Top-left panel for Pairs 1 and 2; top-right panel for Pairs 3 and 4; and bottom panel for Pairs 5 and 6. (Reprinted with permission from Hwangbo et al. [98].)

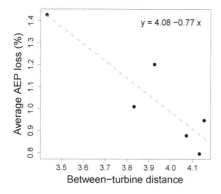
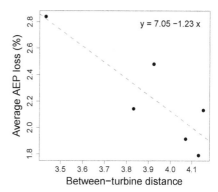

FIGURE 8.9 Relation between AEP losses and turbine spacing. Left panel: for the capacity factor AEP loss; right panel: for the traditional AEP loss. The between turbine distance is expressed as a multiple of the rotor diameter. (Adapted with permission from Hwangbo et al. [98].)

the turbines. The plots show that the northwestern wind, for which Turbine 2 of each pair endures power loss, is more frequent and stronger than the southeastern wind for which Turbine 1 experiences power loss. Unsurprisingly, the AEP loss of Turbine 1 group is usually less than 0.85% (1.67%), whereas the AEP loss for Turbine 2 group is greater than 1.01% (2.36%) and can be as high as 2% (4.04%).

In the literature, it is well known that turbine spacing is a decisive factor affecting the magnitude of wake power loss [13, 129, 192]. Hwangbo et al. [98] conjecture that the variation of the annual power loss between the individual turbine pairs can be explained by the between-turbine distance of each pair. Using the average AEP loss for the six turbine pairs (the fourth row in Table 8.5) and the corresponding between-turbine distances, they fit a simple linear regression model as has been done in [13].

Fig. 8.9 shows the scatter plots and the regression line fitting a respective AEP loss. For the capacity factor AEP loss, the p-values of the intercept and slope estimate are 0.005 and 0.013, respectively. For the traditional AEP loss, the corresponding p-values are 0.006 and 0.022. These results confirm that the turbine spacing indeed by and large explains the pair-wise difference in the AEP losses. An extrapolation based on the fitted regression lines suggests that the wake loss would diminish after the turbine spacing reaches either 5.3 or 5.7 times the rotor diameter, depending on which AEP loss is used in the analysis. Nevertheless, in either circumstance, the 10 times rotor diameter separation used in this study to isolate a particular turbine pair from the rest of turbines appears safe enough to render the turbine pairs free of wake of any other turbines on the wind farm.

Regressing the turbines' inherent production difference (the first row in Table 8.5) on the between-turbine distance, on the other hand, suggests that there is no significant correlation between them. The p-values of the intercept and slope estimate in this case are 0.81 and 0.77, respectively. Unlike the wake effect, the between-turbine production difference does not seem to be affected by the between-turbine distance, much as expected.

GLOSSARY

AEP: Annual energy production

CFD: Computational fluid dynamics

GAM: Generalized additive model

GMRF: Gaussian Markov random field

IEC: International Electrotechnical Commission

MAE: Mean absolute error

MCMC: Markov chain Monte Carlo

RMSE: Root mean squared error

TPRS: Thin plate regression spline

TPRS-N: Thin plate regression spline model with non-negativity constraint

EXERCISES

8.1 In Section 8.6.1, Jensen's model is used in the form of $V_{\text{wake}} = (1 - \kappa_{\text{deficit}}) \cdot V$ with κ_{deficit} being set to 0.075. What if Jensen's model used in the comparison follows Eq. 8.1 instead? Can you update the power difference prediction result in Tables 8.2 and 8.3?

8.2 One modeling strategy mentioned and compared with by both Hwangbo et al. [98] and You et al. [232] is the individual turbine power production model, very much like the power curve model introduced in Chapter 5. The individual turbine model in Hwangbo et al. [98] and You et al. [232] is referred to as IND and follows the model structure of

$$y_t = G(\boldsymbol{x}, \boldsymbol{\beta}_t) + \varepsilon_t,$$

which appears similar to the global term in Eq. 8.17 but here $\boldsymbol{\beta}_t$ is no longer the same for all turbines but tailored to individual turbines. For model performance comparison with IND, please refer to [98, Table 3]. In this exercise, please use the AMK model (Section 5.2.3) as the individual turbine power production model. This is to say, fit the AMK model to two turbines in a pair, respectively, and then compute the power difference. Compare this AMK-based IND model performance with other methods in Tables 8.2 and 8.3.

8.3 In Section 8.6.1, when the data binning approach is applied, the wind speed range used in Step 2 is the whole wind speed spectrum. Please use a narrower wind speed range instead and see what difference this change makes in terms of the power difference prediction errors. Two narrower wind speed range options mentioned in Section 8.3 are: (a) 8.0 ± 0.5 m/s and (b) 5.0–11.0 m/s. Furthermore, please investigate the sensitivity of the data binning approach to the width of wind direction bins. Try and compare the bin-width options of 2.5°, 5° (the current option), 10°, and 15°, in terms of power difference prediction errors.

8.4 In Section 8.6.1, Jensen's model is paired with the IEC binning power curve model to compute the power output. What if the IEC binning power curve model is replaced with the AMK-based power curve model (Section 5.2.3)? Conduct the numerical analysis and see how much it affects the power difference prediction errors.

8.5 In Section 8.6.2, a linear regression model is built to regress the average wake power loss on the between-turbine distances. It was also mentioned there that one can regress the between-turbine difference, $\tilde{\eta}$, on the between-turbine distances and would not find significant correlation between the input and output. Please build the linear regression model and present the scatter plot and the line fit like those in Fig. 8.9.

8.6 Use the `Wake Effect Dataset` and the spline-based single-wake model to investigate the sensitivity of parameter k in that model. Try five different k values: 10, 20, 30, 40, and 50. Please present a plot displaying how the RMSE and MAE values vary with different k values. What conclusion do you draw from the plots?

8.7 Chapter 6 considers the problem of shape-constrained curve fitting, in which an S-shape constraint is imposed. Chapter 8 considers the problem of sign-constrained curve fitting, in which a non-negativity constraint is imposed. In some circumstances, a shape constraint can be expressed as a sign constraint under a functional transformation, and vice versa. Please show that shape constraints like monotonicity, convexity, or concavity can be expressed as a sign constraint of non-negativity or non-positivity. Please state clearly what type of functional transformation is used to make such equivalence possible.

III

Wind Turbine Reliability Management

CHAPTER 9

Overview of Wind Turbine Maintenance Optimization

DOI: 10.1201/9780429490972-9

P art III of this book discusses a few issues related to turbine reliability management. In this part, the data science problem often concerns modeling and estimating $f(z|\boldsymbol{x})$, where z is the mechanical load measured at certain critical spots on a turbine component, such as at the root of the turbine blades. While the conditional density, $f(z|\boldsymbol{x})$, resembles that of $f(y|\boldsymbol{x})$, one unique aspect in Part III is that reliability analysis focuses much more on the tail of $f(z|\boldsymbol{x})$, rather than on the middle region surrounding its mode. In this sense, reliability analysis concentrates on rare events, which take place with a rather small probability. One important branch of data science methodologies pertinent to reliability analysis is random sampling. Chapters 10 and 11 discuss, respectively, the use of Markov chain Monte Carlo methods and importance sampling methods in the context of turbine blade load analysis.

Chapter 12, however, touches upon a different topic relevant to the general theme of reliability management—anomaly detection and fault diagnosis. The data science problem of anomaly detection and fault diagnosis falls into the category of unsupervised learning, in which the class label of a data record is not available. The very purpose of anomaly detection or fault diagnosis is to recover as accurately as possible the class label for that data record, based on observations of explanatory covariates in $\boldsymbol{x}$. While research has been progressing on anomaly detection and fault diagnosis, specialized methods targeting wind turbines are still in great demand. One thing hindering the development on this front, more so than the other data science aspects discussed in this book, is the lack of availability of fault event data resulting from commercial operations to the research community at large. The reason is rather understandable. Whoever owns the reliability or fault event data tends to guard those data diligently, as the implication of liability through reliability data is much more direct than through environmental data (x) or power production data (y). Sometimes, the real owner of this type of data, be it the owner/operator or the turbine's manufacturer, can also become debatable. Even if one party may be willing to share the data with a data science research third party, the other party may not want to divulge. The hydropower plant data used in Chapter 12 is in fact subject to a confidentiality agreement and therefore does not appear among the shared datasets.

Before we proceed with load analysis and anomaly detection in the latter chapters, we start off Part III in the present chapter with a discussion of wind turbine operations and maintenance (O&M), which is in and of itself an important part of reliability management.

9.1　COST-EFFECTIVE MAINTENANCE

Wind turbine O&M plays an important role and accounts for a major portion in the total cost of energy in wind power production. The US Department of Energy's 20%-Wind-Energy-by-2030 report [218, Figure 2-15] shows that while the O&M cost may be as low as 0.5–0.6 cents/kWh in the early years of a turbine's service, it escalates to 1.8–2.3 cents/kWh after 20 years' service. Analysts have also realized that some early estimates of the wind's O&M cost may have been considerably underestimated. In 2011's second issue of the *Wind Stats Report*, William Manganaro, project manager of NAES Corporation, draws a contrast between the 2010 American Wind Energy Association (AWEA)'s estimate of O&M cost at 2.5 cents/kWh and AWEA's initial estimate of 0.5 cents/kWh [181]. Generally speaking, the O&M cost is estimated to account for 20–30 % of the total energy cost in the land-based, utility-scale wind power generation [48], and for offshore wind, the cost portion of O&M is considerably higher. Dr. Fort Felker, former Director of the National Wind Technology Center, stated in a 2009 speech, under the heading of *Critical Elements for 20% Scenario*, that one such critical element is to reduce the O&M costs by 35% from that year's level [62]. Progresses have been made since then but the 35% reduction is still a long way off.

Reducing the cost of O&M for wind power generation appears to be challenging. Under nonstationary loadings due to wind's stochastic nature, a turbine's drive train, especially the gearbox, is prone to failure. Current O&M practice is reactive in nature and depends heavily on what is commonly known as condition-based monitoring or condition-based maintenance (CBM). The concept of CBM is to collect the online monitoring data, and upon diagnosing a turbine's condition, decide whether and when maintenances are needed. A CBM system may ignore the uncertainty in sensory information and respond aggressively to any potential failure signals as is. Doing so of course triggers some of the maintenance actions prematurely. Premature maintenance does not benefit the bottom line of wind operations, and because of that, rarely is such an overconfident approach taken in industrial practices.

On the other side of the spectrum, due to the imperfection of current diagnosis and reliability assessment, compounded with the high cost to undertake a maintenance action, there is always reluctance to take actions until perhaps when it is too late. Scheduled maintenances are rather popular in the wind industry, meaning that the maintenance schedules are fixed *a priori*, e.g., once every six months. Scheduled maintenances are too rigid, so that it either allows too many failures, if the schedules are infrequent, or costs too much, if the schedules are too frequent. The reality is that rigid and reactive maintenance approaches lead to *"a prevalence of unscheduled maintenance"* [182]. It is apparent that neither the overconfident nor the overcautious approach can produce the optimal cost structure for turbine maintenance.

A key in devising a cost-effective maintenance strategy is to handle the uncertainty in sensory information properly and decide the course of action without over-trusting, nor discrediting altogether, the information dynamically collected. In order to compensate for the uncertainty in sensory information, decision makers resort to crew's on-site investigations as well as simulations of wind farm operations, both of which are more expensive, either economically or computationally, than just using online sensors with intermittent analysis. A relevant question in a dynamic maintenance scheduling system is when an on-site investigation should be triggered and how simulation outcomes can be used together with the sensory data.

With this objective in mind, a dynamic, data-driven approach provides a useful school of thought and points to a close coupling between modeling and information-gathering. A data-driven approach could trigger the expensive on-site investigations adaptively, only as needed, and dynamically update the mathematical models by injecting the newly collected data, as appropriate. This chapter discusses, in such a context, two turbine maintenance optimization models, one wind farm simulator, and an approximation strategy allowing optimization and simulation to work together in real time.

Fig. 9.1 illustrates the need for cost-effective maintenance and highlights the pursuit of a better balance in decision making in the presence of uncertainty.

9.2   UNIQUE CHALLENGES IN TURBINE MAINTENANCE

The challenges in wind turbine maintenance are primarily caused by the stochastic nature of turbine operation conditions and the uncertainties in the decision variables induced by the stochasticity. Because of the stochasticity, the condition-based monitoring for wind turbine systems runs into difficulties while attempting to pin down the occurrence and severity of a potential fault or failure. Complicating the matter further are the weather constraints and disruptions, as well as the logistic difficulties such as long lead time and long service time. Not surprisingly, a poorly planned maintenance job contributes substantially to the escalation of O&M costs of wind turbine systems.

Weather does constrain maintenance activities. To maximize power gener-

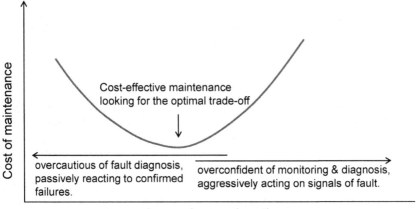

FIGURE 9.1 Plot of cost versus maintenance actions. An overconfident approach, responding to every potential failure signal as if there is no uncertainty, is likely overacting, whereas an overcautious approach, on the other side, likely waits until it is too late to prevent a potential failure.

ation potential, wind farms are built at locations with high wind. But climbing a turbine is not allowed, due to safety concerns, when wind speed is more than 20 m/s. When the speed is higher than 30 m/s, the site becomes inaccessible. The combination of higher failure rates and low accessibility for repairs during windy periods exacerbates the revenue loss because that is the time when wind turbines are supposedly to generate most of its power.

The sheer size of wind turbine components makes it difficult to store spare parts in a warehouse waiting for repairs or replacements. Rather, the large parts are likely ordered and shipped directly from a manufacturer when needed. Doing so leads to long lead time in obtaining parts and can result in costly delays in performing repairs. Pacot et al. [158] state that it may take several weeks for critical parts, such as a gearbox, to be delivered.

The logistic difficulty in turbine maintenance is also caused by the long distances of wind farms from their operation centers, as well as that major repairs require heavy duty equipment, such as a crane or a helicopter, to access the turbine. It certainly takes quite an effort to assemble the maintenance crews and prepare for a major repair. Logistic costs may escalate substantially, depending on the accessibility to the turbine's site, the maintenance strategy, and equipment availability.

The three factors, weather, component size, and distance, can become intertwining with one another, making the already challenging situation even more difficult. For instance, major repair of turbines usually takes weeks to complete due to the physical difficulties of the job (size and distance). The long duration of a repair session in fact increases the likelihood of disruption by adverse weather.

Besides the effect of wind, as mentioned above, other environmental factors can adversely impact a turbine's reliability, such as extreme low temperature, icing, and lighting strikes, or wave and corrosion for offshore turbines. A comprehensive discussion on this topic can be found in [27].

9.3 COMMON PRACTICES

The methods of wind turbine maintenance commence rather naturally with analyzing the historical failure data to elucidate a component's failure probability and then plan maintenance accordingly. These are referred to as the failure statistics-based approaches. In the meanwhile, there have been attempts to establish reliability models, based on first principles and/or stochastic process theories, to provide an understanding of the stochastic aging behavior of turbine components. The recent rapid advancements in microelectronics and sensing technology have allowed real-time measurements and analyses of various characteristics of wind turbines during their operations—this line of approaches is collectively referred to as condition-based maintenance.

9.3.1 Failure Statistics-Based Approaches

Failure statistics-based approaches are purely data-driven. Although often applied to individual components, this type of approach can actually be used for the whole turbine as well. The idea is to use the historical failure data to fit certain probability distributions, which then yield a number of commonly used statistics such as the mean time to failure (MTTF) [82, 227]. The popular distribution here, as in reliability analysis in general, is the Weibull distribution, which has been described in detail in Chapter 2 (Sections 2.2.2–2.2.4). In Chapter 2, the Weibull distribution is used to model the wind speed distribution. It turns out that the Weibull distribution is also a common choice in reliability analysis, because like wind speed, the time to failure is positive and right skewed in distribution. In addition to Weibull distributions, analysts also use the non-homogeneous Poisson process to handle the cases where the number of failures is provided but the time when a failure takes place is missing [80, 210, 211]. Analysts build and run Monte Carlo-based simulation models by making use of the statistics and distributions obtained above in order to study or visualize the stochastic behavior of a turbine's failures. Simulations are furthermore conducted for evaluating the effectiveness of maintenance actions, assessing the impact of turbine reliability on power generation, comparing turbine siting choices, and validating operational strategies [28].

9.3.2   Physical Load-Based Reliability Analysis

It is desirable that a physical relationship between loads and component failures can be established [131], a topic to be discussed in more detail in Chapters 10 and 11. The load-based reliability analysis refers to the studies concerning the fatigue load or extreme load, or the characterization of wake effect. In Chapter 8, we discuss the impact of wake effect on power production. The wake effect also has an adverse impact on turbines in the form of an increased mechanical load. The load-based analysis has mainly concentrated on the structural components of a turbine, i.e., blades and the tower, and are generally carried out offline. National Renewable Energy Laboratory in the US and Risø National Lab in Denmark, the two leading government organizations in wind technology, have developed their respective structural aeroelastic codes and tools to assess the structural loads using computer-based simulations [112, 113, 130]. The IEC publishes design standards [101] for the structural components of a turbine.

9.3.3   Condition-Based Monitoring or Maintenance

CBM involves two necessary analyses: (a) failure modes analysis to understand how the likely failure patterns are associated with the major components in a turbine [177], and (b) analysis of the turbine's operational data acquired by *in situ* sensors for diagnosis and prognosis [230]. Different from the failure statistics-based approach that is done offline, CBM is conducted on a continual, online basis. With the accumulation of online data and advancement in signal analysis, the concept of prognostics and health management has emerged, which is nonetheless within the broad scope of CBM. Although CBM may be used for any components in a wind turbine, the current practice mainly focuses on the drive train, where vibratory responses, acoustic emissions, temperature, and lubrication oil particulate content are the common measurements monitored. Addressing gearbox reliability is one of the major efforts of NREL, who released the Gearbox Reliability Council's report in June 2011 [135].

9.4   DYNAMIC TURBINE MAINTENANCE OPTIMIZATION

The imperfection of today's diagnosis tools is due to the fundamental limitations in our knowledge of the engineering systems, sensing technologies, and data science methods. Uncertainties are not going to disappear from diagnosis and prognosis anytime soon, and as a consequence, analysts need to systematically address the issue of uncertainties in maintenance decision making. General methodologies enabling decision making under uncertainties have seen advancements. One of them is the Markov decision process (MDP) [170], including the partially observable MDP (POMDP) [137, 141] or hidden Markov process [57], for handling the cases when a system's state is not perfectly ob-

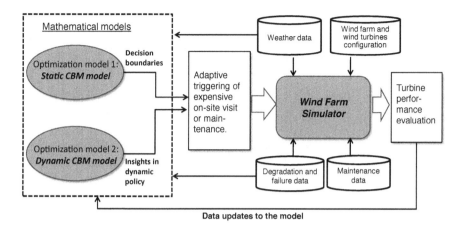

FIGURE 9.2 The framework of the dynamic, data-driven approach for wind turbine maintenance—an integration of models, simulation, and data in a closed loop.

servable due to data or model uncertainties. These methods have been used in traditional power system reliability analysis, like for transformers [110] as well as for wind turbines [25, 26]. The resulting dynamic models consider the stochastic weather constraints and logistic uncertainties as mentioned earlier, and work in an integrated framework with a wind farm simulator to guide model updates and measurement in-taking. A general dynamic maintenance optimization framework is illustrated in Fig. 9.2.

The dynamic maintenance optimization framework includes three important models—two optimization models [25, 26] and a simulation model [28, 161]. The two optimization models come with different flavors: one is a static model whose parameters stay constant, whereas the other is a dynamic model whose parameters are adjusted responding to different seasonal weather patterns. The merit of having both optimization models is that the dynamic model, despite desirable in terms of adaptivity, is expensive to solve, because it is difficult to extract a solution structure due to the complexity of the model. The static model, albeit a simplification of reality, does reveal a solution structure that facilitates solving for maintenance decisions efficiently. The wind farm simulation model is a discrete event-based model, mimicking the operation of a wind farm with hundreds of wind turbines that may degrade along different paths and thus need service at different times. The simulator is specially designed to handle a large number of unorganized random events (e.g., waiting for parts or weather disruptions) and reflect in the simulator's outputs the stochasticity from the operations. The dynamic maintenance optimization framework integrates the optimization models, the simulator, and the data, in a closed loop where sensor data are used to improve the modeling analysis and prediction.

9.4.1 Partially Observable Markov Decision Process

Byon et al. [25, 26] develop maintenance optimization models based on partially observable Markov decision processes. Part of the reason motivating the use of POMDP is that a turbine's SCADA data are noisy and the SCADA data can, at best, provide partial information of the turbine's health status. In the POMDP setting, the system's degrading condition can only be assessed in a probabilistic sense, represented by the information state π. Suppose that the turbine health condition can be discretized into M levels, with Level 1 being the best condition like a new turbine and M being the most deteriorated condition right before failure. Then, the information state can be expressed as $\pi = [\pi_1, \ldots, \pi_M]$, and each element in the π-vector is a probability, indicating the likelihood of the turbine being at that level. This π-vector is used as an important input to the subsequent POMDP model, so that any decision made is based on a probabilistic assessment, rather than the assumption of knowing to which level the turbine has degraded.

Byon et al. [25, 26] consider L number of failure modes. Combining with the M degradation levels, this POMDP model has a total of $M+L$ states. Out of the $M+L$ states, M of them are working states and L of them are failing states. A turbine transitions through the working states and ultimately arrives at one of the failing states, unless some kind of intervention takes place prior to failure. A turbine does not have to exhaust all working states before it fails—it is possible that a turbine can fail at any working state. Once it arrives at one of the failing states, it stays there and will not transition further to other failing states. A turbine can only go from a failing state back to a working state when a repair or replacement operation is carried out.

Transitions between those states are illustrated in the state transition graph in Fig. 9.3, left panel, where P_{ij} is the transition probability from state i to state j. One may notice that there is no state transition between the L failing states but there is a link between any working state to a failing state, per discussion in the earlier paragraph.

Cost-effective maintenance trades multiple choices of action. Besides the options of preventive maintenance (PMT) or continue with SCADA monitoring, i.e., do-it-later (DIL), a possible intermediate action is to conduct on-site visit/investigation (OVI), to find out more certainly how a turbine is operating. This on-site action is much more expensive than purely SCADA-based monitoring, considering the long distance between a service center and a wind farm, but less expensive than a fully blown repair job, which employs expensive heavy-duty equipment and a larger crew. Other than the three options, corrective maintenance (CMT), i.e., the reactive action once a failure has already happened, always needs to be considered. Understandably, corrective maintenance is the most expensive among all options.

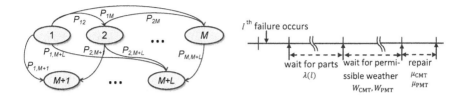

FIGURE 9.3 Maintenance modeling. Left panel: state transition graph and P_{ij} is a transition probability; right panel: uncertainty caused by weather and logistics. (Left panel reprinted with permission from Byon et al. [27].)

The POMDP model takes a dynamic programming form. At time t, let $C_t(\pi_t)$ denote the cost for maintenance until a terminal time (e.g., when the turbine is decommissioned), given that its current state is π_t, and let a_t denote the action options at t. If one considers the three options mentioned earlier, then a_t is chosen from the set of $\{PMT, OVI, DIL\}$. According to the theoretical framework of POMDP [141, 170], the optimal solution regarding which action to take can be decided based on the following optimization problem:

$$C_t(\pi_t) = \min_{a_t} \left\{ c_t(\pi_t, a_t) + \gamma \cdot \sum_{\pi_{t+1} \in S} P(\pi_{t+1} | \pi_t, a_t) \cdot C_{t+1}(\pi_{t+1}) \right\}, \quad (9.1)$$

where a_t is the decision variable to be optimized, c_t is the immediate cost incurred by taking action a_t, γ is the cost discount factor due to the monetary depreciation over time, and

$$S := \{[\pi_1, \cdots, \pi_{M+L}], \quad \text{s.t.} \quad \sum_i \pi_i = 1, \quad \pi_i > 0 \quad \forall i\}$$

is the set of information states.

The above optimization formulation is in recursive form, because the minimum value of C_t depends on that of C_{t+1}. Modeling $C_t(\cdot)$ and $c_t(\cdot)$ for each decision period involves the considerations outlined earlier, such as which action to take and its corresponding cost, when such an action is permissible, and how much revenue loss is incurred due to a turbine's downtime. The solution of the POMDP formulation typically starts from the terminal time, say, at T, when $C_T(\pi_T)$ can be decided since the turbine is supposed to have arrived at one of the failing states. Then the action at $T-1$ can be solved, which gives the value of $C_{T-1}(\pi_{T-1})$. Going further back in time solves for the rest of actions in the turbine's maintenance history.

Byon et al. [25, 26] consider the stochasticity resulting from waiting for permissible weather windows and from repairing, all of which manifest as random events to be included in the decision model; see Fig. 9.3, right panel.

The adverse weather conditions are modeled using two prohibiting probabilities: $W_{\text{CMT}(l)}$ as the probability that the prevailing weather conditions are so adverse that the corrective maintenance required for the lth failure mode, $l \in \{1, \ldots, L\}$, is not allowed, and $W_{\text{PMT}(m)}$ likewise defined for a preventive maintenance that is supposed to restore the system back to the mth degradation level, $m \in \{1, \ldots, M\}$. The waiting periods and repair jobs are characterized through a queuing model having its processing rate, respectively, of $\lambda(l)$, $\mu_{\text{CMT}(l)}$, and $\mu_{\text{PMT}(m)}$.

The costs of different actions are denoted as c_{OVI} for on-site visit/investigation, $c_{\text{PMT}(m)}$ for the mth mode of preventive maintenance repair, and $c_{\text{CMT}(l)}$ for corrective repair upon the lth mode of failure. During the downtime until a repair is completed, a revenue in the amount of τ_t is lost.

With these parameters and settings, the cost and probability items in the dynamic programming formulation in Eq. 9.1 can be uniquely specified. For instance, the consequence associated with $\text{PMT}(m)$ at time t can be expressed as

$$
\begin{aligned}
\text{PMT}_t(m) = {} & W_{\text{PMT}(m)}(\tau_t + \text{PMT}_{t+1}(m)) + \\
& (1 - W_{\text{PMT}(m)})[\tau_t + c_{\text{PMT}(m)} + C_{t+1}(e_m)],
\end{aligned}
\tag{9.2}
$$

where e_m is one of the extreme states of which the mth element in $[\pi_1, \ldots, \pi_M]$ is one and all other elements are zeros. The notation, $\text{PMT}(m)$, denotes the cost associated with the preventive maintenance action restoring a more degraded turbine system to the state e_m. The interpretation of the above equation is that the preventive maintenance cost at t to restore the turbine to state e_m is affected by the weather condition. When the weather condition is not permitting for this type of action, with a probability of $W_{\text{PMT}(m)}$, the preventive maintenance action will then have to wait until the next period. While waiting, the turbine incurs a production loss of τ_t. If the weather condition is permitting, with a probability of $1 - W_{\text{PMT}(m)}$, and the preventive maintenance action is in fact completed, then the costs incurred are the production lost during the repair period, the cost associated with the specific repair action $\text{PMT}(m)$, and the cost of operation for the next period with the turbine in the state of e_m.

9.4.2 Maintenance Optimization Solutions

If one assumes that all the weather-associated parameters in the dynamic programming model in Eq. 9.1 are constant over time, the solution of the dynamic programming model then corresponds to a static maintenance system. One may solve the optimization model offline using the value iteration or policy iteration method [170]. The solution outcome is a decision map, allowing analysts to trade off the three major maintenance actions, namely PMT, DIL, and OVI, based on an online estimation of a turbine's degradation status (coded in π). This maintenance decision process is labeled *static*, because the decision map stays the same, even though the estimation of π changes (move from π_t to π_{t+1} over time). Fig. 9.4 illustrates how the offline optimization and online sensory information work together.

Albeit a simplification of the reality, the static maintenance model does allow analysts to understand the structure of the maintenance policy with clarity. Fig. 9.5 depicts how a maintenance action is selected. Simply put, it depends on the relative costs of the respective actions. The aging status of the turbine system is one of the primary forces driving the cost structure. When a turbine system is relatively new, the cost associated with DIL option is low, as the chance that the system has a catastrophic failure is low in the near future. As a turbine system ages, the chance of failing gets higher, and the consequence of a failure becomes more serious. The choice between doing an on-site investigation versus doing a fully blown maintenance depends on the cost difference between the two options. For older turbines, the choice of full-blown maintenance action is taken most seriously.

The decision maps, as shown in Fig. 9.6, have three action zones. The decision boundary between two decision zones corresponds to the boundary point in Fig. 9.5 between a pair of actions. The two plots in the upper panel of Fig. 9.6 are decision maps produced by a static maintenance model in which the model parameters are kept constant for the entire decision horizon. Under the static model structure, analysts can decide the decision boundaries analytically and populate a decision zone with its respective decision. The difference between the two upper-panel plots is due to the use of different prevailing weather parameters, namely that W_{PMT} and W_{CMT} are under different values. In the left column, $W_{\mathrm{PMT}} = 0.1$ and $W_{\mathrm{CMT}} = 0.3$, whereas in the right column, $W_{\mathrm{PMT}} = 0.5$ and $W_{\mathrm{CMT}} = 0.5$. Apparently, the resulting decision map changes when the weather becomes more prohibiting, as the weather conditions in the left figure turn into that in the right figure. The right figure has a higher probability that an ongoing maintenance may be disrupted, and accordingly, the decision zones of PMT and OVI shrink, whereas the zone of DIL expands, suggesting that one needs a higher confidence of potential failures to undertake more expensive actions.

When the maintenance model uses time-varying parameters, it becomes a dynamic model, which is a better reflection of the reality, as the actual system and the environment constantly change. The difficulty of handling a dynamic model is that the exact solutions will have to be found through iterative numerical procedures. Understandably, these iterative numerical procedures are not computationally cheap to use. The two plots in the lower panel of Fig. 9.6 are decision maps from a dynamic model. As seen in the plots, the dynamic model can produce a path of maintenance policies, rather than identify the decision boundaries delineating the decision zones. To solve for the path of a maintenance policy, Byon and Ding [25] use the backward dynamic programming procedure, as outlined in Algorithm 9.1.

If one looks at the plots in the upper and lower panels of Fig. 9.6, it is not difficult to notice that the policy structure implied by the static maintenance

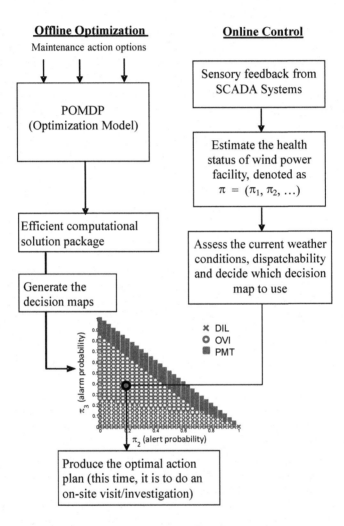

FIGURE 9.4 Offline optimization generates the decision map, while online observations allow an assessment of turbine degradation status. Together they produce the cost-effective maintenance decision.

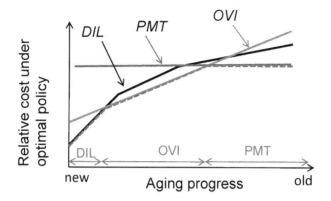

FIGURE 9.5 Maintenance decision structure. When a turbine system is relatively new, the optimal action tends to be do-it-later, as the risk of a catastrophic failure is low. When a turbine system ages, the consequence of a failure gets more serious, so that more aggressive actions are needed. (Adapted with permission from Byon et al. [26].)

Algorithm 9.1 Backward dynamic programming solution procedure.

1. Construct a sample path emanating from π, as well as the extreme sample paths originating from the extreme states e_m, $m = 1, \cdots, M$.

2. Set the turbine's decommission time as T and the decommission value as $C_T(\pi)$, based on a specific system preference.

3. Set $t = T - 1$.

   (a) Update parameters such as $W_{\text{CMT}(l),t}$, $W_{\text{PMT}(m),t}$ and τ_t, $l = 1, \cdots, L$, $m = 1, \cdots, M - 1$.

   (b) Find the optimal decision rule and optimal value at extreme points e_i, $i = 1, \cdots, M$.

   (c) Compute $\text{CMT}_t(l)$ and $\text{PMT}_t(m)$ for each corrective maintenance and preventive maintenance option, as well as that for $\text{DIL}_t(\pi)$ and $\text{OVI}_t(\pi)$, respectively.

   (d) Compute the optimal value function, $C_t(\pi)$, and the corresponding optimal decision rule.

   (e) Set $t = t - 1$, and go back to (a).

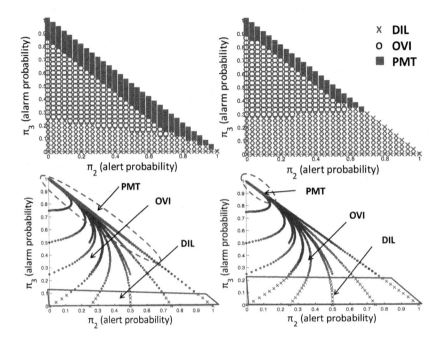

FIGURE 9.6 Static and dynamic maintenance model solutions. Upper panel: static model solutions; lower panel: dynamic model solutions. The two plots in the left column are under the weather condition with $W_{\text{PMT}} = 0.1$ and $W_{\text{CMT}} = 0.3$, whereas the two plots in the right column are under $W_{\text{PMT}} = 0.5$ and $W_{\text{CMT}} = 0.5$.

model are more or less preserved in the dynamic model outputs, although the precise decision boundary would have almost surely shifted to a certain extent, as the model parameters are not exactly the same under the static and dynamic circumstances. Upon this revelation, Byon [24] deems it a sensible approach to solve the dynamic model faster and in real time through the following approach: derive useful approximations in the maintenance policy structure based on the static model, employ the resulting structure to pre-partition the decision map, and invoke the numerical solution procedure only when a decision gets close to the prescribed boundary line.

9.4.3 Integration of Optimization and Simulation

One merit of the POMDP-based maintenance models presented in the preceding sections is that they explicitly model the external environmental conditions under which a turbine system is being operated. The static model, because of its stronger assumptions, may be useful for wind farms that operate in relatively stationary weather conditions, whereas the dynamic model is more adaptive even when there are strong seasonal variations.

Both models nonetheless focus on a single turbine. To the best of our knowledge, using the single-turbine models is the case for the majority of turbine maintenance research [27]. This is not surprising, as building and solving an optimization model for a single turbine is already challenging work, and doing so for a commercial size wind farm housing one hundred plus wind turbines could be analytically intractable.

Maintenance strategies devised based on a single turbine may still be useful to advising operations and decisions on a wind farm. The potential problem is that the randomness in turbine degradation processes cause different turbines to follow different degrading paths, thereby creating a need to adjust maintenance actions accordingly and to resolve the conflicts caused by the multi-turbine environment.

To address this issue, Byon et al. [28] and Pérez et al. [161] develop a discrete event system specification (DEVS) [233]-based simulation platform for a wind farm housing many wind turbines. The specific simulation framework, DEVS, is selected because using it enables an easier modeling of multiscale (time, space and decisions) complex systems. DEVS is a formal modeling and simulation framework based on dynamical systems theory, providing well-defined concepts for coupling atomic and coupled models, hierarchical and modular model construction, and an object-oriented substrate supporting repository reuse. Under the DEVS framework, each turbine is treated as a duplicable module, so that the simulation can be scaled up without difficulty to more than one hundred turbines for a farm. Each turbine module includes a turbine components sub-module, degradation sub-module, power generation sub-module, sensing and maintenance scheduling sub-module. Fig. 9.7 presents a high-level diagram of the DEVS-based wind farm simulation.

After the turbine module is built, one also needs to define the logic that interconnects components in a wind turbine system and mimics the execution and operation in a virtual, cyberspace environment. The simulator is designed to handle a large number of unorganized random events (turbine failures, waiting for parts, weather disruptions) and reflect in the simulator's outputs the stochasticity from the operations.

On the one hand, there are two turbine-based optimization models that can advise on optimal maintenance actions for a single, specific turbine, while on the other hand, there is a wind farm simulator used to evaluate the effectiveness of the resulting maintenance decisions on multiple turbines following different degradation paths, to suggest update and adjustments if necessary. It becomes apparent that one would need to integrate the optimization models and the simulation model in order to materialize the dynamic turbine optimization framework, as highlighted in Fig. 9.2.

Integrating the dynamic MDP model with the wind farm simulator is not trivial, as it creates a computational challenge. Recall that the optimal policy in the dynamic model is solved through a backward dynamic programming

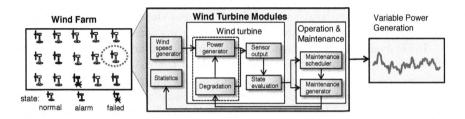

FIGURE 9.7 The hierarchical structure and operation of the DEVS-based wind farm simulator. Each turbine is treated as a duplicable module, which includes sub-modules such as turbine components, degradation, power generation, sensing and maintenance scheduling.

algorithm that is computationally demanding for large-scale problems. When every turbine needs to run this computational procedure in the wind farm simulator for its decision on maintenance, the simulator creeps to a halt.

To alleviate this problem, Byon [24] proposes an approximation approach, taking advantage of an observation made in the previous section, i.e., the decision map from the dynamic model preserves the structure of optimal maintenance policy revealed by the static model. Byon further conjectures that a set of similar decision rules established from a static model may be used to closely *approximate* a dynamic policy. Because this set of rules is rooted in the static policy, they can be expressed in closed form, and their computation leading to decision boundaries can be done in real time.

Fig. 9.8 illustrates how to approximate the dynamic policy based on a set of decision rules derived from a static model. Firstly, Byon [24] adjusts the cost component associated with a maintenance action and makes it updateable with real-time information on weather (characterized by parameters W_{PMT} and W_{CMT}), logistics (characterized by λ), and revenue loss (characterized by τ). Note that for a variable energy source like wind, its revenue loss due to downtime also depends on wind input and may vary over time. Secondly, Byon updates the static decision boundary by plugging in the new dynamic cost obtained. This plug-in is the part of approximation. When an action mismatches with the theoretical optima, the maintenance cost elevates, as compared to the theoretically optimal level. Numerical studies conducted in [24] show, however, that the mismatch rates are around 5% of the decision regions. When costs over different scenarios are considered, the impact of this 5% mismatch is translated to less than 1% increase in terms of maintenance cost. The benefit of using this simple set of approximations reduces the computation remarkably, a 98% reduction as compared to the use of the dynamic programming solution procedure [24]. On balance, Byon states that this approximation approach produces significant benefits and allows the integration of the optimization models and the wind farm simulator in an online decision support fashion.

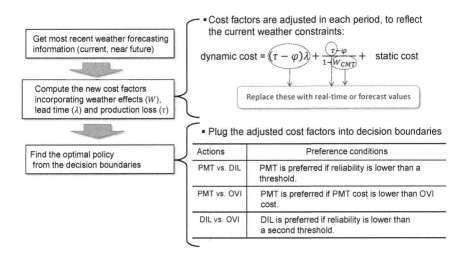

FIGURE 9.8 Approximating the maintenance policy and making it real-time executable. The variable φ indicates the average cost per period.

When executing the integrated framework, real-time weather information or its forecasts in near future are to be used to update the decision boundaries as the simulator runs. This is where wind field analysis and wind forecasting methods in Part I of this book can chime in and benefit the overall wind engineering decision making.

Byon [24] implements this dynamic condition-based maintenance policy for a prototypical wind farm and compares the outcomes with those using scheduled maintenance, a popular practice in industry. The numerical study in [24] shows that using a well-devised dynamic policy leads to a 45% reduction in the annualized failure rate and the number of preventive repairs, and a 23% reduction in the overall maintenance cost. These comparison outcomes are charted in Fig. 9.9.

9.5 DISCUSSION

This chapter presents an overview of a dynamic maintenance optimization paradigm for wind turbine systems. One great challenge in wind turbine maintenance is the non-steady loading and stochasticity in the environment conditions under which turbines operate. These factors drive wind turbines to a fast degradation process, cause the fault diagnosis and condition monitoring to have low specificities, and render any rigid maintenance paradigms costly—all hurt wind energy's market competitiveness. The dynamic, data-driven maintenance approach, with its adaptiveness to a varying environment, strikes a cost-effective balance among different competing cost factors and optimizes for the long-run benefit in the presence of data and model uncertainty.

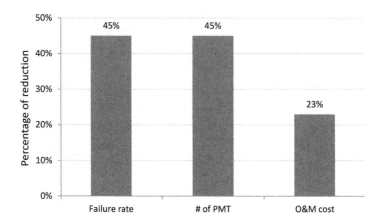

FIGURE 9.9 Benefit of the dynamic turbine maintenance optimization as compared to scheduled maintenance. Shown in the plot is the reduction in terms of the failure rate, number of maintenance actions, and overall O&M cost.

The virtue of a dynamic maintenance framework can be understood in three aspects: (a) It incorporates real-time measurements, learns what the data reveals, and updates the models as needed; (b) it pays full attention to the uncertainty of information, and promotes the use of multi-accuracy information sources (SCADA versus on-site investigation) to compensate for the adverse effect resulting from information uncertainty; and (c) it combines multi-fidelity models (static, dynamic, and simulation) in an integrative fashion, which enables decisions to be made both efficiently and effectively.

Presenting this framework at the beginning of Part III serves as a bridge connecting the wind forecasting work in Part I, the turbine performance analysis work in Part II, and the load analysis and anomaly detection work to be covered in the remaining chapters in Part III.

GLOSSARY

AWEA: American Wind Energy Association

CBM: Condition-based monitoring or condition-based maintenance

CMT: Corrective maintenance

DEVS: Discrete event system specification

DIL: Do it later

IEC: International Electrotechnical Commission

MCMC: Markov chain Monte Carlo

MDP: Markov decision process

MTTF: Mean time to failure

NREL: National Renewable Energy Laboratory

O&M: Operations and maintenance

OVI: On-site visit and investigation

PMT: Preventive maintenance

POMDP: Partially observable Markov decision process

SCADA: Supervisory control and data acquisition

s.t.: Such that

US: United States

EXERCISES

9.1 Consider a turbine system which has only two states, up (operating state) and down (failure state).

   a. What are the values of M and L? What is the dimension of the transition probability matrix $\mathbf{P} = (P_{ij})$? Recall that the transition probability matrix is denoted by $\mathbf{\Pi}$ in Eq. 4.6. Here $\mathbf{P}$ is used in the place of $\mathbf{\Pi}$ because π in this chapter is used to denote the information state, rather than the transition probability.

   b. Can you draw the state transition diagram like in Fig. 9.3, left panel?

   c. Let λ denote the failure rate, which is the probability that the turbine fails and let μ denote the probability with which a failing system can be restored to the operating state. Write down the corresponding transition probability matrix $\mathbf{P}$.

   d. What are the mean time to failure and the mean time to repair for this system?

9.2 Can you write down the costs associated with other action options, i.e., OVI, DIL, and CMT, like what was done for PMT in Eq. 9.2?

9.3 The probability that a system will still operate until the next decision point is called the reliability of the system and denoted by $\Upsilon(\pi) = \sum_{i=1}^{M} \sum_{j=1}^{M} \pi_i P_{ij}$. Express the information state after the next transition, π', in terms of the current state, π, and the reliability of the system, provided that the system has not yet failed.

9.4 The preventive maintenance cost expressed in Eq. 9.2 is for a major repair action. For a minor repair, there are at least two differences. When the weather condition is not permitting, a minor repair can wait, without affecting the power production of the turbine. When the weather condition is permitting, the action of a minor repair is usually assumed to take place fast enough, so that the production loss it causes can be neglected. Given these understandings, how would you write the preventive maintenance cost function for a minor repair?

9.5 Byon [24] identifies that the mismatches between the optimal maintenance policy and the approximating maintenance policy often happen when there is a rapid transition of weather conditions from severe (a higher W_{PMT}) to mild (a lower W_{PMT}). Let us deem 12 m/s the wind speed limit, above which a major preventive maintenance is severely restricted.

a. Use the 10-min data in the Wind Time Series Dataset. Compute the probability of W as the portion of periods that wind speed is above the aforementioned limit in a week.

b. Plot W versus the weeks in a year.

c. Do you observe any time intervals where there is a noticeable transition from a severe weather condition to a mild weather condition?

CHAPTER 10

Extreme Load Analysis

DOI: 10.1201/9780429490972-10

W ind turbines operate under various loading conditions in stochastic weather environments. The increasing size and weight of components of utility-scale wind turbines escalate the loads and thus the stresses imposed on the structure. As a result, modern wind turbines are prone to experiencing structural failures. Of particular interest in a wind turbine system are the extreme events under which loads exceed a threshold, called a *nominal design load* or *extreme load*. Upon the occurrence of a load higher than the nominal design load, a wind turbine could experience catastrophic structural failures. To assess the extreme load, turbine structural responses are evaluated by conducting physical field measurements or performing aeroelastic simulation studies. In general, data obtained in either case are not sufficient to represent various loading responses under all possible weather conditions. An appropriate extrapolation is necessary to characterize the structural loads in a turbine's service life. This chapter focuses on the extreme load analysis based on physical bending moment measurements. Chapter 11 discusses load analysis based on aeroelastic simulations.

10.1 FORMULATION FOR EXTREME LOAD ANALYSIS

Fig. 10.1 presents examples of mechanical loads at different components in a turbine system. The flapwise bending moments measure the loads at the blade roots that are perpendicular to the rotor plane, whereas the edgewise bending moments measure the loads that are parallel to the plane. Shaft- and tower-bending moments measure, in two directions, respectively, the loads on the main shaft connected to the rotor and on the tower supporting the wind power generation system (i.e., blades, rotor, generator etc.).

Same as in the treatment of power response analysis, load response data are arranged in 10-minute intervals in the structural reliability analysis of wind turbines. The basic characteristics of the three inland turbines (ILT) in the **Turbine Bending Moment Dataset** can be found in Table 10.1. The

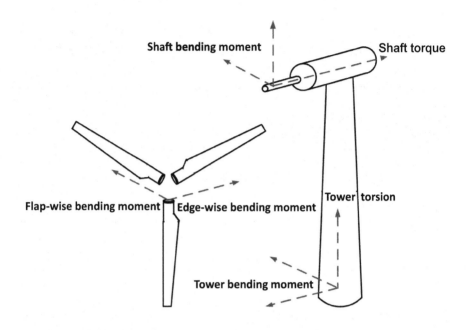

FIGURE 10.1 Illustration of structural loads at different wind turbine components. (Reprinted with permission from Lee et al. [131].)

TABLE 10.1 Specifications of wind turbines in the `Turbine Bending Moment` `Dataset`.

Turbine model (Name of dataset)	NEG-Micon/2750 (ILT1)	Vestas V39 (ILT2)	Nordtank 500 (ILT3)
Hub height (m)	80	40	35
Rotor diameter (m)	92	39	41
Cut-in speed (m/s)	4	4.5	3.5
Cut-out speed (m/s)	25	25	25
Rated speed (m/s)	14	16	12
Nominal power (kW)	2,750	500	500
Control system	Pitch	Pitch	Stall
Location	Alborg, Denmark	Tehachapi Pass, California	Roskilde, Denmark
Terrain	Coastal	Bushes	Coastal

Source: Lee et al. [131]. With permission.

original data are recorded at a much higher frequencies on the three ILTs, as follows:

- ILT1: 25 Hz = 15,000 measurements/10-min;

- ILT2: 32 Hz = 19,200 measurements/10-min;

- ILT3: 35.7 Hz = 21,420 measurements/10-min.

Consider the raw measured variables, which are wind speed, V_{ij}, and load response, z_{ij}, where $i = 1, \ldots, n$ is the index of the 10-minute data blocks and $j = 1, \ldots, m$ is the index of the measurements within a 10-minute block. The load response, z_{ij}, is the bending moment, in millions of Newton-meters, measured at designated locations. Here, m is used to represent the number of measurements in a 10-minute block, equal to 15,000, 19,200, and 21,420 for ILT1, ILT2, and ILT3, respectively, and n is used to represent the total number of the 10-minute intervals in each dataset, taking the value of 1,154, 595, and 5,688, respectively, for ILT1, ILT2, and ILT3. The statistics of the observations in each 10-minute block are calculated as

$$V_i = \frac{1}{m} \sum_{j=1}^{m} V_{ij}, \tag{10.1a}$$

$$s_i = \sqrt{\frac{1}{m-1} \sum_{j=1}^{m} (V_{ij} - V_i)^2}, \quad \text{and} \tag{10.1b}$$

$$z_i = \max \{z_{i1}, z_{i2}, \ldots, z_{im}\}. \tag{10.1c}$$

After this data transformation, z_i denotes the maximum load in the i-th 10-minute block. In this study, we consider only the flapwise bending moments measured at the root of blades. In other words, z in this study is the 10-minute maximum blade-root flapwise bending moment.

Mathematically, an extreme load is defined as an extreme quantile value in a load distribution corresponding to a turbine's service time of T years [202]. Then, the load exceedance probability is defined as

$$P_T = P[z > l_T], \tag{10.2}$$

where P_T is the target probability of exceeding the load level l_T (in the same unit as that of z). The unconditional distribution of z, $f(z)$, is referred to as the *long-term* load distribution and is used to calculate $P[z > l_T]$ in Eq. 10.2.

In Eq. 10.2, the extreme event, $\{z > l_T\}$, takes place with the probability of exceedance (POE), P_T. The waiting time until this event happens should be longer than, or equal to, the service time. Therefore, the exceedance probability, P_T, given the service life T, can be decided via the following way [101, 160]:

$$P_T = \frac{10}{T \times 365.25 \times 24 \times 60}. \tag{10.3}$$

Intuitively, P_T is the reciprocal of the number of 10-minute intervals in T years. For example, when T is 50, P_T becomes 3.8×10^{-7}.

Estimating the extreme load implies finding an extreme quantile l_T in the 10-minute maximum load distribution, given a target service period T, such that Eq. 10.2 is satisfied. Wind turbines should be designed to resist the l_T load level to avoid structural failures during its desired service life.

Since loads are highly affected by wind profiles, analysts consider the marginal distribution of z obtained by using the distribution of z conditional on a wind profile, i.e.,

$$f(z) = \int_{\boldsymbol{x}} f(z|\boldsymbol{x}) f(\boldsymbol{x}) d\boldsymbol{x}. \tag{10.4}$$

This expression is almost identical to that in Eq. 1.1, except that the power response, y, is replaced by the load response, z. The conditional distribution of z given $\boldsymbol{x}$, $f(z|\boldsymbol{x})$, is referred to as the *short-term* load distribution. The long-term load distribution can be computed by integrating out wind characteristics in the short-term load distribution.

The conditional distribution modeling in Eq. 10.4 is a necessary practice in the wind industry. A turbine needs to be assessed for its ability to resist the extreme loads under the specific wind profile at the site where it is installed. Turbine manufacturers usually test a small number of representative turbines at their own testing site, producing $f(z|\boldsymbol{x})$. When a turbine is to be installed at a commercial wind farm, the wind profile at the proposed installation site can be collected and substituted into Eq. 10.4 as $f(\boldsymbol{x})$, so that the site-specific extreme load can be assessed. Without the conditional distribution model, a turbine test would have to be done for virtually every new wind farm; doing so is very costly and hence impractical.

For inland turbines, the wind characteristic vector $\boldsymbol{x}$ in general comprises two elements: a steady-state mean of wind speed and the stochastic variability of wind speed [142, 183]. The first element can be measured by the average wind speed during a 10-minute interval, and the second element can be represented by the standard deviation of wind speed, or the turbulence intensity, also during a 10-minute interval. For offshore turbines, weather characteristics other than wind may be needed, such as wave height [3]. Since the data in the **Turbine Bending Moment Dataset** are all from inland turbines, what is included in $\boldsymbol{x}$ in this chapter is the average wind speed V and the standard deviation of wind speed, s, namely $\boldsymbol{x} := (V, s)$.

10.2 GENERALIZED EXTREME VALUE DISTRIBUTIONS

The 2nd edition of the IEC 61400-1 standard, published in 1999, offers a set of design load cases with *deterministic* wind conditions such as annual average wind speeds, higher and lower turbulence intensities, and extreme wind speeds [100]. In other words, the loads in IEC 61400-1, 2nd edition, are specified as discrete events based on design experiences and empirical models. Veers and Butterfield [223] point out that these deterministic models do not represent the stochastic nature of structural responses, and suggest instead using statistical modeling to improve design load estimates. Moriarty et al. [150] examine the effect of varying turbulence levels on the statistical behavior of a wind turbine's extreme load. They conclude that the loading on a turbine is stochastic at high turbulence levels, significantly influencing the tail of the load distribution.

In response to these developments, the 3rd edition of IEC 61400-1 standard, published in 2005 [101], replaces the deterministic load cases with *stochastic* models, and recommends the use of *statistical* approaches for determining the extreme load level in the design stage. Freudenreich and Argyriadis [67] compare the deterministic load cases in IEC61400-1, 2nd edition, with the stochastic cases in IEC61400-1, 3rd edition, and observe that when statistical approaches are applied, higher extreme load estimates are obtained in some structural responses, such as the blade tip deflection and flapwise bending moment. After the 3rd edition of IEC 61400-1 is published, several research groups devise and recommend statistical approaches for extreme load analysis [3, 65, 67, 149, 153, 174].

According to the classical extreme value theory [38, 200], the short-term distribution, $f(z|\boldsymbol{x})$, can be approximated by a generalized extreme value (GEV) distribution. The pdf of the GEV is

$$f(z) = \begin{cases} \frac{1}{\sigma} \exp\left[-\left(1 + \xi\left(\frac{z-\mu}{\sigma}\right)\right)^{-\frac{1}{\xi}}\right]\left(1 + \xi\left(\frac{z-\mu}{\sigma}\right)\right)^{-1-\frac{1}{\xi}}, & \xi \neq 0, \\ \frac{1}{\sigma} \exp\left[-\frac{z-\mu}{\sigma} - \exp\left(-\frac{z-\mu}{\sigma}\right)\right], & \xi = 0, \end{cases} \tag{10.5}$$

for $\{z : 1 + \xi(z-\mu)/\sigma > 0\}$, where μ is the location parameter, $\sigma > 0$ is the scale parameter, and ξ is the shape parameter that determines the weight of the tail of the distribution. When $\xi > 0$, the GEV corresponds to the Fréchet distribution family with a heavy upper tail, $\xi < 0$ to the Weibull distribution family with a short upper tail and light lower tail, and $\xi = 0$ (or, $\xi \to 0$) to the Gumbel distribution family with a light upper tail.

The cdf of the GEV distribution is

$$F(z) = \begin{cases} \exp\left[-\left(1 + \xi\left(\frac{z-\mu}{\sigma}\right)\right)^{-\frac{1}{\xi}}\right], & \xi \neq 0, \\ \exp\left[-\exp\left(-\frac{z-\mu}{\sigma}\right)\right], & \xi = 0. \end{cases} \tag{10.6}$$

Recall the cdf of Weibull distribution in Eq. 2.5. If modified to be a three-parameter Weibull distribution with a location parameter μ, the cdf reads

$$F(z) = 1 - \exp\left[-\left(\frac{z-\mu}{\eta}\right)^{\beta}\right].$$

Although one can discern that β here is related to ξ in the GEV distribution and η to σ, the two distributions do not appear the same for $\xi \neq 0$. One may wonder why when $\xi < 0$, the GEV distribution is said to be corresponding to the Weibull distribution family. The reason behind the disconnect is that the GEV distribution of $\xi < 0$ corresponds actually to the reverse Weibull distribution, which is the negative of the ordinary Weibull distribution. Recall that a Weibull distribution is used to describe a non-negative random variable with a right-side skewness, like wind speed. The ordinary Weibull distribution deals with the minimum and has a lower bound. The GEV distribution when $\xi < 0$ deals instead with the maximum and sets an upper bound, so that the distribution looks like the mirror image of the ordinary Weibull distribution. The cdf of a reserve Weibull is

$$F(z') = \begin{cases} \exp[-(-z')^{\beta}], & z' < 0, \\ 1, & z' \geq 0. \end{cases} \tag{10.7}$$

If we let $\beta = -1/\xi$ and $z' = -(1 + \xi(z - \mu)/\sigma)$, then one sees that the cdf expression in Eq. 10.6 for $\xi < 0$ and that in Eq. 10.7 are the same.

One of the main focuses of interest in the extreme value theory is to derive the quantile value, given a target probability P_T. The extreme quantile value based on a GEV distribution of z is in fact the design load threshold, l_T, in Eq. 10.2. Given a P_T, l_T can be expressed as a function of the distribution parameters as

$$l_T = \begin{cases} \mu - \frac{\sigma}{\xi}\left[1 - (-\log(1 - P_T))^{-\xi}\right], & \xi \neq 0, \\ \mu - \sigma \log\left[-\log(1 - P_T)\right], & \xi = 0. \end{cases} \tag{10.8}$$

10.3 BINNING METHOD FOR NONSTATIONARY GEV DISTRIBUTION

When using a GEV distribution to model z, one difficulty is that z appears to be a function of input x. Using the three datasets, Lee et al. [131] present a number of scatter plots to illustrate this point. Fig. 10.2 shows the scatter plots between the 10-minute maximum loads and 10-minute average wind speeds. One observes nonlinear patterns between the load and the average wind speed in all three scatter plots, while individual turbines exhibit different response patterns. ILT1 and ILT2 are two pitch-controlled turbines, so when the wind speed reaches or exceeds the rated wind speed, the blades are adjusted to reduce the absorption of wind energy. As a result, the load shows a downward trend after the rated speed. Different from that of ILT1, the load response of ILT2 has a large variation beyond the rated wind speed. This large variation can be attributed to its less capable control system since ILT2 is one of the early turbine models using a pitch control system. ILT3 is a stall controlled turbine, and its load pattern in Fig. 10.2(c) does not have an obvious downward trend beyond the rated speed.

Fig. 10.3 presents the scatter plots between the 10-minute maximum loads and the standard deviations of wind speed during the 10-minute intervals. One

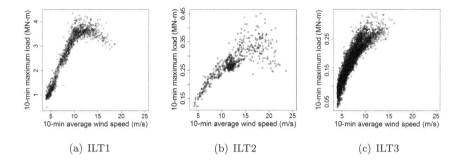

FIGURE 10.2 Scatter plots of 10-minute maximum load versus 10-minute average wind speed. (Reprinted with permission from Lee et al. [131].)

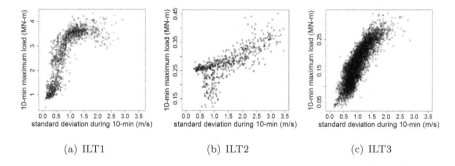

FIGURE 10.3 Scatter plots of 10-minute maximum load versus 10-minute standard deviations of wind speed. (Reprinted with permission from Lee et al. [131].)

also observes nonlinear relationships between them, especially for the newer pitch-controlled ILT1.

Fig. 10.4 shows scatter plots of 10-minute standard deviation versus 10-minute average wind speed. Previous studies [63, 150] suggest that the standard deviation of wind speed varies with the average wind speed, which appears consistent with what is observed in Fig. 10.4.

This type of load response, varying with the input conditions, is known as inhomogeneous or nonstationary response. If a set of constant parameters, $\{\mu, \sigma, \xi\}$, is used in modeling a GEV distribution, the resulting distribution is homogenous or stationary for the entire input domain, which does not match the inhomogeneous reality of the turbine load condition. To address this issue, analysts in the aforementioned extreme load analysis adopt a common framework, referred to as the *binning* approach [131]. The basic idea of the binning method is the same as that used for power curve analysis in Chapter 5, but

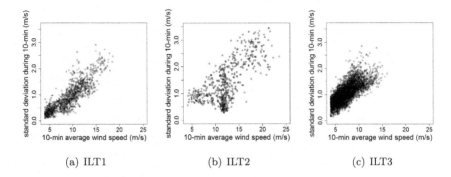

(a) ILT1   (b) ILT2   (c) ILT3

FIGURE 10.4 Scatter plots of 10-minute average wind speed versus 10-minute standard deviations of wind speed. (Reprinted with permission from Lee et al. [131].)

unlike power curve analysis in which a simple sample average is used to model the data in a bin, a GEV distribution is established in each bin in extreme load analysis.

Specifically, the binning approach discretizes the domain of a wind profile vector x into a finite number of bins. One can divide the range of wind speed, from the cut-in speed to the cut-out speed, into multiple bins and set the width of each bin to, say, 2 m/s. Then, in each bin, the conditional short-term distribution, $f(z|x)$, is approximated by a stationary GEV distribution, with the parameters of the distribution fixed at certain constants and estimated by the maximum likelihood method, using the data specific to that bin. The contribution from each bin is summed over all possible bins to determine the final long-term extreme load. In other words, the integration in Eq. 10.4 for calculating the long-term distribution is approximated by the summation of finite elements. The idea of binning is illustrated in Fig. 10.5.

The virtue of the binning method is that by modeling the short-term distribution with a homogeneous GEV distribution, i.e., keeping the GEV distribution parameters constant for a given bin, it provides a simple way to handle the overall nonstationary, inhomogeneous load response across different wind covariates. Assuming the load response stationary or homogenous in a narrow range within a wind speed bin is much more reasonable than assuming stationarity or homogeneity for the entire wind spectrum.

The common binning approach, as Agarwal and Manuel [3] use it to estimate the extreme loads for a 2MW offshore wind turbine, is to use the Gumbel distribution to model the probabilistic behavior of bending moments at critical spots on a turbine structure. To estimate the model parameters, experimental data are collected but only for a short period, say one or two years. As a result, most bins have a small number of data, or sometimes, no data at all. For the bins without data, Agarwal and Manuel estimate the short-term distribution parameters by using a weighted average of all non-empty bins with the weight

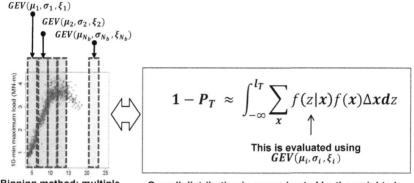

FIGURE 10.5 Binning approach to combine bin-based homogenous GEV distributions for extreme load analysis over the entire wind spectrum. In implementation, ξ_i, $i = 1, \ldots, N_b$, is chosen to be the same constant for all bins.

related to the inverse squared distance between bins. They quantify the uncertainty of the estimated extreme loads using bootstrapping and report the 95% confidence interval for the short-term extreme load given specific weather conditions (i.e., weather bins). Because bootstrapping resamples the existing data for a given weather bin, it cannot precisely capture the uncertainty for those bins with limited data or without data.

Lee et al. [131] present explicit steps to calculate the confidence interval for the binning method. First, they fix ξ to be a constant across all bins. The typical choice of ξ is zero, meaning that a Gumbel distribution is used. But ξ can also be estimated *a priori* and then remains fixed while other parameters are being estimated. After ξ is fixed, the resulting GEV distribution has then two bin-specific parameters, μ and σ. Denote by $\mathbf{\Phi}_c$ the collection of parameters associated with all local GEV distributions, i.e., $\{\mu_1, \sigma_1, \ldots \mu_{N_b}, \sigma_{N_b}, \xi\}$, where N_b is the number of bins, and by $\mathcal{D}_V$ and $\mathcal{D}_s$ the datasets of the observed average wind speeds and the standard deviations. The sampling process is elaborated in Algorithm 10.1.

Despite its popularity, the binning method has obvious shortcomings in estimating extreme loads. A major limitation is that the short-term load distribution in one bin is constructed separately from the short-term distributions in other bins. This approach requires an enormous amount of data to define the tail of each short-term distribution. In reality, the field data can only be collected in a short duration (e.g., one year out of 50 years of service), and consequently, some bins do not have enough data. Then, the binning method may end up with inaccuracy or high uncertainty in the estimates of extreme

---

Algorithm 10.1 Sampling procedure to construct the confidence interval for the binning method. Set $M_w = 1,000$, $M_l = 10,000$, $N_w = 100$, and $N_l = 100$.

---

1. Draw $M_w \times N_w$ samples from the joint distribution $f(\tilde{V}, \tilde{s}|\mathcal{D}_V, \mathcal{D}_s)$ of wind characteristics $(\tilde{V}, \tilde{s})$, where the tilde notation indicates a sampled quantity. Please cross reference Algorithm 10.4 for specific models and steps to draw samples for wind characteristics $(\tilde{V}, \tilde{s})$.

2. Using the data in a bin, employ an MLE method to estimate μ and σ in the GEV, while fixing ξ. Draw a sample of μ and σ for that specific bin from a multivariate normal distributions taking the MLE as its mean and the inverse of the negative of Hessian matrix as its covariance matrix. Not all the bins have data. For those which do not have data, its μ and σ are a weighted average of all non-empty bins with the weight related to the inverse squared distance between bins. Collectively, $\boldsymbol{\Phi}_c$ contains the μ's and σ's from all the bins.

3. Decide which bins the wind characteristic samples $(\tilde{V}, \tilde{s})$'s fall into. Based on the specific bin in which a sample of $(\tilde{V}, \tilde{s})$ falls, the corresponding μ and σ in $\boldsymbol{\Phi}_c$ is chosen. Doing so yields the short-term distribution $f(\tilde{z}|\tilde{V}, \tilde{s}, \boldsymbol{\Phi}_c)$ for that specific bin.

4. Draw N_l samples of $\tilde{z}$ from $f(\tilde{z}|\tilde{V}, \tilde{s}, \boldsymbol{\Phi}_c)$ for each of the total $M_w \times N_w$ samples of $(\tilde{V}, \tilde{s})$. This produces a total of $M_w \times N_w \times N_l$ samples of $\tilde{z}$.

5. One can then compute the quantile value $l_T[\boldsymbol{\Phi}_c]$ corresponding to P_T.

6. Repeat the above procedure M_l times to get the median and confidence interval of l_T.

---

load. In practice, how many bins to use is also under debate, and there is not yet a consensus. The answer to the action of binning sometimes depends on the amount of data—if an analyst has more data, he/she affords to use more bins; otherwise, fewer bins.

The popularity of the binning method in industrial practice is due to the simplicity of its idea and procedure. However, simplicity of a procedure should not be mistaken as simplicity of the resulting model. Suppose that one uses a 6×10 grid to bin the two-dimensional wind covariates and fixes the shape parameter ξ across the bins (a common practice in the industry). The binning method yields 60 local GEV distributions, each of which has two parameters, translating to a total of 121 parameters for the overall model (counting the fixed ξ as well). A model having 121 parameters is not a simple model. The combination of the rigidity of compartmentalization and the unintended high model complexity renders the binning method not scalable and less effective.

10.4 BAYESIAN SPLINE-BASED GEV MODEL

Lee et al. [131] present a Bayesian spline method for estimating the extreme load on wind turbines. The spline method is essentially a method supporting an inhomogeneous GEV distribution to capture the nonlinear relationship between the load response and the wind-related covariates. Such treatment avoids binning the data. The underlying spline model connect all the bins across the whole wind profile, so that load and wind data are pooled together to produce better estimates. The merit of the spline model is demonstrated in Section 10.6 by applying it to three sets of turbine load response data and making comparisons with the binning method.

10.4.1 Conditional Load Model

Recall that in the binning method, a homogeneous GEV distribution is used to model the short-term load distribution in a bin, for it appears reasonable to assume stationarity if the chosen weather bin is narrow enough. A finite number of the homogeneous GEV distributions are then stitched together to represent the nonstationary nature across the entire wind profile; see Fig. 10.5. What Lee et al. [131] propose is to abandon the bins and instead use an inhomogeneous GEV distribution whose parameters are not constant but depend on weather conditions.

Consider 10-minute maximum loads, $z_1, \ldots, z_n$, with corresponding covariate variables $\boldsymbol{x}_1 = (V_1, s_1)$, ..., $\boldsymbol{x}_n = (V_n, s_n)$, as defined in Eq. 10.1. Let us consider modeling z_i conditional on $\boldsymbol{x}$, such that

$$z_i | \boldsymbol{x}_i \sim \text{GEV}(\mu(\boldsymbol{x}_i), \sigma(\boldsymbol{x}_i), \xi), \quad \sigma(\cdot) > 0, \tag{10.9}$$

where the location parameter μ and scale parameter σ in this GEV distribution are a nonlinear function of wind characteristics $\boldsymbol{x}$. The shape parameter ξ is fixed across the wind profile, while its value will still be estimated using the data from a specific wind turbine. The reason behind the fixed ξ is to keep the final model from becoming overly flexible. Too much flexibility could cause complexity in model fitting and parameter estimation.

Let us denote $\mu(\boldsymbol{x}_i)$ and $\sigma(\boldsymbol{x}_i)$ by

$$\mu(\boldsymbol{x}_i) = q(\boldsymbol{x}_i), \tag{10.10}$$

$$\sigma(\boldsymbol{x}_i) = \exp(g(\boldsymbol{x}_i)), \tag{10.11}$$

where an exponential function is used in Eq. 10.11 to ensure the positivity of the scale parameter. To capture the nonlinearity between the load response and the wind-related covariates, Lee et al. [131] model $q(\cdot)$ and $g(\cdot)$ using a Bayesian MARS model [46, 47]. Recall the discussion about splines in Section 5.3.3. A shortcoming of the spline methods is its lack of scalability to model multivariate inputs. One of the methods that addresses this issue is the MARS model [68], which uses an additive model structure, allowing factor interactions to be added through a hierarchical inclusion of interaction terms for the purpose of accomplishing scalability. The Bayesian MARS model is basically a MARS model but includes the number and locations of knots as part of its model parameters and determines these from observed data.

Lee at al. [131] state that they explore simple approaches based on polynomial models for modeling $\mu(\boldsymbol{x})$ and $\sigma(\boldsymbol{x})$. It turns out that polynomial-based approaches lack the flexibility of adapting to the datasets from different types of turbines. Due to the nonlinearity around the rated wind speed and the limited amount of data under high wind speeds, polynomial-based approaches perform poorly in those regions that are generally important for capturing the maximum load. Spline models, on the other hand, appear to work better than a global polynomial model, because they have more supporting points spreading over the input region.

The Bayesian MARS models, i.e., $q(\boldsymbol{x})$ for the location parameter μ and $g(\boldsymbol{x})$ for the scale parameter σ, are represented as a linear combination of the basis functions $B_k^{\mu}(\boldsymbol{x})$ and $B_k^{\sigma}(\boldsymbol{x})$, respectively, such that

$$q(\boldsymbol{x}) = \sum_{k=1}^{K_{\mu}} \beta_k B_k^{\mu}(\boldsymbol{x}), \quad \text{and} \tag{10.12}$$

$$g(\boldsymbol{x}) = \sum_{k=1}^{K_{\sigma}} \theta_k B_k^{\sigma}(\boldsymbol{x}), \tag{10.13}$$

where $\beta_k, k = 1, \ldots, K_{\mu}$ and $\theta_k, k = 1, \ldots, K_{\sigma}$ are the coefficients of the basis functions $B_k^{\mu}(\cdot)$ and $B_k^{\sigma}(\cdot)$, respectively, and K_{μ} and K_{σ} are the number of the respective basis functions. According to Denison et al. [47] who propose the Bayesian MARS method, the basis functions should be specified as

$$B_k(\boldsymbol{x}) = \begin{cases} 1, & k = 1, \\ \prod_{j=1}^{J_k} \left[h_{jk} \cdot (x_{r(j,k)} - t_{jk}) \right]_+, & k = 2, 3, \ldots, K_{\mu} \text{ or } K_{\sigma}. \end{cases} \tag{10.14}$$

Here, $[\cdot]_+ = \max(0, \cdot)$, J_k is the degree of interaction modeled by the basis function $B_k(\boldsymbol{x})$, h_{jk} is the sign indicator, taking the value of either -1 or $+1$, $r(j, k)$ produces the index of the predictor variable which is being split on t_{jk}, whereas t_{jk} is commonly referred to as the knot points.

Lee et al. [131] introduce an integer variable T_k to represent the types of basis functions used in Eq. 10.14. Since two input variables, V and s, are considered for the three inland turbines, there could be three types of basis functions, namely $[\pm(V - *)]_+$ or $[\pm(s - *)]_+$ for the main effect of a respective explanatory variable and $[\pm(V - *)]_+[\pm(s - *)]_+$ for the interactions between them. Let T_k take the integer value of 1, 2, or 3, to represent the three types of basis functions. That is, $[\pm(V - *)]_+$ is represented by $T_k = 1$, $[\pm(s - *)]_+$ represented by $T_k = 2$, and $[\pm(V - *)]_+[\pm(s - *)]_+$ by $T_k = 3$.

To model the location parameter μ for ILT1 and ILT3 data, Lee et al. [131] set $T_k \in \{1, 2, 3\}$, allowing J_k to take either 1 or 2. For ILT2, however, due to its relatively smaller data amount, a model setting $J_k = 2$ produces unstable and unreasonably wide credible intervals. Consequently, Lee et al. set $T_k \in \{1, 2\}$, restricting $J_k = 1$ for ILT2's location parameter μ. For the scale parameter σ, $J_k = 1$ is used for all three datasets. For ITL1 and ILT3, $J_k = 1$ is resulted when setting $T_k \in \{1, 2\}$. For ILT2, again due to its data scarcity, Lee et al. include V as the only input variable in the corresponding scale parameter model; this means $T_k = \{1\}$.

Let $\boldsymbol{\Psi}_a = (\boldsymbol{\Psi}_\mu, \boldsymbol{\Psi}_\sigma, \xi)$ denote all the parameters used in the GEV model in Eq. 10.9, where $\boldsymbol{\Psi}_\mu$ and $\boldsymbol{\Psi}_\sigma$ include the parameters in $q(\cdot)$ and $g(\cdot)$, respectively. These parameters are grouped into two sets: (1) the coefficients of the basis functions in $\boldsymbol{\beta} = (\beta_1, \ldots, \beta_{K_\mu})$ or $\boldsymbol{\theta} = (\theta_1, \ldots, \theta_{K_\sigma})$, and (2) the number of knots, the locations of the knots, and the types of basis function in ϕ_μ or ϕ_σ, as follows,

$$\phi_\mu = \left(K_\mu, \boldsymbol{\Lambda}_2^\mu, \ldots, \boldsymbol{\Lambda}_{K_\mu}^\mu \right), \tag{10.15}$$

where

$$\boldsymbol{\Lambda}_k^\mu = \begin{cases} (T_k^\mu, h_{1k}^\mu, t_{1k}^\mu), & \text{when} \quad T_k^\mu = 1, 2, \\ (T_k^\mu, h_{1k}^\mu, h_{2k}^\mu, t_{1k}^\mu, t_{2k}^\mu), & \text{when} \quad T_k^\mu = 3, \end{cases}$$

and

$$\phi_\sigma = \left(K_\sigma, \boldsymbol{\Lambda}_2^\sigma, \ldots, \boldsymbol{\Lambda}_{K_\sigma}^\sigma \right), \tag{10.16}$$

where

$$\boldsymbol{\Lambda}_k^\sigma = (T_k^\sigma, h_{1k}^\sigma, t_{1k}^\sigma), \quad \text{when} \quad T_k^\sigma = 1, 2.$$

Using the above notations, one can express $\boldsymbol{\Psi}_\mu = (\boldsymbol{\beta}, \phi_\mu)$ and $\boldsymbol{\Psi}_\sigma = (\boldsymbol{\theta}, \phi_\sigma)$.

To complete the Bayesian formulation for the model in Eq. 10.9, priors of the parameters should be specified. Lee et al. [131] use uniform priors on ϕ_μ and ϕ_σ. In the following expressions, we drop the subscript or superscript indicating the association with μ or σ for the sake of notational simplicity, since the priors for both cases are the same. The following priors are used for variables in ϕ:

$$f(K) = \frac{1}{n}, \qquad K \in \{1, \ldots, n\}$$

$$f(T_k) = \begin{cases} 1, & T_k \in \{1\} & \text{for } \phi_\sigma \text{ in ILT2,} \\ \frac{1}{2}, & T_k \in \{1,2\} & \text{for } \phi_\mu \text{ in ILT2 and all other } \phi'_\sigma s, \\ \frac{1}{3}, & T_k \in \{1,2,3\} & \text{for } \phi_\mu \text{ in ILT1 and ILT3,} \end{cases}$$

$$f(h_{\cdot k}) = \frac{1}{2}, \qquad h_{\cdot k} \in \{+1, -1\},$$

$$f(t_{\cdot k}) = \frac{1}{n}, \qquad t_{\cdot k} \in \{V_1, \ldots, V_n\} \text{ or } \{s_1, \ldots, s_n\}.$$

In the above, the dot notation in the expressions of $h_{\cdot k}$ and $t_{\cdot k}$ denotes either 1 or 2.

Given ϕ_μ and ϕ_σ, Lee et al. [131] specify the prior distribution for (β, θ, ξ) as the unit information prior (UIP) [118], which is defined by setting the corresponding covariance matrix to be equal to the Fisher information of one observation. This is accomplished by using a multivariate normal prior distribution with its mean set at the maximum likelihood estimate and its covariance matrix as the inverse of the negative of Hessian matrix.

10.4.2 Posterior Distribution of Parameters

The Bayesian MARS model treats the number and locations of the knots as random quantities. When the number of knots changes, the dimension of the parameter space changes with it. To handle a varying dimensionality in the probability distributions in a random sampling procedure, analysts use a reversible jump Markov chain Monte Carlo (RJMCMC) algorithm developed by Green [78]. The acceptance probability for an RJMCMC algorithm includes a Jacobian term, which accounts for the change in dimension. However, under the assumption that the model space for parameters of varying dimension is discrete, there is no need for a Jacobian. In the turbine extreme load analysis, this assumption is satisfied since only are the probable models over possible knot locations and numbers considered. Instead of using the RJMCMC algorithm, Lee et al. [131] use the reversible jump sampler (RJS) algorithm proposed in [46]. Because the RJS algorithm does not require new parameters to match dimensions between models nor the corresponding Jacobian term in the acceptance probability, it is simpler and more efficient to execute.

To allow for dimensional changes, there are three actions in an RJS algorithm [46, page 53]: BIRTH, DEATH and MOVE, which adds, deletes, or alters a basis function, respectively. Accordingly, the number of knots as well as the locations of some knots change. Denison et al. [46] suggest using equal probability, i.e., 1/3, to propose any of the three moves, and then, use the following acceptance probability, α, while executing a proposed move from a model having k basis functions to a model having k^c basis functions:

$$\alpha = \min\{1, \text{the ratio of marginal likelihood} \times \mathcal{R}\}, \tag{10.17}$$

where $\mathcal{R}$ is a ratio of probabilities defined as:

- For a BIRTH action, $\mathcal{R} = \dfrac{\text{probability of DEATH in model } k^c}{\text{probability of BIRTH in model } k}$;

- For a DEATH action, $\mathcal{R} = \dfrac{\text{probability of BIRTH in model } k^c}{\text{probability of DEATH in model } k}$;

- For a MOVE action, $\mathcal{R} = \dfrac{\text{probability of MOVE in model } k^c}{\text{probability of MOVE in model } k}$.

Lee et al. [131] state that they have $\mathcal{R} = 1$ for most cases, because the probabilities in the denominator and numerator are equal, except when k reaches either the upper or the lower bound.

The marginal likelihood in Eq. 10.17 is expressed as

$$f\left(\mathcal{D}_z|\phi_\mu, \phi_\sigma\right) = \int f\left(\mathcal{D}_z|\beta, \theta, \xi, \phi_\mu, \phi_\sigma\right) f\left(\beta, \theta, \xi|\phi_\mu, \phi_\sigma\right) d\beta d\theta d\xi, \tag{10.18}$$

where $\mathcal{D}_z = (z_1, \ldots, z_n)$ represents a set of observed load data. Since it is difficult to calculate the above marginal likelihood analytically, Lee et al. [131] consider an approximation of $f\left(\mathcal{D}_z|\phi_\mu, \phi_\sigma\right)$. Kass and Wasserman [118] and Raftery [171] show that when UIPs are used, the marginal log-likelihood, i.e., $\log\left(f\left(\mathcal{D}_z|\phi_\mu, \phi_\sigma\right)\right)$, can be reasonably approximated by the Schwarz information criterion (SIC) [197], also known as BIC; please refer to Eq. 2.23.

The SIC is expressed as

$$\text{SIC}_{\phi_\mu, \phi_\sigma} = \log\left(f(\mathcal{D}_z|\hat{\beta}, \hat{\theta}, \hat{\xi}, \phi_\mu, \phi_\sigma)\right) - \frac{1}{2}d_k \log(n), \tag{10.19}$$

where $\hat{\beta}, \hat{\theta}, \hat{\xi}$ are the MLEs of the corresponding parameters obtained conditional on ϕ_μ and ϕ_σ, and d_k is the total number of parameters to be estimated. In this case, $d_k = K_\mu + K_\sigma + 1$ (the inclusion of the last 1 is due to ξ).

Comparing Eq. 10.19 with Eq. 2.23, one may notice that the two expressions are indeed equivalent but differ by a constant of -2. Note that in Chapter 2, a smaller BIC implies a better model fit to data. Here, a larger SIC suggests a better model fit, because of this -2 difference.

There are two dimension-varying states, ϕ_μ and ϕ_σ, in the RJS algorithm. Consequently, two marginal log-likelihood ratios are needed. They are approximated by the corresponding SICs, such as

$$\log \frac{f\left(\mathcal{D}_z|\phi_\mu^c, \phi_\sigma\right)}{f\left(\mathcal{D}_z|\phi_\mu, \phi_\sigma\right)} \simeq \text{SIC}_{\phi_\mu^c, \phi_\sigma} - \text{SIC}_{\phi_\mu, \phi_\sigma}, \tag{10.20}$$

and

$$\log \frac{f\left(\mathcal{D}_z|\phi_\mu, \phi_\sigma^c\right)}{f\left(\mathcal{D}_z|\phi_\mu, \phi_\sigma\right)} \simeq \text{SIC}_{\phi_\mu, \phi_\sigma^c} - \text{SIC}_{\phi_\mu, \phi_\sigma}. \tag{10.21}$$

Then, one uses two acceptance probabilities α_μ and α_σ for accepting or rejecting a new state in ϕ_μ and ϕ_σ, respectively. Using the SICs, α_μ and α_σ are expressed as:

$$\alpha_\mu = \min\left\{1, \exp\left(\mathrm{SIC}_{\phi_\mu^c, \phi_\sigma} - \mathrm{SIC}_{\phi_\mu, \phi_\sigma}\right) \times \mathcal{R}\right\}, \qquad (10.22)$$

and

$$\alpha_\sigma = \min\left\{1, \exp\left(\mathrm{SIC}_{\phi_\mu, \phi_\sigma^c} - \mathrm{SIC}_{\phi_\mu, \phi_\sigma}\right) \times \mathcal{R}\right\}. \qquad (10.23)$$

In order to produce the samples from the posterior distribution of parameters in $\boldsymbol{\Psi}_a$, Lee et al. [131] sequentially draw samples for ϕ_μ and ϕ_σ by using the two acceptance probabilities while marginalizing out $(\boldsymbol{\beta}, \boldsymbol{\theta}, \xi)$, and then, conditional on the sampled ϕ_μ and ϕ_σ, draw samples for $(\boldsymbol{\beta}, \boldsymbol{\theta}, \xi)$ using a normal approximation based on the maximum likelihood estimates and the observed information matrix.

10.4.3 Wind Characteristics Model

To find a site-specific load distribution, the distribution of wind characteristics $f(\boldsymbol{x})$ in Eq. 10.4 needs to be specified. Since a statistical correlation is noticed in Fig. 10.4 between the 10-minute average wind speed, V, and the standard deviation of wind speed, s, the distribution of wind characteristics $f(\boldsymbol{x})$ can be written as a product of the average wind speed distribution $f(V)$ and the conditional wind standard deviation distribution $f(s|V)$.

The probabilistic distribution of wind speed, $f(V)$, is discussed in Chapter 2. At that time, the discussion concentrates on Weibull distribution. The three-parameter Weibull distribution fits the three wind turbine datasets well, as one will see in Section 10.6.1, and is in fact the one used in the case study.

For modeling the 10-minute average wind speed V, the IEC standard suggests using a two-parameter Weibull distribution (W2) or a Rayleigh distribution (RAY) [101]. These two distributions are arguably the most widely used ones for this purpose. But analysts [31, 134] note that under different wind regimes other distributions may fit wind speed data better, including the three-parameter Weibull distribution (W3), three-parameter log-normal distribution (LN3), three-parameter Gamma distribution (G3), and three-parameter inverse-Gaussian distribution (IG3).

What Lee et al. [131] suggest to do is to take the total of six candidate distribution models for average wind speed (W2, W3, RAY, LN3, G3, IG3) and conduct a Bayesian model selection to choose the best distribution fitting a given average wind speed dataset. Lee et al. assume UIP for the parameters involved in the aforementioned models, and as such, the Bayesian model selection is again based on maximizing the SIC. The chosen best wind speed model is denoted by $\mathcal{M}_V$. Then, the distribution of 10-minute average wind speed V is expressed as

$$V_i \sim \mathcal{M}_V(\boldsymbol{\nu}), \qquad (10.24)$$

where $\boldsymbol{\nu}$ is the set of parameters specifying $\mathcal{M}_V$. For instance, if $\mathcal{M}_V$ is W3, then $\nu = (\nu_1, \nu_2, \nu_3)$, where ν_1, ν_2, and ν_3 represent, respectively, the location, scale, and shape parameter of the three-parameter Weibull distribution.

For modeling the standard deviation of wind speed s, given the average wind speed V, the IEC standard [101] recommends using a two-parameter truncated normal distribution (TN2), which appears to be what analysts have commonly used [63]. The distribution is characterized by a location parameter η and a scale parameter δ. In the literature, both η and δ are treated as a constant. But Lee et al. [131] observe that datasets measured at different sites have different relationships between the average wind speed V and the standard deviation s. Some of the V-versus-s scatter plots show nonlinear patterns.

Motivated by this observation, Lee et al. [131] employ a Bayesian MARS model for modeling η and δ, similar to what is done in Section 10.4.1 for the conditional load model. The standard deviation of wind speed s, conditional on the average wind speed V, can then be expressed as

$$s_i | V_i \sim \text{TN2}(\eta(V_i), \delta(V_i)), \tag{10.25}$$

where $\eta(V_i) = q_\eta(V_i)$ and $\delta(V_i) = \exp(g_\delta(V_i))$, like their counterparts in Eq. 10.10 to Eq. 10.13, are linear combinations of the basis functions taking the general form as in Eq. 10.14. Notice that both of the functions have only one input variable, which is the average wind speed.

Let $\boldsymbol{\Psi}_\eta = (\boldsymbol{\beta}_\eta, \boldsymbol{\phi}_\eta)$ and $\boldsymbol{\Psi}_\delta = (\boldsymbol{\theta}_\delta, \boldsymbol{\phi}_\delta)$ denote the parameters in $q_\eta(\cdot)$ and $g_\delta(\cdot)$. Since the basis functions for q_η and g_δ have a single input variable, only one type of basis function is needed, i.e., $T_k = 1$. For this reason, $\boldsymbol{\phi}_\eta$ and $\boldsymbol{\phi}_\delta$ are much simpler than $\boldsymbol{\phi}_\mu$ and $\boldsymbol{\phi}_\sigma$, their counterparts in Eq. 10.15 and Eq. 10.16, and are expressed as follows:

$$\boldsymbol{\phi}_\eta = \left(K_\eta, \boldsymbol{\Lambda}_2^\eta, \ldots, \boldsymbol{\Lambda}_{K_\eta}^\eta \right),$$
$$\text{where } \boldsymbol{\Lambda}_k^\eta = (T_k^\eta, h_{1k}^\eta, t_{1k}^\eta) \quad \text{and} \quad T_k^\eta = 1; \tag{10.26}$$

and

$$\boldsymbol{\phi}_\delta = \left(K_\delta, \boldsymbol{\Lambda}_2^\delta, \ldots, \boldsymbol{\Lambda}_{K_\delta}^\delta \right),$$
$$\text{where } \boldsymbol{\Lambda}_k^\delta = (T_k^\delta, h_{1k}^\delta, t_{1k}^\delta) \quad \text{and} \quad T_k^\delta = 1. \tag{10.27}$$

Lee et al. [131] choose the prior distribution for $(\boldsymbol{\beta}_\eta, \boldsymbol{\theta}_\delta)$ as UIP, the prior for $(\boldsymbol{\phi}_\eta, \boldsymbol{\phi}_\delta)$ as uniform distribution, and set $f(T_k) = 1$ because T_k is always 1. They solve this Bayesian MARS model using an RJS algorithm, as in the preceding two sections.

The predictive distributions of the average wind speed $\tilde{V}$ and the standard deviation $\tilde{s}$ are

$$f(\tilde{V} | \mathcal{D}_V) = \int f(\tilde{V} | \boldsymbol{\nu}, \mathcal{D}_V) f(\boldsymbol{\nu} | \mathcal{D}_V) d\boldsymbol{\nu}, \tag{10.28}$$

and

$$f(\tilde{s} | \tilde{V}, \mathcal{D}_V, \mathcal{D}_s) = \int \int f(\tilde{s} | \tilde{V}, \boldsymbol{\Psi}_\eta, \boldsymbol{\Psi}_\delta, \mathcal{D}_V, \mathcal{D}_s) f(\boldsymbol{\Psi}_\eta, \boldsymbol{\Psi}_\delta | \mathcal{D}_V, \mathcal{D}_s) d\boldsymbol{\Psi}_\eta d\boldsymbol{\Psi}_\delta. \tag{10.29}$$

10.4.4 Posterior Predictive Distribution

Analysts are interested in getting the posterior predictive distribution of the quantile value l_T, based on the observed load and wind data $\mathcal{D} :=$ $(\mathcal{D}_z, \mathcal{D}_V, \mathcal{D}_s)$. Under a Bayesian framework, one draws samples, $\tilde{z}$'s, from the predictive distribution of the maximum load, $f(\tilde{z}|\mathcal{D}, \boldsymbol{\Psi}_a)$, which is

$$f(\tilde{z}|\mathcal{D}, \boldsymbol{\Psi}_a) = \int \int f(\tilde{z}|\tilde{V}, \tilde{s}, \boldsymbol{\Psi}_a, \mathcal{D}) f(\tilde{V}, \tilde{s}|\mathcal{D}_V, \mathcal{D}_s) d\tilde{V} d\tilde{s}, \quad (10.30)$$

where $f(\tilde{V}, \tilde{s}|\mathcal{D}_V, \mathcal{D}_s)$ can be expressed as the product of Eq. 10.28 and Eq. 10.29.

To calculate a quantile value of the load for a given P_T, one goes through the steps in Algorithm 10.2. The predictive mean and Bayesian credible interval of the extreme load level, l_T, are obtained when running the RJS algorithm. The RJS runs through M_l iterations, and at each iteration, one obtains a set of samples of the model parameters, $\boldsymbol{\Psi}_a$, and calculates an $l_T[\boldsymbol{\Psi}_a]$. Once the M_l values of $l_T[\boldsymbol{\Psi}_a]$ are obtained, the mean and credible interval of l_T can then be numerically computed.

Algorithm 10.2 Sampling procedure to obtain the posterior predictive distribution of load response z. Set $M_w = 1,000$, $M_l = 10,000$, $N_w = 100$, and $N_l = 100$.

---

1. Draw $M_w \times N_w$ samples from the joint posterior predictive distribution $f(\tilde{V}, \tilde{s}|\mathcal{D}_V, \mathcal{D}_s)$ of wind characteristics $(\tilde{V}, \tilde{s})$. This is realized by employing Algorithm 10.4;

2. Draw a set of samples from the posterior distribution of model parameters $\boldsymbol{\Psi}_a = (\boldsymbol{\Psi}_\mu, \boldsymbol{\Psi}_\sigma, \xi)$. This is realized by employing the RJS algorithm in Section 10.4.2 (or Steps 1–11 of Algorithm 10.3);

3. Given the above samples of wind characteristics and model parameters, one calculates (μ, σ, ξ) that are needed in a GEV distribution. This yields a short-term distribution $f(\tilde{z}|\tilde{V}, \tilde{s}, \boldsymbol{\Psi}_a)$;

4. Integrate out the wind characteristics $(\tilde{V}, \tilde{s})$, as implied in Eq. 10.30, to obtain the long-term distribution $f(\tilde{z}|\mathcal{D}, \boldsymbol{\Psi}_a)$.

5. Draw $N_l \times M_w \times N_w$ samples from $f(\tilde{z}|\mathcal{D}, \boldsymbol{\Psi}_a)$ and compute the quantile value $l_T[\boldsymbol{\Psi}_a]$ corresponding to P_T.

6. Repeat the above procedure M_l times to get the median and confidence interval of l_T.

10.5 ALGORITHMS USED IN BAYESIAN INFERENCE

In this section, more details of the implementation procedures are provided to facilitate the Bayesian inference.

The procedure consists of two main parts: Algorithm 10.3, which is to construct the posterior predictive distribution of the extreme load level l_T, and Algorithm 10.4, which is to obtain the posterior predictive distribution of wind characteristics (V, s). The main algorithms use the RJS subroutine for the location parameter μ and the scale parameter σ. These two subroutines are separately listed in Algorithms 10.5 and 10.6. The two subroutines look the same but differ in terms of the specific variables and parameters used therein.

Algorithm 10.2 in Section 10.4.4 carries out the same task as Algorithm 10.3 does. The difference is that Algorithm 10.2 outlines the main steps, whereas Algorithm 10.3 presents more detailed steps.

10.6 CASE STUDY

This section presents numerical analysis of extreme loads recorded in the **Turbine Bending Moment Dataset** and discusses the difference between the spline-based approach and the binning-based approach.

10.6.1 Selection of Wind Speed Model

The first task is to select a model, out of the six candidate models mentioned in Section 10.4.3, for the average wind speed. This model selection is done using the SIC.

Table 10.2 presents the SIC values of the six candidate average wind speed models using a respective ILT dataset. The boldfaced values indicate the largest SIC for a given dataset, and accordingly, the corresponding model is chosen for that dataset.

Regarding the average wind speed model, all candidate distributions except RAY provide generally a good model fit for ILT1 with a similar level of fitting quality, but W3 outperforms others slightly. For the ILT2 data, W2, W3, LN3 and G3 produce similar SIC values. In the ILT3 data, W3, LN3, G3 and IG3 perform similarly. Again W3 is slightly better. For this reason, W3 is chosen as the average wind speed model.

10.6.2 Pointwise Credible Intervals

As a form of checking the conditional maximum load model, Lee et al. [131] produce the 95% pointwise credible intervals of the load response under different wind speeds and standard deviations. The resulting credible intervals are presented in Figs. 10.6 and 10.7.

To generate these figures, Lee et al. [131] take a dataset and fix V or s at one specific speed or standard deviation at a time and then draw the posterior

---

Algorithm 10.3 Construct the posterior predictive distribution of the extreme load level using the Bayesian spline models. Set $M_w = 1,000$, $M_l = 10,000$, $N_w = 100$, and $N_l = 100$.

---

1. Set $t = 0$ and the initial $\phi_\mu^{(t)}$ and $\phi_\sigma^{(t)}$ both to be a constant scalar.

2. At iteration t, K_μ and K_σ are equal to the number of basis functions specified in $\phi_\mu^{(t)}$ and $\phi_\sigma^{(t)}$. Find the MLEs of $\beta^{(t)}, \theta^{(t)}, \xi^{(t)}$ and the inverse of the negative of Hessian matrix, given $\phi_\mu^{(t)}$ and $\phi_\sigma^{(t)}$.

3. Generate u_μ^1 uniformly on $[0,1]$ and choose a move in the RJS procedure. Denote by $b_{K_\mu}, r_{K_\mu}, m_{K_\mu}$ the proposal probabilities associated with a move type; they are all set as $\frac{1}{3}$. Call Algorithm 10.5 to execute the RJS procedure.

4. Find the MLEs $(\beta^*, \theta^*, \xi^*)$ and the inverse of the negative of Hessian matrix, given ϕ_μ^* and $\phi_\sigma^{(t)}$.

5. Generate u_μ^2 uniformly on $[0,1]$ and compute the acceptance ratio α_μ in Eq. 10.22, using the results from Step 2 and Step 4.

6. Accept ϕ_μ^* as $\phi_\mu^{(t+1)}$ with probability $\min(\alpha_\mu, 1)$. If ϕ_μ^* is not accepted, let $\phi_\mu^{(t+1)} = \phi_\mu^{(t)}$.

7. Generate u_σ^1 uniformly on $[0,1]$ and choose a move in the RJS procedure. Denote by $b_{K_\sigma}, r_{K_\sigma}, m_{K_\sigma}$ the proposal probabilities associated with a move type; they are all set as $\frac{1}{3}$. Call Algorithm 10.6 to execute the RJS procedure.

8. Find the MLEs $(\beta^*, \theta^*, \xi^*)$ and the inverse of the negative of Hessian matrix, given $\phi_\mu^{(t+1)}$ and ϕ_σ^*.

9. Generate u_σ^2 uniformly on $[0,1]$ and compute the acceptance ratio α_σ in Eq. 10.23, using the results from Step 4 and Step 8.

10. Accept ϕ_σ^* as $\phi_\sigma^{(t+1)}$ with probability $\min(\alpha_\sigma, 1)$. If ϕ_σ^* is not accepted, let $\phi_\sigma^{(t+1)} = \phi_\sigma^{(t)}$.

11. After initial burn-ins (set to 1,000 samples), draw a posterior sample of $(\beta^{(t+1)}, \theta^{(t+1)}, \xi^{(t+1)})$ from the approximated multivariate normal distribution at the maximum likelihood estimates and the inverse of the negative of Hessian matrix. Depending on the acceptance or rejection that happened in Step 6 and Step 10, the MLEs to be used are obtained from either Step 2, Step 4, or Step 8.

12. Take the posterior sample of Ψ_a, obtained in Step 6, Step 10, and Step 11, and calculate a sample of μ and σ using Eq. 10.10 and Eq. 10.11, respectively, for each pair of the $M_w \times N_w$ samples of (V, s) obtained in Algorithm 10.4. This generates $M_w \times N_w$ samples of μ and σ.

13. Draw N_l samples for the 10-minute maximum load $\tilde{z}$ from each GEV distribution with μ_i, σ_i, and ξ_i, $i = 1, \ldots, M_w \times N_w$, where μ_i and σ_i are among $M_w \times N_w$ samples obtained in Step 12, and ξ_i is always set as $\xi^{(t+1)}$.

14. Get the quantile value (that is, the extreme load level $l_T[\Psi_a]$) corresponding to $1 - P_T$ from the $M_w \times N_w \times N_l$ samples of $\tilde{z}$.

15. To obtain a credible interval for l_T, repeat Step 2 through Step 14 M_l times.

Algorithm 10.4 Obtain the posterior predictive distribution of wind characteristics (V, s). Set $M_w = 1,000$ and $N_w = 100$.

1. Find the MLEs of $\boldsymbol{\nu}$ for all candidate distributions listed in Section 10.4.3.

2. Use the SIC to select the "best" distribution model for the average wind speed V. The chosen distribution model is used in the subsequent steps to draw posterior samples.

3. Draw a posterior sample of $\boldsymbol{\nu}$ from the approximated multivariate normal distribution at the MLEs and the inverse of the negative of Hessian matrix.

4. Draw N_w samples of $\tilde{V}$ using the distribution chosen in Step 2 with the parameter sampled in Step 3.

5. Implement the RJS algorithm again, namely Step 1 through Step 11 in Algorithm 10.3, to get one posterior sample of $\boldsymbol{\Psi}_\eta = (\boldsymbol{\beta}_\eta, \boldsymbol{\phi}_\eta)$ and $\boldsymbol{\Psi}_\delta = (\boldsymbol{\theta}_\delta, \boldsymbol{\phi}_\delta)$.

6. Take the posterior sample of $\boldsymbol{\Psi}_\eta$ and $\boldsymbol{\Psi}_\delta$, obtained in Step 5, and calculate a sample of η and δ using Eq. 10.25 for each sample of $\tilde{V}$. This generates N_w samples of η and δ.

7. Draw a sample for the standard deviation of wind speed $\tilde{s}$ from each truncated normal distribution with η_i, δ_i, $i = 1, \ldots, N_w$. Using the N_w samples of η and δ obtained in Step 6, one obtains N_w samples of $\tilde{s}$.

8. To get $M_w \times N_w$ samples of $\tilde{V}$ and $\tilde{s}$, repeat Step 3 through Step 7 M_w times.

Algorithm 10.5 Three types of move in the RJS for location parameter μ.

1. If $u_\mu^1 \leq b_{K_\mu}$, then go to BIRTH step, denoted by $\phi_\mu^* =$ BIRTH-proposal($\phi_\mu^{(t)}$), which is to augment $\phi_\mu^{(t)}$ with a $\Lambda_{K_\mu+1}^\mu$ that is selected uniformly at random;

2. Else if $b_{K_\mu} \leq u_\mu^1 \leq b_{K_\mu} + r_{K_\mu}$,
   then go to DEATH step, denoted by $\phi_\mu^* =$ DEATH-proposal($\phi_\mu^{(t)}$), which is to remove from $\phi_\mu^{(t)}$ with a Λ_k^μ where $2 \leq k \leq K_\mu$ that is selected uniformly at random;

3. Else go to MOVE step, denoted by $\phi_\mu^* =$ MOVE-proposal($\phi_\mu^{(t)}$), which first do $\phi_\mu^\dagger =$ DEATH-proposal($\phi_\mu^{(t)}$) and then do $\phi_\mu^* =$ BIRTH-proposal($\phi_\mu^\dagger$).

Algorithm 10.6 Three types of move in the RJS for scale parameter σ.

---

1. If $u_\sigma^1 \leq b_{K_\sigma}$, then go to BIRTH step, denoted by $\phi_\sigma^* = $ BIRTH-proposal($\phi_\sigma^{(t)}$), which is to augment $\phi_\sigma^{(t)}$ with a $\Lambda_{K_\sigma+1}^\sigma$ that is selected uniformly at random;

2. Else if $b_{K_\sigma} \leq u_\sigma^1 \leq b_{K_\sigma} + r_{K_\sigma}$,
   then go to DEATH step, denoted by $\phi_\sigma^* = $ DEATH-proposal($\phi_\sigma^{(t)}$), which is to remove from $\phi^{(t)}$ with a Λ_k^σ where $2 \leq k \leq K_\sigma$ that is selected uniformly at random;

3. Else go to MOVE step, denoted by $\phi_\sigma^* = $MOVE-proposal($\phi_\sigma^{(t)}$), which first do $\phi_\sigma^\dagger = $ DEATH-proposal($\phi_\sigma^{(t)}$) and then do $\phi_\sigma^* = $ BIRTH-proposal($\phi_\sigma^\dagger$).

TABLE 10.2   SIC for the average wind speed models.

Distributions	ILT1	ILT2	ILT3
W2	-2,984	-1,667	-12,287
W3	**-2,941**	**-1,663**	**-11,242**
RAY	-3,120	-1,779	-13,396
LN3	-2,989	-1,666	-11,444
G3	-2,974	-1,666	-11,290
IG3	-2,986	-2,313	-11,410

Source: Lee et al. [131]. With permission.

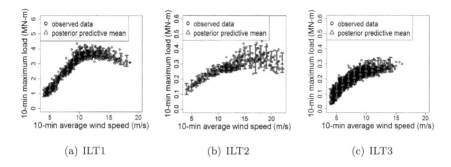

(a) ILT1   (b) ILT2   (c) ILT3

FIGURE 10.6 The 95% pointwise credible intervals of the load response against wind speeds. (Reprinted with permission from Lee et al. [131].)

samples for $\tilde{z}$ from the posterior predictive distribution of conditional maximum load, $f(\tilde{z}|\boldsymbol{x})$. Suppose that one wants to generate the credible interval at wind speed V_* or standard deviation s_*. The posterior predictive distributions are computed as follows:

$$f(\tilde{z}|(V,s) \in \mathcal{D}_{V_*}, \mathcal{D}_z) = \int f(\tilde{z}|(V,s) \in \mathcal{D}_{V_*}, \boldsymbol{\Psi}_a)f(\boldsymbol{\Psi}_a|\mathcal{D}_z)d\boldsymbol{\Psi}_a, \quad (10.31)$$

and

$$f(\tilde{z}|(V,s) \in \mathcal{D}_{s_*}, \mathcal{D}_z) = \int f(\tilde{z}|(V,s) \in \mathcal{D}_{s_*}, \boldsymbol{\Psi}_a)f(\boldsymbol{\Psi}_a|\mathcal{D}_z)d\boldsymbol{\Psi}_a, \quad (10.32)$$

where $\mathcal{D}_{V_*}$ and $\mathcal{D}_{s_*}$ are subsets of the observed data such that $\mathcal{D}_{V_*} = \{(V_i, s_i) : V_* - 0.5 < V_i < V_* + 0.5, \text{ and}, (V_i, s_i) \in \mathcal{D}_{V,s}\}$ and $\mathcal{D}_{s_*} = \{(V_i, s_i) : s_* - 0.05 < s_i < s_* + 0.05, \text{ and}, (V_i, s_i) \in \mathcal{D}_{V,s}\}$. Given these distributions, samples for $\tilde{z}$ are drawn to construct the 95% credible interval at V_* or s_*. The result is shown as one vertical bar in either a V-plot in Fig. 10.6 or an s-plot in Fig. 10.7. To complete these figures, the process is repeated in the V-domain with 1 m/s increment and in the s-domain with 0.2 m/s increment. These figures show that the variability in data are reasonably captured by the spline method.

10.6.3 Binning versus Spline Methods

In the procedure of estimating the extreme load level, two different distributions of maximum load z are involved—one is the conditional maximum load distribution $f(z|\boldsymbol{x})$, namely the short-term distribution, and the other is the unconditional maximum load distribution $f(z)$, namely the long-term distribution. Using the observed field data, it is difficult to assess the estimation accuracy of the extreme load levels in the long-term distribution,

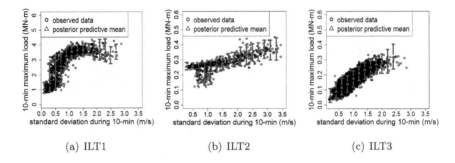

(a) ILT1   (b) ILT2   (c) ILT3

FIGURE 10.7 The 95% pointwise credible intervals of the load response against standard deviations. (Reprinted with permission from Lee et al. [131].)

because of the relatively small amount of observation records. For this reason, this section evaluates a method's performance of estimating the tail of the short-term distribution $f(z|\boldsymbol{x})$. Doing so makes sense, as the short-term distribution underlies the difference between the Bayesian spline method and the IEC standard procedure based on binning. In Section 10.6.5, a simulation is employed to generate a much larger dataset, allowing to compare the performance of the two methods in estimating the extreme load level in the long-term distribution.

To evaluate the tail part of a conditional maximum load distribution, Lee et al. [131] compute a set of upper quantile estimators and assess their estimation qualities using the generalized piecewise linear (GPL) loss function [73]. A GPL is defined as follows:

$$
S_{\tau,b}(\hat{l}(\boldsymbol{x}_i), z(\boldsymbol{x}_i)) =
$$
$$
\begin{cases}
\left(\mathbb{1}(\hat{l}(\boldsymbol{x}_i) \geq z(\boldsymbol{x}_i)) - \tau \right) \frac{1}{|b|} ([\hat{l}(\boldsymbol{x}_i)]^b - [z(\boldsymbol{x}_i)]^b), & \text{for } b \neq 0, \\
\left(\mathbb{1}(\hat{l}(\boldsymbol{x}_i) \geq z(\boldsymbol{x}_i)) - \tau \right) \log\left(\frac{\hat{l}(\boldsymbol{x}_i)}{z(\boldsymbol{x}_i)} \right), & \text{for } b = 0,
\end{cases} \quad (10.33)
$$

where $\hat{l}(\boldsymbol{x}_i)$ is the τ-quantile estimation of $f(z|\boldsymbol{x}_i)$ for a given $\boldsymbol{x}_i$, $z(\boldsymbol{x}_i)$ is the observed maximum load in the test dataset, given the same $\boldsymbol{x}_i$, b is a power parameter, and $\mathbb{1}$ is the indicator function. The power parameter b usually ranges between 0 and 2.5. When $b = 1$, the GPL loss function is the same as the piecewise linear (PL) loss function.

For the above empirical evaluation, Lee et al. [131] randomly divide a dataset into a partition of 80% for training and 20% for testing. They use the training set to establish a short-term distribution $f(z|\boldsymbol{x})$. For any $\boldsymbol{x}_i$ in the test set, the τ-quantile estimation $\hat{l}(\boldsymbol{x}_i)$ can be computed using $f(z|\boldsymbol{x})$. And then, the GPL loss function value is taken as the average of all $S_{\tau,b}$ values

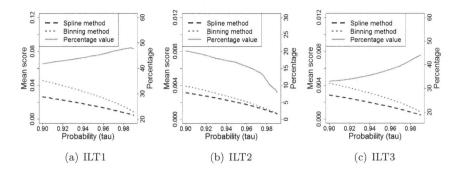

(a) ILT1   (b) ILT2   (c) ILT3

FIGURE 10.8 Comparison of piecewise linear loss function: the left vertical axis represents the mean score values and the right vertical axis
represents a percentage value, which is the reduction in the mean
scores when the spline method is compared with the binning method.
(Reprinted with permission from Lee et al. [131].)

over the test set, as follows:

$$\overline{S}_{\tau,b} = \frac{1}{n_t} \sum_{i=1}^{n_t} S_{\tau,b}(\hat{l}(\boldsymbol{x}_i), z_i), \tag{10.34}$$

where n_t is the number of data points in a test set, and z_i is the same as $z(\boldsymbol{x}_i)$.
Apparently, $\overline{S}_{\tau,b}$ is a mean score. The training/test procedure is repeated 10
times, and the final mean score is the average of the ten mean scores. For
notational simplicity, this final mean score is still called the mean score and
represented by $\overline{S}_{\tau,b}$, as long as its meaning is clear in the context.

In this comparison, two methods are used to establish the short-term distribution: the binning method and the Bayesian spline method. In the sampling
algorithms outlined in Sections 10.3 and 10.5, $N_l = 100$ samples are drawn
from the short-term distribution. As such, one can evaluate the quality of
quantile estimations of the short-term distribution for a τ up to 0.99.

First, let us take a look at the comparisons in Fig. 10.8, which compares
the PL loss (i.e., $b = 1$) of both methods as τ varies in the above-mentioned
range. The left vertical axis shows the values of the mean score of the PL loss,
whereas the right axis is the percentage value of the reduction in mean scores
when the spline method is compared with the binning method. For all three
datasets, the spline method maintains lower mean scores than the binning
method.

When τ is approaching 0.99 in Fig. 10.8, it looks like that the PL losses
of the spline and binning methods are getting closer to each other. This is
largely due to the fact that the PL loss values are smaller at a higher τ, so
that their differences are compressed in the figure. If one looks at the solid
line in the plots, which represents the percentage of reduction in the mean

TABLE 10.3   Mean scores of GPL/PL for the 0.9-quantile estimators.

Power parameter	ILT1		ILT2		ILT3	
	Binning	Spline	Binning	Spline	Binning	Spline
$b = 0$	0.0185	0.0108	0.0129	0.0103	0.0256	0.0171
$b = 1$	0.0455	0.0265	0.0040	0.0031	0.0042	0.0028
$b = 2$	0.1318	0.0782	0.0013	0.0010	0.0008	0.0005

Source: Lee et al. [131]. With permission.

TABLE 10.4   Mean scores of GPL/PL for the 0.99-quantile estimators.

Power parameter	ILT1		ILT2		ILT3	
	Binning	Spline	Binning	Spline	Binning	Spline
$b = 0$	0.0031	0.0018	0.0022	0.0020	0.0045	0.0027
$b = 1$	0.0086	0.0045	0.0007	0.0006	0.0008	0.0005
$b = 2$	0.0270	0.0135	0.0003	0.0002	0.0002	0.0001

Source: Lee et al. [131]. With permission.

score, the spline method's advantage over the binning method is more evident in the cases of ILT1 and ILT3 datasets. When τ gets larger, the spline method produces a significant improvement over the binning method, with a reduction of PL loss ranging from 33% to 50%. The trend is different when using the ILT2 dataset. But still, the spline method can reduce the mean scores of the PL loss from the binning method by 8% to 20%. Please note that ILT2 dataset is the smallest set, having slightly fewer than 600 data records. The difference observed over the ILT2 case is likely attributable to the scarcity of data.

Lee et al. [131] compute the mean scores of the GPL loss under three different power parameters $b = 0, 1, 2$ for each method. Table 10.3 presents the results under $\tau = 0.9$, whereas Table 10.4 is for $\tau = 0.99$. In Table 10.3, the spline method has a mean score 20% to 42% lower than the binning method. In Table 10.4, the reductions in mean scores are in a similar range.

In order to understand the difference between the spline method and binning method, Lee et al. [131] compare the 0.99 quantiles of the 10-minute maximum load conditional on a specific wind condition. This is done by computing the difference in the quantile values of conditional maximum load from the two methods for different weather bins. The wind condition of each bin is approximated by the median values of V and s in that bin. Fig. 10.9 shows the standardized difference of the two 0.99 quantile values in each bin. The darker the color is, the bigger the difference. Lee et al. exclude comparisons in the weather bins with very low likelihood, which is the bins of low wind speed and high standard deviation or high wind speed and low standard deviation.

One can observe that the two methods produce similar results at the bins having a sufficient number of data points, which are mostly weather bins in the central area. The results are different when the data are scarce—this tends to

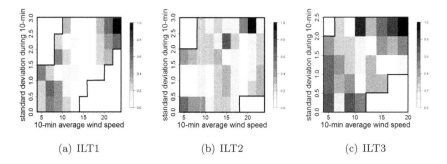

(a) ILT1   (b) ILT2   (c) ILT3

FIGURE 10.9 Comparison of the 0.99-quantiles between binning method and spline method. (Reprinted with permission from Lee et al. [131].)

TABLE 10.5 Estimates of extreme load levels ($l_T, T = 20$ years), unit: MN-m.

Datasets	Binning method	Spline method
ILT1	6.455 (6.063, 7.092)	4.750 (4.579, 4.955)
ILT2	0.752 (0.658, 0.903)	0.576 (0.538, 0.627)
ILT3	0.505 (0.465, 0.584)	0.428 (0.398, 0.463)

Source: Lee et al. [131]. With permission.

happen at the two ends of the average wind speed and standard deviation. This echoes the point made earlier that without binning the weather conditions, the spline method is able to make better use of the available data and overcome the problem of limited data for rare weather events.

Lee et al. [131] also note that the spline method, although conceptually and procedurally more involved, produces an overall model with fewer parameters. To see this, consider the following—for the three ILT datasets, the average $(K_\mu + K_\sigma)$ from the RJS algorithm is between 12 and 18. The number of model parameters d_k in Eq. 10.19 is generally less than 20, a number far smaller than the number of parameters used in the binning method. As explained in Section 10.3, when one uses a 6×10 grid to bin the two-dimensional wind covariates, the binning method in fact uses a total of 121 parameters for the overall model. Evidently, the spline method uses a sophisticated procedure to find a simpler model that is more capable.

10.6.4 Estimation of Extreme Load

Tables 10.5 and 10.6 show the estimates of the extreme load levels l_T, corresponding to $T = 20$ and $T = 50$ years, respectively. The values in parentheses are the 95% credible (or confidence) intervals.

One can observe that the extreme load levels, l_T, obtained by the binning

TABLE 10.6 Estimates of extreme load level (l_T, $T = 50$ years), unit: MN-m.

Datasets	Binning method	Spline method
ILT1	6.711 (6.240, 7.485)	4.800 (4.611, 5.019)
ILT2	0.786 (0.682, 0.957)	0.589 (0.547, 0.646)
ILT3	0.527 (0.480, 0.621)	0.438 (0.405, 0.476)

Source: Lee et al. [131]. With permission.

method are generally higher than those obtained by the spline method. This should not come as a surprise. As one pushes for a high quantile, more data would be needed in each weather bin but the amounts in reality are limited due to the binning method's compartmentalization of data. The binning method also produces a wider confidence interval than the spline method, as a result of the same rigidity in data handling. The procedure of computing the binning method's confidence interval is explained in Algorithm 10.1.

10.6.5 Simulation of Extreme Load

In this section, a simulation study is undertaken to assess the estimation accuracy of extreme load level in the long-term distribution. The simulations use one single covariate x, mimicking the wind speed, and a dependent variable z, corresponding to the maximum load. Algorithm 10.7 is used to generate the simulated data. A set of simulated data thus generated is included in the **Turbine Bending Moment Dataset** and ready to use, but interested readers are welcome to generate the simulated load response data by themselves.

Once the training dataset $\mathcal{D}_{TR}$ is simulated, both the binning method and spline method are used to estimate the extreme load levels l_T corresponding to two probabilities: 0.0001 and 0.00001. This estimation is based on drawing samples from the long-term distribution of z, as described in Section 10.4.4, which produces the posterior predictive distribution of l_T. To assess the estimation accuracy of the extreme quantile values, Lee et al. [131] also generate 100 additional simulated datasets, by repeating Step 1 through Step 3 in Algorithm 10.7, each of which consists of $100,000$ data points. For each dataset, one can compute the observed quantile values $l_{0.0001}$ and $l_{0.00001}$. Using the 100 simulated datasets, one can obtain 100 different samples of these quantiles.

Fig. 10.10(a) shows a scatter plot of the simulated x's and z's in $\mathcal{D}_{TR}$, which resembles the load responses observed. Figs. 10.10(b) and (c) present, under the two selected probabilities, the extreme load levels estimated by the two methods as well as the observed extreme quantile values. One notices that the binning method tends to overestimate the extreme quantile values and yields wider confidence intervals than the spline method. Furthermore, the degree of overestimation appears to increase as the probability corresponding

---

Algorithm 10.7 Simulated data generation to mimic wind speed and load response for assessing the long-term distribution.

---

1. Generate a sample x_i from a three-parameter Weibull distribution. Then sample x_{ij}, $j = 1, \ldots, 1,000$, from a normal distribution having x_i as its mean and a unit variance. The set of x_{ij}'s represents the different wind speeds within a bin.

2. Draw samples, z_{ij}, from a normal distribution with its mean as μ_{ij}^s and its standard deviation as σ_{ij}^s, which are expressed as follows:

$$\mu_{ij}^s = \begin{cases} \frac{1.5}{[1+48\times\exp(-0.3\times x_{ij})]}, & \text{if } x_i < 17, \\ \frac{1.5}{[1+48\times\exp(-0.3\times x_{ij})]} + [0.5 - 0.0016 \times (x_i + x_i^2)], & \text{if } x_i \geq 17, \end{cases}$$

$$\tag{10.35}$$

$$\sigma_{ij}^s = 0.1 \times \log(x_{ij}). \tag{10.36}$$

The above set of equations is used to create a z response resembling the load data. The parameters used in the equations are chosen through trials so that the simulated z looks like the actual mechanical load response. While many of the parameters used above do not have any physical meaning, some of them do; for instance, the "17" in "$x_i < 17$" bears the meaning of the rated wind speed.

3. Find the maximum value $z_i = \max\{z_{i,1}, \ldots, z_{i,1000}\}$ and treat z_i as the maximum load response corresponding to x_i.

4. Repeat Step 1 through Step 3 for $i = 1, ..., 1,000$ to produce the training dataset with $n = 1,000$ data pairs, and denote this dataset by $\mathcal{D}_{\mathrm{TR}} = \{(x_1, z_1), \ldots, (x_{1000}, z_{1000})\}$.

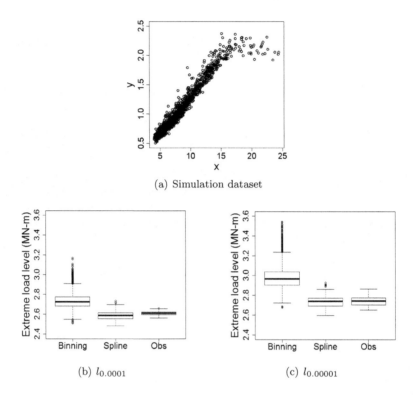

(a) Simulation dataset

(b) $l_{0.0001}$ (c) $l_{0.00001}$

FIGURE 10.10 Simulation dataset, estimated and observed extreme quantile values: (a) an example of the simulated dataset, (b) and (c) boxplots of the binning estimate, of the Bayesian spline estimate, and of the respective sample quantile across 100 simulated datasets. (Reprinted with permission from Lee et al. [131].)

to an extreme quantile value becomes smaller. This observation confirms what is observed in Section 10.6.4 using the field data.

GLOSSARY

BIC: Bayesian information criterion

cdf: Cumulative distribution function

G3: Three-parameter Gamma distribution

GEV: Generalized extreme value

GPD: Generalized Pareto distribution

GPL: Generalized piecewise linear (loss function)

IEC: International Electrotechnical Commission

IG3: Three-parameter inverse-Gaussian distribution

ILT: Inland turbine

LN3: Three-parameter log-normal distribution

MARS: Multivariate adaptive regression spline

MCMC: Markov chain Monte Carlo

MLE: Maximum likelihood estimation

pdf: Probability density function

PL: Piecewise linear (loss function)

POE: Probability of exceedance

POT: Peak over threshold

RAY: Rayleigh distribution

RJMCMC: Reserve jump Markov chain Monte Carlo

RJS: Reserve jump sampler

SIC: Schwarz information criterion

TN2: Two-parameter truncated normal distribution

UIP: Unit information prior

W2: Two-parameter Weibull distribution

W3: Three-parameter Weibull distribution

EXERCISES

10.1 In R, the package evd has a set of functions related to the reverse Weibull distribution. To generate values for the probability density function of reverse Weibull distribution, one can use the function drweibull(x, loc, scale, shape, log = FALSE), where loc, scale, and shape are the three parameters to be specified. Their default values are 0, 1, and 1, respectively. Please use this function to plot a pdf curve for the reverse Weibull distribution by setting loc = 0. Compare the reverse Weibull pdf curve with the Weibull distribution pdf plot under the same values of their respective scale and shape parameters. For computing the Weibull pdf, please use the dweibull function in the stats package.

10.2 Plot the cdf and pdf curves for GEV distribution when, respectively, $\xi = 1$, $\xi = 0$, and $\xi = -1$. Make a note of the pattern of upper and lower tails under respective ξ values.

10.3 Understand the sensitivity of design load, l_T, to the parameters in a GEV distribution. Let $z \sim \text{GEV}(\mu, \sigma, \xi)$, where $\mu = 0$, $\sigma = 1$, and $\xi = 1$.

  a. Compute l_T in Eq. 10.8 for $P_T = 3.8 \times 10^{-7}$.

  b. Keeping $\sigma = 1$ and $\xi = 1$, change the location parameter such that the change in l_T is doubled (or halved). Quantify the change in the location parameter.

  c. Keeping $\mu = 0$ and $\xi = 1$, change the scale parameter such that the change in l_T is doubled (or halved). Quantify the change in the scale parameter.

  d. Keeping $\mu = 0$ and $\sigma = 1$, change the shape parameter such that the change in l_T is doubled (or halved). Quantify the change in the shape parameter.

  e. Repeat the same exercise when the initial GEV distribution has the same μ and σ but $\xi = -1$.

  f. Repeat the same exercise when the initial GEV distribution has the same μ and σ but $\xi = 0$.

10.4 To understand the binning method's lack of scalability, consider the following scenarios. Suppose that one has a full year of 10-minute data pairs, $\{x_i, z_i\}$, with no missing data at all.

  a. How many data pairs are there in this one-year dataset?

  b. If x is one dimensional, and analysts use 10 bins for this variable, how many data points are there, on average, per bin?

  c. What if x is three dimensional and each variable uses 10 bins, how many data points are there, on average, per bin? What if x is six dimensional?

  d. Suppose that in order to adequately fit a GEV distribution with three constant parameters, one would need 25 data points. If we want to have sufficient amount of data points per bin to fit a GEV distribution for every single bin, what is the highest dimensionality of the input space that the binning method can serve?

10.5 Using ILT1 data in the **Turbine Bending Moment Dataset** and the binning method to estimate l_T corresponding to P_T of both 20-year service and 50-year service. This time, instead of fixing the bin-based GEV distribution as the Gumbel distribution, please treat ξ as the third parameter in each bin and estimate based on the data. For the bins that do not have data or sufficient amount of data, follow the treatment in Algorithm 10.1. Compare the estimated $l_{20\text{-yr}}$ and $l_{50\text{-yr}}$ with its counterpart in Tables 10.5 and 10.6.

10.6 Suppose that we use a two-parameter Weibull distribution, instead of the three-parameter Weibull distribution, to model the average wind speed (recall that we used the two-parameter Weibull distribution to model the wind speed in Chapter 2). Reproduce the 95% credible interval plots in Fig. 10.6 and see if there is any noticeable difference.

10.7 Consider a TN2 model with constant η and δ, and, a LN2 model (two-parameter log-normal distribution) with constant distribution parameters, as alternatives for modeling wind speed standard deviation s. Use the SIC as the criterion to select the modeling option that produces the best model fit to the three turbines in the **Turbine Bending Moment Dataset**. Is the best model the TN2 with functional η and δ, TN2 with constant η and δ, or LN2 with constant parameters?

10.8 An alternative method to evaluate Eq. 10.2 is through a peak-over-threshold (POT) approach. The POT approach is to model the unconditional distribution of z directly without accounting for the effect of wind covariates in $\boldsymbol{x}$. First, select a threshold u for the z values in a dataset. Use the z data points above this threshold (that is where the name comes from) to estimate a generalized Pareto distribution (GPD). Assume that the extreme load z follows this GPD and estimate l_T for the corresponding P_T based on the estimated GPD. For the three turbine datasets, use this POT approach to estimate their 20-year and 50-year l_T and compare the outcome with those in Tables 10.5 and 10.6. When using the POT approach, set u as the 95-percentile value of the respective dataset.

10.9 Apply the POT method to the simulated training dataset, again with the threshold, u, set as the 95-percentile value of that dataset. Estimate $l_{0.0001}$ and $l_{0.00001}$, as what has been done while using the binning and spline methods. Compare the POT outcome with those in Fig. 10.10.

CHAPTER 11

Computer Simulator-Based Load Analysis

DOI: 10.1201/9780429490972-11

The principal challenge in reliability assessment of wind turbines is rooted in the fact that a small tail probability, $f(z > l_T)$, in the order of 10^{-7} for $T = 50$ years, needs to be estimated. To accurately estimate this type of small probability requires a sufficient number of high-z load response values. If one opts to collect enough high-z values from physical turbine systems, it takes tens of years, as the high-z values, by definition, are rare events. Adding to the challenge is that hardly have any commercial wind turbines been installed with strain sensors, due to cost concerns. Physically measured bending moments are typically obtained on a few test turbines and only for a short duration of time, which is the reason behind the need for an extrapolation and the modeling of the conditional load density, as explained in Chapter 10. Wind engineers have been developing aeroelastic simulators that can produce reasonably trustworthy bending moments response under a wind force input. The availability of these simulators lends a degree of convenience to load analysis, as a simulator can be steered, at least in principle, towards the region of high load responses so as to produce more high-z data points. For this reason, using the aeroelastic simulators could expedite and enhance the estimation of extreme load distribution and facilitate the reliability assessment of wind turbines. Of course, running aeroelastic turbine load simulators can be computationally expensive. Data science methods are much needed to make the simulator-based load analysis efficient and practical.

11.1 TURBINE LOAD COMPUTER SIMULATION

11.1.1 NREL Simulators

Two NREL simulators are popularly used in the study to generate the structural load response on a turbine—one is TurbSim [112] and the other is FAST [113]. To simulate the structural load response on a turbine, it takes two steps: first, TurbSim generates an inflow wind profile in front of a wind turbine, and second, FAST takes the inflow profile as input and simulates structural and mechanical load responses at multiple turbine components. Please refer to Fig. 10.1 for an illustration of load responses on turbine components.

More than a single-point turbulence computed based on the hub height wind speed, TurbSim simulates a full-field stochastic inflow turbulence environment in front of a turbine, "*reflect[ing] the proper spatiotemporal turbulent velocity field relationships seen in instabilities associated with nocturnal boundary layer flows*" [120]. The input to TurbSim is the hub height wind speed, either measured or simulated, and the output is the full-field inflow environments to be used to drive a downstream load simulator. FAST is the aeroelastic dynamic load simulator, and uses wind inflow data and solves for the rotor wake effects and blade-element loads.

The data in the `Simulated Bending Moment Dataset` are simulated using the two simulators [37]. According to Choe et al. [37], the 10-minute average wind speed is simulated using a Rayleigh distribution. Recall that as mentioned in Section 10.4.3, the IEC standard recommends using a Rayleigh distribution to model the 10-minute average wind speed, although the numerical studies in Section 10.6.1 show that other distributions may fit the actual wind speed data better. TurbSim and FAST are used to simulate a turbine's 10-minute operations, given the average wind speed. The maximum load responses at a blade root are recorded as the output. Two load types, the edgewise and flapwise bending moments, are simulated and their respective maximum values in a 10-minute interval are recorded in the `Simulated Bending Moment Dataset`. Choe et al. [37] define a simulation replication or one simulator run as a single execution of the 10-minute turbine operation simulation that generates a 10-minute maximum edgewise and flapwise load. The *simulated* maximum load is still denoted by z, same as the notation used for the physical maximum load.

Fig. 11.1 illustrates the load responses simulated from TurbSim and FAST, following the procedure discussed in [149].

11.1.2 Deterministic and Stochastic Simulators

Not only does the wind industry use computer simulators to complement physical experiments or make up data deficiency in physical measurements, the use of computer simulators, sometimes referred to as *computer experiments*, is popular and common in many other engineering applications [124, 193].

Many computer simulators used in engineering applications are based on

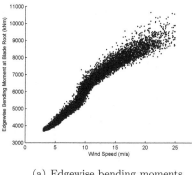

(a) Edgewise bending moments

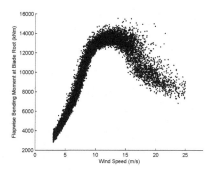

(b) Flapwise bending moments

FIGURE 11.1 Simulated blade root load responses. (Reprinted with permission from Choe et al. [37].)

solving a set of partial differential equations, or a mix of differential and algebraic equations, derived from physical laws and principles. Finite element analysis in mechanics is a frequently mentioned example of this type of computer simulators. These computer simulators are referred to as *deterministic* computer simulators, and numerical analyses run on such simulators are called deterministic computer experiments. They are called "deterministic" because for the same given input, the simulator's output remains the same, no matter how many times one re-runs the same simulator. Let us denote by $g(\cdot)$ the function of the black-box simulator, and by $z(x) := g(x)$ the output of the simulator, given the input at x. Note that z here refers to a generic output, although it could be, but not necessarily, the load response. For a deterministic computer simulator, $z(x)$ does not change, as long as x stays the same.

The development of turbine load simulators is indeed based on aerodynamic and aeroelastic physical principles. But TurbSim and FAST are not deterministic simulators, because for a given input, x, the load response is not guaranteed to be the same. Rather, the simulator response exhibits randomness, resembling the characteristics of noisy physical measurements. This is because the turbine load simulators embed a large number of uncontrollable variables inside the black-box simulator. These variables take different values, produced from certain random number generators, at individual runs of the same simulator, so that even if the input, x, remains the same, the output, z, could be, and is almost surely different. The turbine load simulators are therefore known as the *stochastic* computer simulators, and numerical analyses run on them are stochastic simulations or stochastic computer experiments, mimicking physical experiments. For the stochastic simulators, their g function should include two types of inputs—x that can be set prior to running the simulator and ϵ that is not controlled explicitly but takes its values from random number generators. In other words, the simulator response z is such that $z(\boldsymbol{x}) = g(\boldsymbol{x}, \boldsymbol{\epsilon})$.

When simulating the turbine load response, $\boldsymbol{x}$ is considered a stochastic input, and its marginal distribution, $f(\boldsymbol{x})$, is either known *a priori*, or practically, estimated from the historical data. Random samples of $\boldsymbol{x}$ are drawn from $f(\boldsymbol{x})$ and used to drive TurbSim and FAST simulators. For a deterministic simulator, its inputs can be stochastic and drawn from its own marginal distribution. Analysts may question whether a stochastic simulator may become deterministic, if one treats the combined set of variables, $(\boldsymbol{x}, \boldsymbol{\epsilon})$, as a new input. This is to say, let us specify the joint probability distribution, $f(\boldsymbol{x}, \boldsymbol{\epsilon})$, for both $\boldsymbol{x}$ and $\boldsymbol{\epsilon}$ and then draw samples from $f(\boldsymbol{x}, \boldsymbol{\epsilon})$ to drive the stochastic simulator. Given a specific value of $(\boldsymbol{x}, \boldsymbol{\epsilon})$, the simulator response, $g(\boldsymbol{x}, \boldsymbol{\epsilon})$, remains the same, no matter how many times the same value of $(\boldsymbol{x}, \boldsymbol{\epsilon})$ is used to re-run the same simulator.

Technically, this is correct. In fact, nothing is exactly uncontrolled in computer simulations—even the random numbers generated are, rigorously speaking, pseudo random numbers. But practically, there are too many random variables embedded inside the load response simulators to be specified with a joint probability distribution. According to Choe et al. [37], $\boldsymbol{\epsilon}$ in the NREL simulators has over eight million elements. By contrast, elements in $\boldsymbol{x}$ are far fewer—its number is generally in a single digit. It is thus practical to specify a joint distribution only for $\boldsymbol{x}$ and draw samples from it, while leaving $\boldsymbol{\epsilon}$ to be individually handled by its own random number generator. A computer simulator is stochastic in the sense that $\boldsymbol{\epsilon}$ is left uncontrolled.

Another branch of stochastic simulators are commonly found in discrete event simulations. One of such simulations is mentioned in Section 9.4.3, which is the DEVS-based simulation platform for a wind farm. In that wind farm simulator, a number of inputs or parameters, such as the number and locations of wind turbines, can be specified by analysts running the simulation, but there are many more random variables left to be individually handled by a respective random number generator, such as the degradation path for a turbine component. In the end, even under a fixed $\boldsymbol{x}$, the wind farm simulator changes its response when it is re-run.

11.1.3 Simulator versus Emulator

Running computer simulators is to reduce cost by not conducting too many physical experiments, either too expensive, or too time consuming, or unrealistic. But running computer simulators incurs its own cost, in the form of computational expense. Depending on the fidelity of a computer simulator, the time to run one simulation replication ranges from a couple of minutes (low-fidelity ones) to hours or even days (high-fidelity ones).

Analysts therefore develop efficient, or computationally cheap, mathematical surrogates of computer simulators and hope to rein in the computational expense by running a small number of computer simulations but a large number of the surrogate models. The surrogate models are models of models, because computer simulators are themselves mathematical models of a physical reality, rather than the physical reality itself. For this reason, a surrogate model is known as a meta-model. They are also called *emulators*, to be differentiated from the simulators, and the surrogate models do mean to emulate the behavior of a respective simulator.

A popular branch of emulators is based on Gaussian process regression, or the kriging model, as introduced in Section 3.1.3. To model simulator responses, the location input, $\mathbf{s}$, used in the spatial modeling in Section 3.1.3, is to be replaced by a generic input, $\boldsymbol{x}$. While Gaussian process regression used in Section 3.1.3 for spatial modeling has an input dimension of two, the same modeling approach can be easily extended to general applications of more than two inputs, without changing much of the formulations and solution procedures as outlined in Section 3.1.3.

When modeling a stochastic simulator response, Eq. 3.8 or Eq. 3.17 can be directly used, as the simulator response is treated as if it were a physical response. A training dataset, collected from running the stochastic simulators at different $\boldsymbol{x}$'s, is needed to estimate the parameters in the Gaussian process model. The resulting model, if we express it by $\hat{g}(\boldsymbol{x})$, is a meta-model or an emulator.

When modeling a deterministic simulator response, the main difference is to use Eq. 3.8 or Eq. 3.17 without the nugget effect, i.e., remove ε in the respective equation. This is because a deterministic simulator returns the same response for the same input, so that an emulator is supposed to produce the precise response at the same input value of $\boldsymbol{x}$. It can be shown that a Gaussian regression model without the nugget effect interpolates precisely through the training data points, known as its interpolating property (see Exercise 3.3).

The popularity of the Gaussian process model as an emulator arises from its modeling of deterministic computer simulators. When deterministic computer simulators become common, analysts realize its difference from physical experiments, particularly the aspect of having noise-free responses, and therefore seek a different modeling approach. Sacks and his co-authors adopt the Gaussian process model from spatial statistics to model computer experiments [189, 190] and note the interpolating property; their effort launched the field of design and analysis of computer experiments.

But Gaussian process models are not the only emulator choice, especially when it comes to modeling the stochastic computer simulators. Recall that the response of a stochastic computer simulator looks more like physical measurements. Many data science methods introduced in this book, employed to model various types of physical responses, can be used to model the response of a stochastic computer simulator and hence be an emulator. As we will see in Section 11.4, the emulator used in the turbine load analysis is not a Gaussian process model.

11.2 IMPORTANCE SAMPLING

Let us first consider the use of deterministic computer simulator in reliability analysis. Given a wind condition $\boldsymbol{x}$, the deterministic computer simulator produces a load response, $z = g(\boldsymbol{x})$. This output can be compared with the design load, or a turbine system's resistance level, l, to see if the turbine structure may fail under the simulated load response z. For reliability assessment, analysts are interested in knowing the failure probability $P(z > l)$, which was expressed in Eq. 10.1 with a subscript T. Here we drop the subscript for the simplicity of notation. Relying on the response of a deterministic computer simulator, this failure probability can be expressed as

$$P(z > l) = \int \mathbb{1}(g(\boldsymbol{x}) > l) f(\boldsymbol{x}) d\boldsymbol{x} = \mathbb{E}[\mathbb{1}(g(\boldsymbol{x}) > l)], \qquad (11.1)$$

where $\mathbb{1}(\cdot)$ is the indicator function.

11.2.1 Random Sampling for Reliability Analysis

Computer simulators, including the turbine load simulators, are considered black boxes because an output is numerically computed by going through thousands of lines of computer codes. It is impractical to analytically evaluate the failure probability $P(z > l)$ in Eq. 11.1. It is, however, rather straightforward to evaluate the failure probability empirically through random sampling. The simplest method is the plain version Monte Carlo method, also known as the crude Monte Carlo (CMC), which is to draw random samples, $\{\boldsymbol{x}_1, \ldots, \boldsymbol{x}_{N_T}\}$, from $f(\boldsymbol{x})$, where N_T is the number of the random samples. Each one of the samples is used to drive the computer simulator and produce a corresponding load output. As such, N_T is also the number of simulation runs.

The simulated load response is then compared with l. If $g(\boldsymbol{x}) > l$, a failure occurs and the indicator function, $\mathbb{1}(g(\boldsymbol{x}_i) > l)$, returns a one; otherwise, no failure occurs and the indicator function, $\mathbb{1}(g(\boldsymbol{x}_i) > l)$, returns a zero. The failure probability is empirically estimated by

$$\hat{P}(z > l) = \frac{1}{N_T} \sum_{i=1}^{N_T} \mathbb{1}(g(\boldsymbol{x}_i) > l). \qquad (11.2)$$

The estimate is simply counting how many times, among the N_T runs, the simulator output exceeds the design load level l.

The crude Monte Carlo method is easy to use and applies to almost any applications. Its main shortcoming is the inefficiency for reliability assessment. Heidelberger [89] presents the following example to stress the point. Let us denote the probability in Eq. 11.1 by P and the estimate in Eq. 11.2 by $\hat{P}_{\text{CMC}}$. It is not difficult to show (see Exercise 11.1) that the expectation and variance of $\hat{P}_{CMC}$ are, respectively,

$$\mathbb{E}[\hat{P}_{CMC}] = P, \quad \text{and}$$
$$Var[\hat{P}_{CMC}] = \frac{1}{N_T}P(1-P). \tag{11.3}$$

If using a normal approximation, the $100(1 - \alpha)\%$ confidence interval for P is $\hat{P}_{CMC} \pm z_{\alpha/2}\sqrt{P(1-P)/N_T}$. A similar treatment is used in Eq. 2.8. The expectation expression in Eq. 11.3 also means that the crude Monte Carlo estimate is unbiased.

Heidelberger [89] asks that how many random samples, or equivalently, how many simulator runs, are required in order to estimate the 99% confidence interval of P to be within 10% of the true probability. To accomplish the desired estimation accuracy, it requires that $z_{\alpha/2}\sqrt{P(1-P)/N_T} \leq 0.1P$ for $\alpha = 0.01$, or equivalently,

$$2.58\sqrt{\frac{(1-P)}{P} \cdot \frac{1}{N_T}} \leq 0.1,$$

so that

$$N_T \geq 666 \times \frac{1-P}{P}. \tag{11.4}$$

For a well-designed product, its failure probability P is small, suggesting $1 - P \approx 1$, so that N_T is roughly of $666/P$, which is going to be large for a small P. Suppose the target failure probability is at the level of $P = 10^{-5}$. To have an accurate enough estimate of this small probability, the sample size or the number of simulation runs required is 6.7×10^7. Even if a single run of the computer simulator takes only one second, 6.7×10^7 seconds still translate to more than two years. The essence of reliability assessment is to capture and characterize the behavior of rare events. While attempting to come up with enough samples of the rare events, the inefficiency of the crude Monte Carlo leads to a high computational demand.

11.2.2 Importance Sampling Using Deterministic Simulator

Importance sampling is to introduce another density function, $q(\boldsymbol{x})$, to draw samples of $\boldsymbol{x}$, where $q(\boldsymbol{x})$ is referred to as the importance sampling density. We explain later where the name comes from.

While using $q(\boldsymbol{x})$, the failure probability expression in Eq. 11.1 can be written differently, i.e.,

$$P(z > l) = \int \mathbb{1}(g(\boldsymbol{x}) > l)\frac{f(\boldsymbol{x})}{q(\boldsymbol{x})}q(\boldsymbol{x})d\boldsymbol{x} = \mathbb{E}\left[\mathbb{1}(g(\boldsymbol{x}) > l)\frac{f(\boldsymbol{x})}{q(\boldsymbol{x})}\right]. \tag{11.5}$$

By multiplying and dividing $q(\boldsymbol{x})$ in the integrand, the above probability expression remains equivalent to that in Eq. 11.1.

Denote by

$$\mathcal{L}(\boldsymbol{x}) = \frac{f(\boldsymbol{x})}{q(\boldsymbol{x})}$$

the likelihood ratio between the two density functions. Eq. 11.5 can be expressed as

$$P(z > l) = \mathbb{E}[\mathbb{1}(g(\boldsymbol{x}) > l)\mathcal{L}(\boldsymbol{x})].$$

The empirically estimated failure probability based on importance sampling (IS) density is then

$$\hat{P}_{\text{IS}}(z > l) = \frac{1}{N_T} \sum_{i=1}^{N_T} \mathbb{1}(g(\boldsymbol{x}_i) > l)\mathcal{L}(\boldsymbol{x}_i), \qquad (11.6)$$

where the samples, $\{\boldsymbol{x}_1, \ldots, \boldsymbol{x}_{N_T}\}$, are drawn from $q(\boldsymbol{x})$.

Technically, any valid density function can be used as $q(\boldsymbol{x})$ in importance sampling, and $\hat{P}_{\text{IS}}$ is an unbiased estimator of P, as long as $q(\boldsymbol{x}) = 0$ implies that $\mathbb{1}(g(\boldsymbol{x}) > l)f(\boldsymbol{x}) = 0$ for any $\boldsymbol{x}$, which means that a non-zero feasible sample under the old density $f(\cdot)$ with $g(\boldsymbol{x}) > l$ must also be a non-zero feasible sample under the new density $q(\cdot)$. However, this does not mean that an arbitrary choice of $q(\boldsymbol{x})$ can help address the computational inefficiency problem of the crude Monte Carlo method. To understand the choice for an optimal importance sampling density, we first provide an intuitive understanding how importance sampling works.

The condition to be verified for failures, $g(\boldsymbol{x}) > l$, defines the events of interest (EOI) for a reliability assessment. But the concentration of $f(\boldsymbol{x})$ does not coincide with the EOI. The region of $\boldsymbol{x}$, whose corresponding response belongs to the EOI, is referred to as the *critical region*. By the nature that the EOI in reliability analysis are rare, random sampling from $f(\cdot)$ has a low hit rate on the critical region. An importance sampling can help if the density so chosen, $q(\cdot)$, steers the sample concentration towards the critical region. This means that while $f(\cdot)$ is small over the critical region, $q(\cdot)$ needs to be large on that region, so as to make the EOI likely to occur. The name, "importance," is given to the sampling approach because the new density is supposed to place the right importance on the critical region, or the new density concentrates on the region of importance.

This intuition is realized through variance reduction in random sampling. To see this, consider the following. The variance of the importance sampling estimator in Eq. 11.6 can be expressed as

$$Var[\hat{P}_{\text{IS}}] = \frac{1}{N_T^2} \sum_{i=1}^{N_T} \mathbb{E}_q \left[(\mathbb{1}(g(\boldsymbol{x}_i) > l)\mathcal{L}(\boldsymbol{x}_i))^2 \right] + C$$

$$= \frac{1}{N_T} \mathbb{E}_f \left[\mathbb{1}(g(\boldsymbol{x}) > l)\mathcal{L}(\boldsymbol{x}) \right] + C, \qquad (11.7)$$

where the subscript placed on the expectation operator is to make explicit which probability measure the expectation is taken with respect to and C is a constant not depending on the sampling action. The first equality in the above question means that reducing the variance of the importance sampling estimator corresponds to selecting a $q(\boldsymbol{x})$ that reduces the second moment of $\mathbb{1}(g(\boldsymbol{x}) > l)\mathcal{L}(\boldsymbol{x})$.

Let us take a look at the likelihood ratio, which is $\mathcal{L}(\boldsymbol{x}) = f(\boldsymbol{x})/q(\boldsymbol{x})$. For importance sampling, following the intuition above, $f(\boldsymbol{x})$ is small in the critical region where $\mathbb{1}(g(\boldsymbol{x}_i) > l) = 1$, while $q(\boldsymbol{x})$ should be large. As such, the likelihood ratio, $\mathcal{L}(\boldsymbol{x})$, is small. Consequently, the variance of the importance sampling is small, according to Eq. 11.7. A proper choice of the importance sampling density is thereby to reduce the likelihood ratio, which in turn makes the samples less spread out (small variance). Together with the unbiasedness property of $\hat{P}_{\text{IS}}$, a variance-reduced importance sampling is able to concentrate on the critical region to sample. For the derivation of Eq. 11.7, please see Exercise 11.2.

The theoretically optimal importance sampling density is

$$q_{\text{IS}}^* = \frac{\mathbb{1}(g(\boldsymbol{x}) > l)f(\boldsymbol{x})}{P(z > l)}, \tag{11.8}$$

because this q_{IS}^* leads to a failure probability estimate that has a zero (and hence the smallest) variance, and one sample from it gives us the unconditional POE, $P(z > l)$, exactly. Practically, this q_{IS}^* is not implementable. The probability $P(z > l)$ is unknown and precisely what analysts want to estimate using the simulators and random samples. Moreover, the critical region, implied by $g(\boldsymbol{x}) > l$, is not known, either, before the simulator is run on the random samples of $\boldsymbol{x}$.

De Boer et al. [45] present a cross-entropy-based approximation to implement the idea of importance sampling. Consider the case that the density function, $f(\boldsymbol{x})$, can be parameterized by a vector $\boldsymbol{u}$. To make this parametrization explicit, let us express it as $f(\boldsymbol{x}; \boldsymbol{u})$. Suppose that the importance sampling density takes the same function form but uses different parameters, i.e., $q(\boldsymbol{x}) := f(\boldsymbol{x}; \boldsymbol{v})$. The likelihood ratio can be expressed as

$$\mathcal{L}(\boldsymbol{x}; \boldsymbol{u}, \boldsymbol{v}) = \frac{f(\boldsymbol{x})}{q(\boldsymbol{x})} = \frac{f(\boldsymbol{x}; \boldsymbol{u})}{f(\boldsymbol{x}; \boldsymbol{v})}. \tag{11.9}$$

The cross-entropy algorithm is iterative in nature. When the algorithm starts, it attempts to find an event not so rare, by setting a probability, say $\kappa = 0.01$, so that there are almost surely EOI produced from the simulator. Let t be the iteration index and N_t be the sample size at the t-th iteration. When the N_t samples are evaluated using the simulator at the t-th iteration, the responses are labeled as $\{g_1^{(t)}, \ldots, g_{N_t}^{(t)}\}$. Without ambiguity, the superscript (t) is often dropped. We order the simulator response from smallest to largest, such that $g_{(1)} \le g_{(2)} \le \cdots \le g_{(N_t)}$, where $g_{(j)}$ is the j-th order-statistic of the sequence $\{g(\boldsymbol{x}_1), \ldots, g(\boldsymbol{x}_{N_t})\}$.

The iterative cross-entropy algorithm constructs a sequence of reference parameters $\{v_t, t \geq 0\}$ and design load thresholds $\{l_t, t \geq 1\}$. It starts with $v_0 = u$ and updates both v_t and l_t by steering the sampling action towards the critical region. The specific steps are outlined in Algorithm 11.1. The optimization formulation in Step 3 is based on the minimization of the Kullback-Leibler distance between the optimal importance sampling density, q_{IS}^* in Eq. 11.8, and the actual importance sampling density to be used for the next iteration, $q^{(t+1)}(x) := f(x; v_t)$. The Kullback-Leibler distance is also termed the *cross entropy* between the two density functions of interest (see Exercise 11.3).

Algorithm 11.1 Iterative cross-entropy approximation for importance sampling.

---

1. Set $\hat{v}_0 = u$ and $t = 1$.

2. Draw samples, $\{x_1, \ldots, x_{N_t}\}$, from the density $q^{(t)}(x) := f(x; v_{t-1})$. Compute the $(1 - \kappa)N_t$-th order-statistic of $\{g(x_1), \ldots, g(x_{N_t})\}$ and set that as the estimate of l_t, i.e.,

$$\hat{l}_t = g_{(\lceil (1-\kappa)N_t \rceil)}.$$

   If $\hat{l}_t \geq l$, then let $\hat{l}_t = l$.

3. Use the same samples drawn in Step 2, $\{x_1, \ldots, x_{N_t}\}$, to solve the following optimization problem and get an update of v_t. Denote the solution by $\hat{v}_t$.

$$\max_{v} \frac{1}{N_t} \sum_{i=1}^{N_t} \mathbb{1}(g(x_i) \geq \hat{l}_t)) \mathcal{L}(x_i; u, \hat{v}_{t-1}) \ln f(x_i; v). \qquad (11.10)$$

4. If $\hat{l}_t < l$, set $t = t + 1$ and reiterate from Step 2. Else proceed to Step 5.

5. Estimate the failure probability by using Eq. 11.6, re-written below as

$$\hat{P}_{IS}(z > l) = \frac{1}{N_T} \sum_{i=1}^{N_T} \mathbb{1}(g(x_i) > l) \mathcal{L}(x_i; u, \hat{v}_T).$$

   where T is the final number of iterations.

Dubourg et al. [53] present a different approximation approach, which is based on the use of a meta-model. The idea is simple. First, draw a small number of samples of x, say a couple of hundreds, and use the computer simulator to generate the corresponding structural responses. Using this small set of simulator-generated samples, Dubourg et al. [53] build a Gaussian process emulator, which can run more efficiently and be used to generate a much larger sample set, say several thousands or even tens of thousands. The importance sampling estimate in Eq. 11.6, instead of relying on the simulator function $g(\cdot)$, now uses the emulator function, $\hat{g}(\cdot)$.

One challenge faced by this meta-model-based approach is that with the initial small number of samples, the chance of having a sufficient number of EOI is low. The subsequent Gaussian process emulator is therefore unlikely able to gain a good accuracy in the tail probability estimation when there are very few quality samples to build the meta-model in the first place. Like the cross-entropy approach, an iterative procedure appears unavoidable for the meta-model-based approach, which gradually steers the sampling action towards the critical region.

11.3 IMPORTANCE SAMPLING USING STOCHASTIC SIMULA-TORS

The importance sampling described in Section 11.2 relies on the use of a deterministic computer simulator. This is reflected in the failure verification function, $\mathbb{1}(g(\boldsymbol{x}) > l)$. Due to the deterministic nature of the simulator used, $g(\boldsymbol{x})$ is a constant for a given $\boldsymbol{x}$, so that $g(\boldsymbol{x}) > l$ is either true or false, meaning $\mathbb{1}(g(\boldsymbol{x}) > l)$ is either one or zero, once $\boldsymbol{x}$ is fixed. This is no longer true for stochastic simulators, because $g(\boldsymbol{x})$ varies even for the same $\boldsymbol{x}$. The verification condition, $g(\boldsymbol{x}) > l$, compares in fact a random variable with a threshold, and for this reason, the indicator function is no longer appropriate to be used to capture the failure verification outcome. Rather, a probability should be assessed of this condition, namely $P(g(\boldsymbol{x}) > l)$.

In the context of stochastic simulators, the crude Monte Carlo estimate is changed to

$$
\hat{P}_{\mathrm{CMC}}(z > l) = \frac{1}{M} \sum_{i=1}^{M} \hat{P}(g(\boldsymbol{x}_i) > l)
$$

$$
= \frac{1}{M} \sum_{i=1}^{M} \left(\frac{1}{N_i} \sum_{j=1}^{N_i} \mathbb{1}(g_j(\boldsymbol{x}_i) > l) \right),
$$

(11.11)

where $\{\boldsymbol{x}_1, \ldots, \boldsymbol{x}_M\}$ are M random samples from $f(\cdot)$ and M is called the *input sample size*. At each input $\boldsymbol{x}_i$, the simulator is run N_i times to produce N_i outputs, $g_1(\boldsymbol{x}_i), \ldots, g_{N_i}(\boldsymbol{x}_i)$, each of which is a realization of a stochastic process and can then be compared with the design threshold l in a deterministic manner. The number of simulations per input, N_i, is called the *allocation size*. The total number of simulator runs is then $N_T = \sum_{i=1}^{M} N_i$.

Apparently, the inclusion of the inner summation in Eq. 11.11 is the major difference between the failure probability estimate using a stochastic simulator and that using a deterministic simulator. When using a deterministic simulator, N_i is set simply one, so that $N_T = M$. When using a stochastic simulator, the sample average of N_i simulator responses under the same $\boldsymbol{x}_i$ is used to approximate the probability, $\hat{P}(g(\boldsymbol{x}_i) > l)$.

Importance sampling based on deterministic simulators can be explicitly referred to as the deterministic importance sampling (DIS), whereas importance sampling based on stochastic simulators is referred to as the stochastic importance sampling (SIS). In the sequel, some of the "IS" subscripts used previously is replaced by "DIS." For instance, q_{IS}^* in Eq. 11.8 is expressed as q_{DIS}^* from this point onwards.

Choe et al. [37] develop two versions of the stochastic importance sampling method, referred to as SIS1 and SIS2, respectively, which are to be explained in the sequel.

11.3.1 Stochastic Importance Sampling Method 1

Noticing the difference between Eq. 11.2 and Eq. 11.11, when introducing an importance sampling density to the stochastic simulators, Eq. 11.6 should be written as

$$\hat{P}_{\mathrm{SIS1}}(z > l) = \frac{1}{M} \sum_{i=1}^{M} \hat{P}(g(\boldsymbol{x}_i) > l) \mathcal{L}(\boldsymbol{x}_i) = \frac{1}{M} \sum_{i=1}^{M} \left(\frac{1}{N_i} \sum_{j=1}^{N_i} \mathbb{1}(g_j(\boldsymbol{x}_i) > l) \right) \mathcal{L}(\boldsymbol{x}_i),$$

(11.12)

where the samples, $\{\boldsymbol{x}_1, \ldots, \boldsymbol{x}_M\}$, are drawn from $q(\boldsymbol{x})$. Here, $P(g(\boldsymbol{x}_i) > l)$ is the probability of exceedance, conditioned on input $\boldsymbol{x}_i$. Let us denote this conditional POE by

$$S(\boldsymbol{x}) := P(g(\boldsymbol{x}) > l). \tag{11.13}$$

In Eq. 11.12, the conditional POE is estimated by the sample mean of successes.

In SIS1, Choe et al. [37] state that N_T and M are assumed given and the goal is to find the optimal allocation, N_i, and the optimal importance sampling density function, $q_{\mathrm{SIS1}}(\cdot)$.

Recall the intuition behind importance sampling described in Section 11.2.2. The optimal importance sampling density is supposed to minimize the variance of the failure probability estimate. For $\hat{P}_{\mathrm{SIS1}}(z > l)$, Choe et al. [37] obtain

$$
\begin{aligned}
Var[\hat{P}_{\mathrm{SIS1}}] &= Var\left[\frac{1}{M} \sum_{i=1}^{M} \hat{S}(\boldsymbol{x}_i)\mathcal{L}(\boldsymbol{x}_i) \right] \\
&= \frac{1}{M^2} \mathbb{E}\left[Var\left\{ \sum_{i=1}^{M} \hat{S}(\boldsymbol{x}_i)\mathcal{L}(\boldsymbol{x}_i) \right\} \right] + \frac{1}{M^2} Var\left[\mathbb{E}\left\{ \sum_{i=1}^{M} \hat{S}(\boldsymbol{x}_i)\mathcal{L}(\boldsymbol{x}_i) \right\} \right] \\
&= \frac{1}{M^2} \mathbb{E}\left[\sum_{i=1}^{M} \frac{1}{N_i} S(\boldsymbol{x}_i)(1 - S(\boldsymbol{x}_i))(\mathcal{L}(\boldsymbol{x}_i))^2 \right] + \frac{1}{M} Var\left[S(\boldsymbol{x})\mathcal{L}(\boldsymbol{x}) \right].
\end{aligned}
$$

(11.14)

Choe et al. further prove that the following allocation sizes and importance sampling density function make $\hat{P}_{\text{SIS1}}$ an unbiased estimator and minimize the variance of the failure probability estimate in Eq. 11.14:

$$q^*_{\text{SIS1}}(\boldsymbol{x}) = \frac{1}{C_{q1}} f(\boldsymbol{x}) \sqrt{\frac{1}{N_T} S(\boldsymbol{x})(1 - S(\boldsymbol{x})) + S(\boldsymbol{x})^2}, \qquad (11.15a)$$

$$N^*_i = \frac{N_T \sqrt{\frac{N_T(1-S(\boldsymbol{x}_i))}{1+(N_T-1)S(\boldsymbol{x}_i)}}}{\sum_{j=1}^{M} \sqrt{\frac{N_T(1-S(\boldsymbol{x}_j))}{1+(N_T-1)S(\boldsymbol{x}_j)}}}, \qquad (11.15b)$$

where C_{q1} is a normalizing constant such that

$$C_{q1} = \int f(\boldsymbol{x}) \sqrt{\frac{1}{N_T} S(\boldsymbol{x})(1 - S(\boldsymbol{x})) + S(\boldsymbol{x})^2} \, d\boldsymbol{x}.$$

When using the above formula for N_i, N_i is rounded to the nearest integer. If the rounding yields a zero, Choe et al. suggest using one in its place in order to ensure unbiasedness in the failure probability estimation.

The importance sampling density is a re-weighted version of the original density for $\boldsymbol{x}$. It gives more weight to the critical region when EOI are more likely to occur, and less weight to the region when EOI do not happen as often, so as to refocus the sampling effort on the critical region.

The allocation size is roughly proportional to $\sqrt{1 - S(\boldsymbol{x}_i)}$, after we approximate $1 + (N_T - 1)S(\boldsymbol{x}_i)$ by one for a small $S(\boldsymbol{x}_i)$. This allocation policy says that for a smaller failure probability, one needs a larger size of samples. This result may sound counter-intuitive at first, because one would expect the smaller failure probability area to be accompanied by a smaller sample size. While sampling from the critical region where EOI are more likely to occur, the optimal importance sampling density, q^*_{SIS1}, concentrates more resources on the region where $g(\cdot)$ is close to l, rather than on the region where $g(\cdot)$ is much greater than l. This turns out to be a good strategy because for the region where $g(\cdot)$ is much greater than l, the certainty is high, foreclosing the need for large sample sizes. In summary, among the important input conditions under which a system can possibly fail, SIS1's allocation strategy finds it a more judicious use of the simulation resources by allocating a larger (smaller) number of replications in the region with a relatively small (large) $S(\boldsymbol{x})$.

The q^*_{SIS1} reduces to q^*_{DIS} in Eq. 11.8 (where it was called q^*_{IS} then) when the stochastic simulator is replaced by a deterministic simulator. Under a deterministic simulator, $N_i = 1$, $N_T = M$, and $S(\boldsymbol{x}) = \mathbb{1}(g(\boldsymbol{x}) > l)$. The last expression means that under a deterministic simulator, the conditional POE deteriorates to an indicator function, taking either zero or one. As such, $S(\boldsymbol{x})(1 - S(\boldsymbol{x})) = 0$, so that the density function becomes

$$q^*_{\text{SIS1}} = \frac{S(\boldsymbol{x})f(\boldsymbol{x})}{\int S(\boldsymbol{x})f(\boldsymbol{x})d\boldsymbol{x}} = \frac{\mathbb{1}(g(\boldsymbol{x}) > l)f(\boldsymbol{x})}{\int \mathbb{1}(g(\boldsymbol{x}) > l)f(\boldsymbol{x})d\boldsymbol{x}} = \frac{\mathbb{1}(g(\boldsymbol{x}) > l)f(\boldsymbol{x})}{P(z > l)}.$$

11.3.2 Stochastic Importance Sampling Method 2

Choe et al. [37] propose an alternative stochastic importance sampling-based estimator that restricts N_i to one, such that

$$\hat{P}_{\text{SIS2}}(z > l) = \frac{1}{N_T} \sum_{i=1}^{N_T} \mathbb{1}(g(\boldsymbol{x}_i) > l)\mathcal{L}(\boldsymbol{x}_i). \tag{11.16}$$

Although the right-hand side of Eq. 11.16 looks the same as that in Eq. 11.6, a profound difference is that $g(\cdot)$ function here is not deterministic. As a result, q_{DIS} cannot be used as q_{SIS2}. Choe et al. [37] present the optimal density function as

$$q^*_{\text{SIS2}}(\boldsymbol{x}) = \frac{1}{C_{q2}} f(\boldsymbol{x})\sqrt{S(\boldsymbol{x})}, \tag{11.17}$$

where C_{q2} is another normalizing constant such that

$$C_{q2} = \int f(\boldsymbol{x})\sqrt{S(\boldsymbol{x})}d\boldsymbol{x}.$$

The importance sampling density, q^*_{SIS2}, also reduces to q^*_{DIS} when the stochastic simulator is replaced by a deterministic simulator. Again, under a deterministic simulator, $S(\boldsymbol{x}) = \mathbb{1}(g(\boldsymbol{x}) > l)$, i.e., an indicator function taking either zero or one. Therefore, $\sqrt{S(\boldsymbol{x})} = S(\boldsymbol{x})$, so that the density function becomes

$$q^*_{\text{SIS2}} = \frac{f(\boldsymbol{x})S(\boldsymbol{x})}{\int f(\boldsymbol{x})S(\boldsymbol{x})d\boldsymbol{x}} = \frac{\mathbb{1}(g(\boldsymbol{x}) > l)f(\boldsymbol{x})}{P(z > l)}.$$

11.3.3 Benchmark Importance Sampling Method

Choe et al. [37] mimic the deterministic importance sampling density function by replacing the failure-verifying indicator function in Eq. 11.6 with the conditional POE, and call the resulting importance sampling density function the benchmark importance sampling (BIS) density, i.e.,

$$q^*_{\text{BIS}}(\boldsymbol{x}) = \frac{S(\boldsymbol{x})f(\boldsymbol{x})}{P(z > l)} = \frac{S(\boldsymbol{x})f(\boldsymbol{x})}{\int S(\boldsymbol{x})f(\boldsymbol{x})d\boldsymbol{x}},$$

and use Eq. 11.6 as the failure probability estimator. To be consistent with the notations used in q^*_{SIS1} and q^*_{SIS2}, we denote by C_{qB} the normalizing constant in the above density function, i.e.,

$$q^*_{\text{BIS}}(\boldsymbol{x}) = \frac{1}{C_{qB}} f(\boldsymbol{x})S(\boldsymbol{x}), \tag{11.18}$$

where,

$$C_{qB} = \int f(\boldsymbol{x})S(\boldsymbol{x})d\boldsymbol{x}.$$

11.4   IMPLEMENTING STOCHASTIC IMPORTANCE SAMPLING

In the stochastic importance sampling densities, described in the preceding section, two pieces of detail need to be sought out for their implementation. The first is about modeling the conditional POE, $S(x)$, and the other is how to sample from a resulting importance sampling density without necessarily computing the normalized constant in the denominator.

11.4.1   Modeling the Conditional POE

Choe et al. [37] suggest using a meta-modeling approach to establish an approximation for $S(x)$, but argue that using the Gaussian process model is appropriate when the modeling focus is on the part around the mode of a probability density function. For extreme load and failure probability analysis, the focus is instead on the extreme quantiles and the tail probability of a skewed distribution. Unlike Dubourg et al. [53] who use the Gaussian process-based approach, Choe et al. use a generalize additive model for location, scale, and shape (GAMLSS) [179].

In Chapter 10, a GEV distribution is used to model the extreme load on critical turbine components. The GEV distribution has three distribution parameters: location, scale, and shape. Section 10.4 presents an inhomogeneous GEV distribution, in which the location parameter and the scale parameter are modeled as a function of the input x using MARS models. The approach in Section 10.4 falls under the broad umbrella of GAMLSS.

In [37], Choe et al. still use a GEV distribution and also model its location and shape parameter as a function of the input, while keeping the shape parameter fixed, the same approach as used in Section 10.4. Choe et al. choose to include only the wind speed in x, so that the functions for the location and shape parameters are univariate. For this reason, Choe et al. use a smoothing spline to model both functions, rather than the MARS function used in Section 10.4. Recall that a smoothing spline handles a univariate input well but does not scale very well in higher input dimensions. MARS is one popular multivariate spline-based models handling multi-dimensional inputs. Please visit Section 5.3.3 for more details on smoothing splines and spline-based regression.

Choe et al. [37] obtain a training dataset using the NREL simulators. The training dataset consists of 600 observation pairs of $\{x_i, y_i\}$, $i = 1, 2, \ldots, 600$. The x is the wind speed sampled from a Rayleigh distribution but truncated between the cut-in wind speed at 3 m/s and the cut-out wind speed at 25 m/s. The y is the corresponding load response obtained by running the NREL simulators. Slightly different from the smoothing spline formulation in Eq. 5.22, here are two smoothing splines, one for the location parameter and the other for the scale parameter, to be estimated simultaneously. Following the GAMLSS framework, Choe et al. maximize an objective function regularized by both smoothing splines. Let $\mu(x)$ be the location function, $\sigma(x)$ be

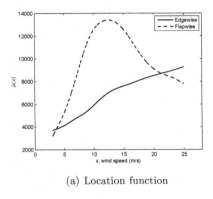

(a) Location function

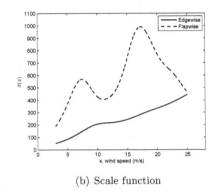

(b) Scale function

FIGURE 11.2 Location and scale parameter functions for both bending moments responses. (Reprinted with permission from Choe et al. [37].)

the scale function, and γ_μ and γ_σ be the two respective penalty parameters. The objective function is then

$$\min \left\{ \text{log-lik} - \gamma_\mu \int \mu''(t)^2 dt - \gamma_\sigma \int (\log \sigma(t)'')^2 dt \right\}, \qquad (11.19)$$

where log-lik refers to the log-likelihood function using the training dataset.

Fig. 11.2 presents the estimated functions for the location and scale parameters using the 600 data pairs in the training set. The shape parameter, kept constant in the above modeling process, is estimated at -0.0359 for the edgewise bending moments response and at -0.0529 for the flapwise bending moments response. In both cases, the resulting GEV distribution exhibits the pattern of a reverse Weibull distribution.

11.4.2 Sampling from Importance Sampling Densities

The three importance sampling densities, q^*_{SIS1}, q^*_{SIS2}, and q^*_{BIS} in Section 11.3, all have a normalizing constant in the denominator of their respective expression. Let us refer to this normalizing constant generically as C_q. Specifically, $C_q = C_{q1}$ in q^*_{SIS1}, $C_q = C_{q2}$ in q^*_{SIS2}, and $C_q = C_{qB}$ in q^*_{BIS}. In order to compute the failure probability estimate using Eq. 11.12 or Eq. 11.16, these constants need to be numerically evaluated. All the constants involve the integration of one known function, $f(x)$, and a meta-model function, $S(x)$, so that a numerical integration routine can compute these constants. In their study [37], Choe et al. use the MATLAB function quadgk for the numerical integration whose input is the univariate wind speed. If one has multiple inputs in x and needs to use a numerical integrator for multivariate inputs, Choe et al. recommend using mcint.

For drawing samples from the importance density functions, Choe et

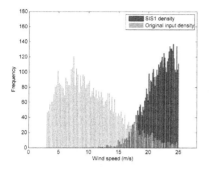 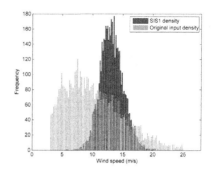

(a) Edgewise moments with $l = 9,300$ kNm (b) Flapwise moments with $l = 14,300$ kNm

FIGURE 11.3 Empirical SIS1 importance sampling density for both bending moments responses, overlaid on top of the density function of wind speed $f(x)$. (Reprinted with permission from Choe et al. [37].)

al. [37] skip the computing of these normalizing constant. They advocate using an acceptance-rejection algorithm to sample from the respective importance sampling density. The acceptance-rejection algorithm samples a u from a uniform distribution over the interval of $[0, f(\boldsymbol{x})]$ and then compares u with $C_q \cdot q^*(\boldsymbol{x})$. If u is smaller, then accept $\boldsymbol{x}$ as a valid sample; otherwise, reject this sample and repeat the sampling action and check again.

Note that the acceptance-rejection condition, $u \leq C_q \cdot q^*(\boldsymbol{x})$, does not involve computing C_q, because $C_q \cdot q^*(\boldsymbol{x})$ can be determined based on $f(\boldsymbol{x})$ and $S(\boldsymbol{x})$, according to Eqs. 11.15a, 11.17, and 11.18.

11.4.3 The Algorithm

Choe et al.'s algorithm to execute the importance sampling using stochastic simulators is summarized in Algorithm 11.2. Fig. 11.3 presents the empirical importance sampling densities of both bending moments responses, overlaid on top of the original wind speed density $f(x)$. The importance sampling densities in Fig. 11.3 are obtained by using q^*_{SIS1}. Similar results can be obtained by using either q^*_{SIS2} or q^*_{BIS}. One can observe from Fig. 11.3 that the distribution of samples over the wind spectrum is different under the importance sampling density versus that under the original wind distribution. Where the high mass of samples appears depends on the physical mechanism governing the bending moments response and exhibits close correlation with the trend shown in the respective plot in Fig. 11.1.

Algorithm 11.2 Importance sampling algorithm using stochastic simulators.

1. Approximate the conditional POE, $S(\boldsymbol{x})$, with a meta-model. In the case of turbine load response, estimate $S(\boldsymbol{x})$ using a small training dataset and fit an inhomogeneous GEV distribution model.

2. Select one of the stochastic importance sampling densities and obtain the set of samples, $\{\boldsymbol{x}_i, \ i = 1, \ldots, M\}$, based on the following acceptance-rejection procedure:

   (a) Sample $\boldsymbol{x}$ from the original input distribution, $f(\boldsymbol{x})$.

   (b) Sample a u from the uniform distribution over $[0, f(\boldsymbol{x})]$.

   (c) If $u \leq C_q \cdot q^*(\boldsymbol{x})$, return $\boldsymbol{x}$ as an sample drawn from the respective importance sampling density; otherwise, discard the sample and draw a new sample of $\boldsymbol{x}$ from $f(\cdot)$.

   (d) Repeat the acceptance-rejection check and sampling action until the prescribed sample size M is reached.

3. For SIS1, determine the allocation size, N_i^*, using Eq. 11.15b, for each $\boldsymbol{x}_i$. For SIS2 and BIS, $N_i^* = 1$.

4. Run the stochastic simulator N_i^* times at each $\boldsymbol{x}_i, \ i = 1, 2, \ldots, M$.

5. Estimate the failure probability using Eq. 11.12 or Eq. 11.16.

11.5   CASE STUDY

Choe et al. [37] present both a numerical analysis, illustrating various aspects of the stochastic importance sampling method, and a case study, using the NREL simulator's responses to estimate the failure probability and to demonstrate the computational benefit of using the importance sampling method.

11.5.1   Numerical Analysis

In the numerical analysis, Choe et al. [37] use the following data generating mechanism

$$x \sim \mathcal{N}(0,1),$$
$$y|x \sim \mathcal{N}(\mu(x), \sigma^2(x)), \tag{11.20}$$

where $\mu(x)$ and $\sigma(x)$ in the distribution of y are functions of input x. Specifically, $\mu(x)$ and $\sigma(x)$ are chosen as

$$\mu(x) = 0.95\delta x^2(1 + 0.5\cos(5x) + 0.5\cos(10x)), \quad \text{and}$$
$$\sigma^2(x) = 1 + 0.7|x| + 0.4\cos(x) + 0.3\cos(14x). \tag{11.21}$$

To use the stochastic importance sampling densities in Section 11.3, Choe et al. [37] specify the meta-models used for $\mu(x)$ and $\sigma(x)$, respectively, as

$$\hat{\mu}(x) = 0.95\delta x^2(1 + 0.5\rho\cos(5x) + 0.5\rho\cos(10x)), \quad \text{and}$$
$$\hat{\sigma}(x) = 1 + 0.7|x| + 0.4\rho\cos(x) + 0.3\rho\cos(14x), \tag{11.22}$$

which are nearly the same as the location and scale functions in Eq. 11.21 but with a ρ inserted to control the accuracy of meta-modeling between $\mu(x)$ and $\hat{\mu}(x)$ and between $\sigma(x)$ and $\hat{\sigma}(x)$. Both $\mu(x)$ and $\hat{\mu}(x)$ also include a δ to control the similarity between the importance sampling density and the original density function of x.

The simulation parameters are set as $N_T = 1,000$ and $M = 300$ when using SIS1 or simply $N_T = 1,000$ for using SIS2 and BIS. To assess the uncertainty of the failure probability estimates, the numerical experiment is repeated 500 times so as to compute the standard error of a failure probability estimate. The computational efficiency is measured by the relative computational ratio of $N_T/N_T^{(CMC)}$, where $N_T^{(CMC)}$ is the total number of simulation runs required by a crude Monte Carlo method to achieve a standard error comparable to that achieved by using the importance sampling method.

The first numerical analysis sets $\rho = 1$, while choosing $\delta = 1$ or $\delta = -1$, and running for three failure probabilities, $P = 0.1, 0.05,$ or 0.01. The analysis outcome is presented in Table 11.1. Choe et al. [37] observe that the computational benefit of using the stochastic importance sampling, as indicated by a small relative computational ratio, is more pronounced when the target probability is smaller, which is a desired property for the importance sampling method.

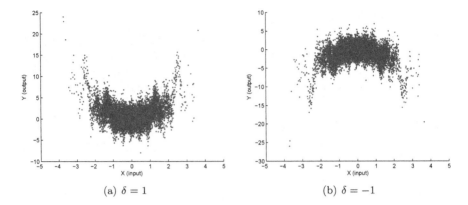

(a) $\delta = 1$ (b) $\delta = -1$

FIGURE 11.4 Sample scatter plots under different δ's. (Reprinted with permission from Choe et al. [37].)

The parameter δ affects the critical region where the importance sampling density function is supposed to draw its samples. In this simulation study, the critical region is where the large positive y values can be found. When $\delta = 1$, the critical region is where $|x|$ is large, i.e., at both ends of the input area and far away from the origin. This choice of δ thus makes the importance sampling density different from the original density of x, as the original density centers at zero. The choice of $\delta = -1$ flips the spread of samples vertically. Consequently, the critical region under $\delta = -1$ is around the origin, so that the resulting importance sampling density function has a great overlap with the original density function of x. In other words, $\delta = -1$ makes the importance sampling density less different from the original density. To appreciate this effect, please see the sample scatter plots in Fig. 11.4, drawn with $\delta = 1$ and $\delta = -1$, respectively. Note how much in each case, or how much less, the positive tails overlap with the area around the origin.

When the importance sampling density is different from the original density, the computational gain by using the importance sampling method is supposed to be more substantial. This is confirmed by the analysis outcome in Table 11.1, where the computational benefit is greater when $\delta = 1$ than that when $\delta = -1$.

In the second analysis, Choe et al. [37] vary ρ in $\hat{\mu}(x)$ and $\hat{\sigma}(x)$ so that the meta-model may deviate from the respective true function. Table 11.2 presents the analysis result for $\rho = 1$, 0.50, and 0. The standard error of the failure probability estimate does increase as ρ decreases, but the rate of increase for SIS1 and SIS2 is slower than that for BIS. The slowest increase is witnessed in the case of SIS2, whose standard error increases about 67% from a perfect meta-model (when $\rho = 1$) to a meta-model substantially different from the original model (when $\rho = 0$), whereas the standard error increases three times

TABLE 11.1 Estimates of the failure probability and the associated standard errors ($\rho = 1$).

		$\delta = 1$		
		$P = 0.10$	$P = 0.05$	$P = 0.01$
SIS1	Average estimate	0.1004	0.0502	0.0100
	Standard error	0.0068	0.0039	0.0005
	$N_T/N_T^{(CMC)}$	51%	32%	2.5%
SIS2	Average estimate	0.0999	0.0501	0.0100
	Standard error	0.0069	0.0042	0.0006
	$N_T/N_T^{(CMC)}$	53%	37%	3.6%
BIS	Average estimate	0.1002	0.0505	0.0101
	Standard error	0.0089	0.0068	0.0014
	$N_T/N_T^{(CMC)}$	88%	97%	20%
CMC	Average estimate	0.1005	0.0506	0.0100
	Standard error	0.0092	0.0070	0.0030

		$\delta = -1$		
		$P = 0.10$	$P = 0.05$	$P = 0.01$
SIS1	Average estimate	0.1001	0.0500	0.0100
	Standard error	0.0090	0.0062	0.0026
	$N_T/N_T^{(CMC)}$	90%	81%	68%
SIS2	Average estimate	0.1001	0.0500	0.0099
	Standard error	0.0086	0.0064	0.0028
	$N_T/N_T^{(CMC)}$	82%	86%	79%
BIS	Average estimate	0.1009	0.0503	0.0101
	Standard error	0.0095	0.0067	0.0031
	$N_T/N_T^{(CMC)}$	100%	95%	97%
CMC	Average estimate	0.1005	0.0498	0.0100
	Standard error	0.0096	0.0071	0.0031

Source: Choe et al. [37]. With permission.

TABLE 11.2   Effect of ρ on failure probability estimate
($\delta = 1$ and $P = 0.01$).

		ρ		
		1.00	0.50	0
SIS1	Average estimate	0.0100	0.0100	0.0101
	Standard error	0.0005	0.0008	0.0017
SIS2	Average estimate	0.0100	0.0101	0.0100
	Standard error	0.0006	0.0007	0.0010
BIS	Average estimate	0.0101	0.0100	0.0102
	Standard error	0.0014	0.0018	0.0063
CMC	Average estimate	0.0099	0.0099	0.0099
	Standard error	0.0030	0.0030	0.0030

Source: Choe et al. [37]. With permission.

in the case of SIS1 and four and a half times in the case of BIS. Choe et al. state that SIS2 is less sensitive to the quality of meta-modeling, making SIS2 a robust, and thus favored, choice in the applications of importance sampling. While the standard errors of SIS1 and SIS2 remain substantially smaller than that of CMC even when $\rho = 0$, the standard error of BIS grows exceeding, and in fact, more than doubling, that of CMC at $\rho = 0$, indicating that the approach disregarding the stochasticity in a stochastic simulator's response has serious drawbacks.

Recall that the deterministic importance sampling density in Eq. 11.8 leads to a failure probability estimate of zero variance. It is also mentioned in Sections 11.3.1 and 11.3.2 that when the response of a stochastic simulator becomes less variable under a given input, the two stochastic importance sampling densities reduce to a deterministic importance sampling density. Putting the two pieces of information together, one expects to see failure probability estimates of much smaller standard errors when SIS1 and SIS2 are used on less variable stochastic simulators.

To show this effect, Choe et al. [37] devise a numerical experiment in their third analysis, in which they change $\sigma(x)$ in Eq. 11.21 to

$$\sigma^2(x) = \tau^2, \tag{11.23}$$

while keeping $\mu(x)$ unchanged. Choe et al. vary τ to control the variability in the response of the simulator. Fig. 11.5 visualizes the variability in response under three values of τ. Comparing the spread of data samples for a given x value demonstrates that the variability in the response when $\tau = 0.5$ is much smaller than that when $\tau = 8$.

Table 11.3 presents the failure probability estimates and the associated

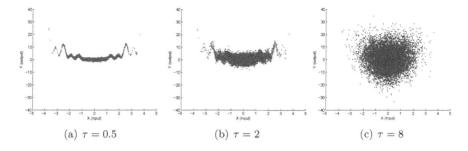

(a) $\tau = 0.5$ (b) $\tau = 2$ (c) $\tau = 8$

FIGURE 11.5 Sample scatter plots under different τ's. (Reprinted with permission from Choe et al. [37].)

TABLE 11.3 Effect of randomness in the simulator's response on the failure probability estimate ($\rho = 1$, $\delta = 1$ and $P = 0.01$).

		τ				
		0.50	1.00	2.00	4.00	8.00
SIS1	Average estimate	0.0102	0.0101	0.0101	0.0102	0.0100
	Standard error	0.0001	0.0001	0.0005	0.0021	0.0028
SIS2	Average estimate	0.0102	0.0101	0.0101	0.0104	0.0100
	Standard error	0.0001	0.0002	0.0006	0.0023	0.0028

Source: Choe et al. [37]. With permission.

standard errors. It is evident that when τ gets smaller, the standard errors of the failure probability estimates, resulting from both stochastic importance sampling methods, get close to zero quickly.

11.5.2 NREL Simulator Analysis

Choe et al. [37] employ the stochastic importance sampling method to estimate the failure probability using the NREL simulators. Both edgewise bending moments and flapwise bending moments are studied. There are two design load levels used, which are $l = 8,600$ kNm and $l = 9,300$ kNm for edgewise bending moments, and $l = 13,800$ kNm and $l = 14,300$ kNm for flapwise bending moments. The two load levels are chosen so that they correspond roughly to the failure probability of $P = 0.05$ and $P = 0.01$, respectively. The total computational runs set for the two design levels are, respectively, $N_T = 1,000$ and $N_T = 3,000$ for the edgewise bending moments response and $N_T = 2,000$ and $N_T = 9,000$ for the flapwise bending moments response. When using the same number of computational runs, the average estimates of the failure probability by the three importance sampling methods are comparable but their standard errors are different. Using SIS1 leads to the smallest standard error, whereas using BIS sees a sizeable increase in the resulting standard error.

To assess the computation required for the crude Monte Carlo method to attain the same level of estimation accuracy, one could run the simulators a sufficient number of times, as one has run the simulator under the importance sampling method. Running the NREL simulator takes about one minute, not much for a single run. The difficulty is that a crude Monte Carlo method needs sometimes more than 60,000 runs of simulation to attain the same level of estimation accuracy as the importance sampling method does. Sixty thousand NREL simulator runs would take more 40 days to complete, too time consuming to be practical. For this reason, $N_T^{(\mathrm{CMC})}$ is computed by using Eq. 11.3 without actually running the NREL simulators under CMC. To compute N_T, one plugs in the standard error attained by the importance sampling method and the target probability value, P (note that in Eq. 11.3, $Var[\hat{P}_{\mathrm{CMC}}]$ is the square of the standard error). Taking the edgewise bending moments as an example, CMC needs about 11,000 runs to attain the same estimation accuracy attained in 1,000 runs for $l = 8,600$ kNm by the importance sampling method using SIS2, or about 51,000 runs for $l = 9,300$ kNm, as compared to 3,000 runs needed by the importance sampling method using SIS2. When compared with the importance sampling method using SIS1, the two run numbers become 18,000 and 61,000, respectively.

Tables 11.4 and 11.5 present, respectively, the failure probability estimates for edgewise and flapwise bending moments. In the tables, the standard error is computed by repeating the computer experiments 50 times. The 95% confidence intervals of the standard error are obtained by running a bootstrap resampling and using the bootstrap percentile interval [55]. In general, SIS1 performs the best but SIS2 performs rather comparably. Both SIS1 and SIS2 outperform BIS by a noticeable margin. Both SIS1 and SIS2 use only a fraction of simulation runs that would be needed by CMC in the case of estimating the failure probability for edgewise bending moments. The computational benefit in the case of flapwise bending moments is not as pronounced as in the case of edgewise bending moments, primarily because the importance sampling densities are not as much different from the original density $f(\cdot)$ in the case of flapwise bending moments. Still, even for flapwise bending moments, the computation needed by SIS1 is only about one-third of what is needed for CMC.

It is interesting to observe the appreciable difference between the stochastic importance sampling methods and BIS, especially between SIS2 and BIS. Looking at Eqs. 11.17 and 11.18, one notices that the density functions are rather similar. The normalizing constants are different, but that difference does not affect the sampling process outlined in Algorithm 11.2. The essential difference is in the numerator, where SIS2 uses a $\sqrt{S(\boldsymbol{x})}$, while BIS uses $S(\boldsymbol{x})$ without taking the square root. That simple action apparently makes a profound difference, as SIS2 is more efficient and requires fewer simulation runs than BIS does, for achieving a comparable standard error. Comparisons

TABLE 11.4   Estimates of the failure probability and the associated standard errors for edgewise bending moments.

| Method | $l = 8,600, N_T = 1,000$ | | |
	Average estimate	Standard error (95% bootstrap CI)	$N_T/N_T^{(CMC)}$
SIS1	0.0486	0.0016 (0.0012, 0.0020)	5.5%
SIS2	0.0485	0.0020 (0.0016, 0.0024)	8.7%
BIS	0.0488	0.0029 (0.0020, 0.0037)	18%

| Method | $l = 9,300, N_T = 3,000$ | | |
	Average estimate	Standard error (95% bootstrap CI)	$N_T/N_T^{(CMC)}$
SIS1	0.00992	0.00040 (0.00032, 0.00047)	4.9%
SIS2	0.01005	0.00044 (0.00036, 0.00051)	5.9%
BIS	0.00995	0.00056 (0.00042, 0.00068)	9.6%

Source: Choe et al. [37]. With permission.

TABLE 11.5 Estimates of the failure probability and the associated standard errors for flapwise bending moments.

Method	$l = 13,800, N_T = 2,000$		
	Average estimate	Standard error (95% bootstrap CI)	$N_T/N_T^{(\mathrm{CMC})}$
SIS1	0.0514	0.0028 (0.0022, 0.0033)	32%
SIS2	0.0527	0.0032 (0.0025, 0.0038)	42%
BIS	0.0528	0.0038 (0.0030, 0.0044)	59%
Method	$l = 14,300, N_T = 9,000$		
	Average estimate	Standard error (95% bootstrap CI)	$N_T/N_T^{(\mathrm{CMC})}$
SIS1	0.01070	0.00061 (0.00047, 0.00074)	32%
SIS2	0.01037	0.00063 (0.00046, 0.00078)	34%
BIS	0.01054	0.00083 (0.00055, 0.00110)	59%

Source: Choe et al. [37]. With permission.

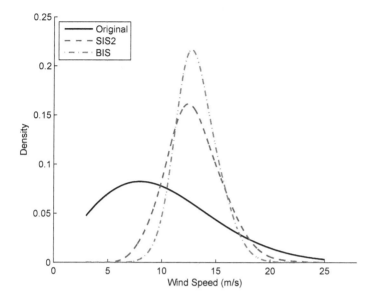

FIGURE 11.6 Comparison of three density functions: the original density function of wind speed, $f(\cdot)$, the SIS2 density function, and the BIS density function.

presented in Table 11.2 also show that SIS2 is more robust than BIS against, or less sensitive to, the meta-model's misspecification.

Fig. 11.6 presents a comparison of the two density functions. The original density function of wind speed, $f(\cdot)$, is also shown in Fig. 11.6. Not surprisingly, the concentration of $f(\cdot)$ does not coincide with the critical region. It turns out that BIS is able to focus on the correct sampling region. As compared with SIS2, however, BIS's focus is a bit too narrow and that action back fires. The square-root operation used in SIS2 appears crucial to attain the right balance in the bias (where to focus) versus variance (how narrowly to focus) tradeoff.

One more note is regarding the level of the target failure probability used in the case study. In Section 11.2.1, we cite a target failure probability at the level of 10^{-5}, but in the case study, the target probability is at or larger than 0.01. The reason that a smaller probability is not used is because doing so demands many more simulation runs than a numerical analysis could tolerate. Consider an importance sampling method that uses one percent of the runs required by CMC, for achieving the desired estimation accuracy for a target failure probability of $P = 10^{-5}$. As calculated in Section 11.2.1, CMC would need 6.7×10^7 simulation runs, and one percent of the CMC simulations is to run the simulator 6.7×10^5 times. When each simulator run takes one minute, as opposed to one second, it takes more than a year to run the simulator that many times. In practice, in order to estimate the small failure probability for a turbine's 20-year or 50-year service, an iterative procedure, in the spirit of Algorithm 11.1, together with parallel computation taking advantage of multiple CPU cores, is inevitable. These solution approaches are in fact being actively pursued in ongoing research.

GLOSSARY

BIS: Benchmark importance sampling

CI: Confidence interval

CMC: Crude Monte Carlo

CPU: Central processing unit

DEVS: Discrete event system specification

DIS: Deterministic importance sampling

EOI: Events of interest

GAMLSS: Generalized additive model for location, scale, and shape

GEV: Generalized extreme value

IEC: International Electrotechnical Commission

IS: Importance sampling

MARS: Multivariate adaptive regression splines

NREL: National renewable energy laboratory

POE: Probability of exceedance

SIS: Stochastic importance sampling

EXERCISES

11.1 Prove the expectation and variance formulas in Eq. 11.3, which are about a crude Monte Carlo method's ability to estimate a failure probability.

11.2 Derive Eq. 11.7, the variance expression for the importance sampling method using a deterministic computer simulator.

11.3 The Kullback-Leibler distance between a pair of density functions, $g(\cdot)$ and $h(\cdot)$, is defined as

$$\mathcal{D}(g, h) = \mathbb{E}_g \left[\ln \frac{g(\boldsymbol{x})}{h(\boldsymbol{x})} \right] = \int g(\boldsymbol{x}) \ln g(\boldsymbol{x}) d\boldsymbol{x} - \int g(\boldsymbol{x}) \ln h(\boldsymbol{x}) d\boldsymbol{x}.$$
(P11.1)

The cross-entropy between the same two density functions is defined as

$$\mathcal{H}(g, h) = \mathbb{E}_g\left[-\ln h(\boldsymbol{x})\right] = -\int g(\boldsymbol{x})\ln h(\boldsymbol{x})d\boldsymbol{x}. \qquad \text{(P11.2)}$$

The entropy function of $g(\cdot)$ is the cross-entropy between $g(\cdot)$ and itself, i.e., $\mathcal{H}(g, g)$.

a. Express the Kullback-Leibler distance using an entropy and a cross-entropy.

b. In Algorithm 11.1, our objective is to minimize the distance between $q^*(\boldsymbol{x})$ and $f(\boldsymbol{x}, \boldsymbol{v})$, in order to choose $\boldsymbol{v}$, where q^* is the optimal importance sampling density to be solved for. Show that the minimization of $\mathcal{D}(q^*, f)$ is the same as maximizing $-\mathcal{H}(q^*, f)$, the negative cross-entropy between the two density functions. This is why the algorithm is referred to as a cross-entropy method.

c. Prove that Eq. 11.10 is meant to minimize the $\mathcal{D}(q^*, f)$ or maximize $-\mathcal{H}(q^*, f)$ (through their empirical counterparts).

11.4 Derive Eq. 11.14, the variance expression of the failure probability estimate, $\hat{P}_{\text{SIS1}}$.

11.5 Derive the optimal density, q^*_{SIS1}, and the optimal allocation, N_i^*, in Eq. 11.15.

11.6 Prove that the variance of $\hat{P}_{\text{SIS2}}$ takes the following expression.

$$Var\left[\hat{P}_{\text{SIS2}}\right] = \frac{1}{N_T}\left(\mathbb{E}_f\left[S(\boldsymbol{x})\mathcal{L}(\boldsymbol{x})\right] - P(z > l)^2\right). \qquad \text{(P11.3)}$$

11.7 Derive the optimal density, q^*_{SIS2}, in Eq. 11.17.

11.8 Using the data pairs in the training set, build a kriging-based meta-model, $S(\boldsymbol{x})$. For this purpose, please use the ordinary kriging model in Eq. 3.8 without the nugget effect. Establish a kriging meta-model for the edgewise bending moments response and another for the flapwise bending moments response.

11.9 Using the meta-models created in Exercise 11.8 and draw wind speed samples using the importance sampling density functions. Plot the empirical distribution of the resulting samples and overlay them on top of the original wind speed samples; the same is done in Fig. 11.3. Observe the empirical distributions and compare them with their counterparts in Fig. 11.3.

11.10 Let us modify the location function and the associated meta-model in Eqs. 11.21 and 11.22 to the following:

$$\mu(x) = 0.95x^2(1 + 0.5\cos(10\nu x) + 0.5\cos(20\nu x)), \quad \text{and}$$
$$\hat{\mu}(x) = 0.95\beta x^2(1 + 0.5\cos(10\nu x) + 0.5\cos(20\nu x)),$$

(P11.4)

where β is the scaling difference between the two functions, while ν controls the roughness of the location function. When $\nu = 0$, the location function and its meta-model reduce to a quadratic function of x.

a. Set the target failure probability $P = 0.01$ and the roughness parameter $\nu = 0.5$. Investigate the effect of β on the failure probability estimate. Try for $\beta = 0.90, 0.95, 1.00, 1.05,$ and 1.10. For each of the β values, produce the average and standard error of the failure probability estimate for four methods, SIS1, SIS2, BIS, and CMC.

b. Set $P = 0.01$ and $\beta = 1$. Investigate the effect of ν on the failure probability estimate. Try for $\nu = 0, 0.50,$ and 1.00. Same as in part (a), produce the average and standard error of the failure probability estimate for four methods, SIS1, SIS2, BIS, and CMC.

CHAPTER 12

Anomaly Detection and Fault Diagnosis

DOI: 10.1201/9780429490972-12

L oad assessment, as introduced in Chapters 10 and 11, definitely plays an important role in wind turbine reliability management. But load assessment addresses a specialized category of problems and faults, which happen as a result of excessive mechanical load. A wind turbine generator is a complex system, comprising a large number of electro-mechanical elements. Many other types of operational anomalies and faults could happen and do take place. This is the reason that we dedicate the last chapter to the general topic of anomaly detection and fault diagnosis. Anomaly detection techniques are supposed to identify anomalies from loads of seemingly homogeneous data and lead analysts and decision makers to timely, pivotal and actionable information. It bears a high relevance with the mission of reliability management for wind turbines.

In this chapter, we could not run the case study using wind turbine fault data. Instead, the methods introduced in the chapter are demonstrated using a group of publicly accessible benchmark datasets, plus a hydropower plant dataset.

12.1 BASICS OF ANOMALY DETECTION

12.1.1 Types of Anomalies

Loosely speaking, anomalies, also referred to as outliers, are data points or a cluster of data points which lie away from the neighboring points or clusters and are inconsistent with the overall pattern of the data. A universal definition of anomaly is difficult to come by, as what constitutes an anomaly often depends on the context.

Goldstein and Uchida [77] illustrate a few different types of anomalies; please see Fig. 12.1. Points A_1 and A_2 are referred to as the global point-

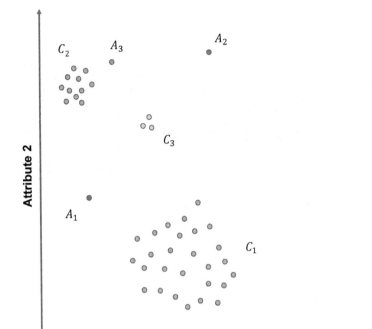

FIGURE 12.1 Illustration of different types of anomalies. (Source: Goldstein and Uchida [77].)

wise anomalies, as they are far away from the existing data points and data clusters. Point A_3 is referred to as a local pointwise anomaly, because, globally, this point is closer to the data cluster C_2 than to many other data points and data clusters, but locally and relative to C_2, it is away from the rest of data points in that cluster. The data clusters, C_2 (including A_3) and C_3, are considered as the collective anomalies or anomalous clusters, referring to the situation when a whole set of data behaves differently from the other regular clusters or data points. Of course, one can argue that should we deem C_2 to represent the normal operation conditions, the cluster, C_1, is the anomalous cluster. That is certainly possible. Absent strong prior knowledge suggesting otherwise, however, the anomalies are always treated as the minority cases in a dataset. Such treatment makes intuitive sense for especially engineering systems, because if the faults and anomalies become more numerous than the supposedly normal operation conditions, the said system is definitely in need of a redesign or an overhaul.

The complexity in defining anomalies translates to the challenges faced in anomaly detection. Methods for detecting anomalous clusters have a strong connection with the research in the field of clustering [86, Section 14.3] and community detection [66, 154]. But in this chapter, our main focus is on the pointwise anomalies.

A branch of research relevant to anomaly detection is the field of statistical process control (SPC) or statistical quality control (SQC) [147]. SPC considers a time series response from a process that has a natural, underlying randomness and aims to detect a change that is deemed substantially greater than the inherent fluctuation of the underlying process. SPC methods are usually manifested in a control chart, a run chart with the horizontal axis representing the time and the vertical axis representing the response or a test statistic computed out of the response.

Fig. 12.2 presents two types of anomalies often encountered in the SPC literature. The left-hand plot of Fig. 12.2 shows a spike type of change. The right-hand plot shows a sustained mean shift—in this particular case, the process response increases substantially at time t_0 and stays there for the next period of time, until a new change happens in the future. This time instance, t_0, is called a *change point*, and for this reason, SPC methods are part of the change-point detection methods.

The two types of changes correspond naturally to the pointwise anomalies and the clustered anomalies. Apparently, the spike type of change is a pointwise anomaly, whereas the sustained mean shift partitions the dataset into two clusters of different operational conditions, one being normal and the other being anomalous. To represent both types of change detection, the term anomaly detection, also labeled as novelty detection or outlier detection, is often used together with the term change detection, creating the expression *change and anomaly detection*. The subtle difference between change detection and anomaly detection lies in the different types of changes or anomalies to be detected.

To detect a change or an anomaly, the SPC approaches rely on a statistical hypothesis test to decide if a mean shift or a spike comes from a statistically different distribution. The distribution in question is typically assumed to be, or approximated by, a Gaussian distribution. The detection mission can thus be reduced to detecting whether the key distribution parameter, either the mean or the variance (or both), is different, with a degree of statistical significance, from that of the baseline process. A control chart runs this statistical hypothesis test repeatedly or iteratively over time, comparing the newly arrived observations with the control limits, i.e., the decision limits, that are established according to the chosen degree of statistical significance, and triggering an anomaly alarm when something exceeds the control limits.

12.1.2   Categories of Anomaly Detection Approaches

Goldstein and Uchida [77] categorize anomaly detection approaches in three broad branches, depending on the availability and labeling of the data in a training set.

The first category is *supervised anomaly detection*, when one has appropri-

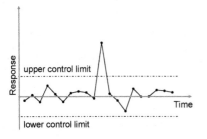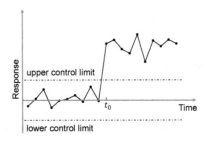

FIGURE 12.2 Control charts, change-point detection, and two types of anomalies.

ately labeled training data in advance, both normal and anomalous instances, so that analysts can train a model based on these labeled data and use the model to decide the labels of future data. This is in fact the classical two-class classification approach, and many of the data science methods introduced in the early part of this book can be used for this purpose, including SVM, ANN, kernel regression and classification, MARS, CART, among others. While using the two-class approaches for anomaly detection, analysts need to be mindful of the imbalance nature of the data instances in the training set. Understandably, the anomalies are far fewer than the normal instances. This data imbalance issue has received considerable attention and is still an active research topic [29, 35, 167].

The second category is known as *semi-supervised anomaly detection*, in which one has only the normal instances and no anomalous data. The idea is to employ the normal data to train a model and create a decision boundary enclosing the normal data. The approach classifies future observations as anomalies if they fall outside the decision boundaries. In other words, the semi-supervised anomaly detection is to define what constitutes the normalcy and treats anything that deviates from the normalcy as anomalies. One-class SVM [196] falls under this category. Park et al. [159] develop a non-parametric semi-supervised anomaly detection method which is proven to be asymptotically optimal. Park et al. show that under the Gaussianity assumption, their optimal detection method reduces to a Hotelling's T^2, a popular method used in SPC [147, Section 11.3].

The most difficult scenario is the absence of any labels for the data or the inability to assume that all data points are normal. As a result, it is not possible to conduct a supervised training. One therefore has to rely entirely on the structure of the dataset and to detect the anomalies in an unsupervised manner. This last category is known as *unsupervised anomaly detection*.

The SPC methods are commonly considered as a method of semi-supervised anomaly detection, as the control limits used in the control charts are based on the normal condition data, known as the *in-control* data in the SPC vernacular. But an SPC procedure usually starts with a dataset for which one cannot guarantee all data instances are normal. This creates the desire of separating change and anomaly detection into two stages: Phase I and Phase II. Phase I is to separate the anomalous cases from the normal majority, whereas Phase II uses the normal majority to establish a decision boundary, or control limits, to flag an incoming observation if it is deemed anomalous. In this sense, Phase I in an SPC procedure is unsupervised, while Phase II is semi-supervised.

Our focus in this chapter is unsupervised anomaly detection. The relevance of an unsupervised anomaly detection is evident not only to the wind industry but to many engineering systems, instrumented with various types of sensors on many components or subsystems. When a service and maintenance engineer suspects that there is a malfunction in a turbine, she/he extracts a dataset from the control system that contains the collected sensor data for that turbine for a selected period of time (weeks, months, or even years), and then stores the data in a relational database or simply in a CSV file for further analysis. Staring at the spreadsheet of data, a service and maintenance engineer often wonders if there is an automated, efficient way to isolate the anomalies from the rest of the data. The historical data in the spreadsheet have almost surely both normal condition data and anomalies. It is just that the service and maintenance engineers do not know which is which. An unsupervised anomaly detection is meant to answer the call.

12.1.3   Performance Metrics and Decision Process

To assess the performance of an anomaly detection method, the usual type-I error versus type-II error trade-off applies. The type-I error, also known as false alarms or false positives, is when the underlying truth of the instance is normal, but the method nonetheless flags it as an anomaly. The type-II error, on the other hand, is when the underlying truth of the instance is an anomaly, but the method fails to flag it. The type-II error is also referred to as missed detections or false negatives. The trade-off between the two types of error says that with all other conditions and parameters held unchanged, one type of error can only be reduced at the expense of increasing the other type of error. Of course, it is possible to reduce both types of errors, but doing so calls for more capable methods or more resources like a larger sample size.

In the mission of anomaly detection, the desire to have a higher detection capability, or equivalently, a smaller type-II error, often triumphs a small type-I error. The fundamental reason is because an anomaly detection method is useful only if it can detect something. A method that rarely detects is utterly useless no matter how nice a property it has in terms of the false positive rate. In the meanwhile, if a detection method triggers too many false alarms, it will eventually become a nuisance and will be turned off in practice.

One common practice in maintaining a healthy trade-off between these two errors for anomaly detection is to set a cut-off threshold, say, N_o, and let an anomaly detection method rank the data from being most likely anomalous to being least likely so. The top N_o ranked data instances are flagged as anomalies, whereas the rest are treated as normal. Once N_o is given, a commonly used performance metric is the precision at N_o ($P@N_o$) [30], defined as the proportion of correct anomalies identified in the top N_o ranks, such as

$$P@N_o = \frac{\#\{o_i \in \mathcal{O} \mid \text{rank}(o_i) \leq N_o\}}{N_o}, \tag{12.1}$$

where $\mathcal{O}$ is the set of true anomalies and o_i is the i-th element in the ranked dataset, according to their likelihood of being an anomaly. A small rank value implies a higher likelihood, so the most likely instance has a $\text{rank}(o_i) = 1$.

Under a given N_o, the goal is to have as high a $P@N_o$ as possible. When N_o is the number of true anomalies, the number of false positives or false alarms is simply $N_o - N_o \times P@N_o$. In reality, the number of true anomalies is not known. Still, a high detection rate at N_o strongly implies a lower false positive rate. For this reason, one does not always present the false positive rate separately.

Without knowing the number of true anomalies, one practical problem is how to set the cut-off threshold N_o. A good practice is to set N_o to be larger than the perceived number of anomalies but small enough to make the subsequent identification operations feasible. The rationale behind this choice lies in the fact that the false positive rate for anomaly detection problems is generally high, especially compared to the standard used for supervised learning methods. Despite a relatively high proportion of false positives, anomaly detection methods can still be useful, particularly used as a pre-screening tool. By narrowing down the candidate anomalies, it helps human experts a great deal to follow up with each circumstance and decide how to take a proper action or deploy a countermeasure. A fully automated anomaly detection is not yet realistic, due to the challenging nature of the problem. Therefore, a useful pre-screening tool, as the current anomaly detection offers, is valuable in filling the void, while analysts strive for the ultimate, full automation goal.

Not only is the number of true anomalies not known in practice, which data instance is a genuine anomaly is also unknown, as the dataset itself is unlabeled and finding out the anomalous instances is precisely what the method intends to do. Verifying the detection accuracy has to rely on another layer of heightened scrutiny, be it a more expensive and thus more capable detection instrument or method or a laborious and time-consuming human examination. In Section 12.6, we use a group of 20 benchmark datasets for which the true anomalies are known, plus a hydropower data for which the anomalies are verified manually by domain experts.

12.2 BASICS OF FAULT DIAGNOSIS

Detecting an anomaly is an important first step to inform proper actions to respond. Sometimes, the response or countermeasure needed could be obvious, once the nature of the anomaly is revealed, but oftentimes, the anomaly just reveals the symptom of the problem. Yet multiple root causes may lead to the same symptom, so that a diagnostic follow-up is inevitable. This is very much analogous to medical diagnosis. A high body temperature and sore throat are anomalous symptoms on a healthy person. But a large number of diseases can cause these symptoms. Deciding what specific pathogen causes the symptoms is necessary before a proper medicine can be administrated to cure the illness.

Diagnosis of engineering systems relies heavily on the knowledge of the systems and know-how of their operations. The diagnostic process can hardly be fully automatic. Rather it is almost always human experts driven and could be labor intensive. But data science methods can facilitate the diagnostic process. For instance, a data science method can help find out which variables contribute to the anomalies and provide a pointed interpretation of each anomaly, thus aiding the domain expert to verify the root causes and fix them, if genuine. In this section, we present two commonly used diagnosis-aiding approaches: diagnosis based on supervised learning and visualization, and diagnosis based on signature matching.

12.2.1   Tree-Based Diagnosis

One immediate benefit of anomaly detection is that the outcomes of the detection can be used to convert the original unsupervised learning problem into a supervised learning problem. Suppose that the anomaly detection method does an adequate job, analysts can then label the data instances in the training set, according to their respective detection outcome. With the labeled dataset, many supervised learning methods can be used to extract rules or find out process variable combinations leading to the anomalous conditions.

The application of supervised learning methods is rather straightforward. While various methods can be used for this purpose, tree-based methods, like CART, are popular, due to its ability to visualize what leads to the anomalous outcomes. CART produces the learning outcomes in the fashion of mimicking a human-style decision-making process, which is another reason behind engineers' fondness of using this tool.

In the hydropower plant case, to be discussed in Section 12.6.2, after the anomalies are detected, a CART is built to facilitate the diagnosis process. While the bulk detail of that case study is to be discussed later, let us present the CART's learning outcome in Fig. 12.3.

From the resulting tree in Fig. 12.3, one can see that using the variables *Oil Temperature of Bearing 4*, *Air Pressure*, *Turbine Vertical Vibration* and *Delta Oil temp - Air Temp of Bearing 1* can correctly classify 25 anomalies based on the proper combination of their conditions. One such condition is when the oil temperature of bearing 4 is less than 27.216 degrees Celsius, the turbine generator almost surely behaves strangely. This condition consistently leads to eleven anomalous observations. Such specific information can certainly help domain experts go to the right components and subsystems

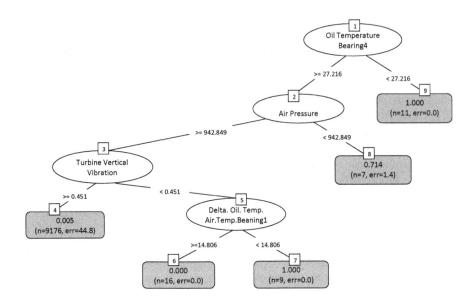

FIGURE 12.3 CART model based on anomaly detection and to be used to facilitate the diagnostic process. (Reprinted with permission from Ahmed et al. [4].)

to perform a follow-up and authenticate the root cause. The pertinent process conditions revealed by the CART model expedite the diagnostic process because the domain experts or the maintenance engineers save the effort of sifting through the large number of variables and data records to find and locate such conditions.

12.2.2 Signature-Based Diagnosis

The idea of signature-based diagnosis is intuitive. A signature library is built to store unique signatures of certain forms that have been associated with specific root causes or faults. If the data collected from ongoing operations can reveal the fault signature lurking in the current process, comparing the estimated signature with the ones in the signature library leads naturally to the identification of the responsible root cause, fulfilling the task of fault diagnosis.

While an intuitive idea, the specifics behind how to build the signature library and how to estimate the signature for ongoing operations can become involved. There is no universal definition of what constitutes a signature. A particular harmonics in the vibration signal resulting from gearbox rotations or the trace of a type of metallic ingredient in the lubricant oil can be signatures sought after. On the other hand, a signature may not be in plain sight but needs to be worked out by building a mathematical model first. Invariably, the signature library building process is performed offline, while the signature estimation is conducted online, in a fashion similar to Fig. 9.4, in which one can replace "offline optimization" with "offline signature library" and "online control" by "online signature estimation."

The signature-based diagnosis approach has been successfully applied in many other industries, such as in the automotive assembly process [49]. The approach is general enough. With a similar model built for a wind turbine system, the method is applicable to the wind energy applications.

Let us briefly explain how the model-based signature matching works. For the sake of simplicity, let us consider a system whose input and output can be linked through a linear model, such as

$$y = \mathbf{H}x + \varepsilon, \tag{12.2}$$

where y and x are the outputs and inputs, ε is assumed to be a Gaussian noise, and $\mathbf{H}$ is the system matrix capturing the input-output relationship. This equation in fact looks similar to the observation equation in Eq. 2.37, except that the one in Eq. 2.37 has a univariate output, whereas the model above has a multivariate response.

Assume $\varepsilon \sim \mathcal{N}(\mathbf{0}, \sigma_\varepsilon^2 \mathbf{I})$. Then, taking the variance of both sides of Eq. 12.2 and furthermore assuming independence between x and ε, one has

$$\mathbf{\Sigma}_y = \mathbf{H}\mathbf{\Sigma}_x \mathbf{H}^T + \sigma_\varepsilon^2 \mathbf{I}. \tag{12.3}$$

Suppose that one of the elements in x, say, x_i, is malfunctioning. As a result, x_i creates a substantially large variation source of the magnitude of σ_i^2. Assume that all other elements in x are properly functioning and thus have zero variance, or a variance so small relative to σ_i^2 that it can be approximated by zero. As such,

$$\mathbf{\Sigma}_x = \begin{pmatrix} 0 & & & & \\ & \ddots & & & \\ & & \sigma_i^2 & & \\ & & & \ddots & \\ & & & & 0 \end{pmatrix}. \tag{12.4}$$

Substituting the above $\mathbf{\Sigma}_x$ into Eq. 12.3, one gets

$$\mathbf{\Sigma}_y = \sigma_i^2 \mathbf{h}_i \mathbf{h}_i^T + \sigma_\varepsilon^2 \mathbf{I}, \tag{12.5}$$

where $\mathbf{h}_i$ is the i-th column of $\mathbf{H}$.

With the presence of a systematic fault, the magnitude of the background noise, measured by σ_ε^2, is supposed to be much smaller than that of the fault, σ_i^2; otherwise, the fault may not be a real fault, or it may not be detectable. Aware of this, let us approximate Eq. 12.5 by dropping the term of the background noise. So the approximation reads

$$\mathbf{\Sigma}_y \simeq \sigma_i^2 \mathbf{h}_i \mathbf{h}_i^T. \tag{12.6}$$

What Eq. 12.6 implies is that $\mathbf{h}_i$ is an eigenvector of $\mathbf{\Sigma_y}$. To see this, applying $\mathbf{\Sigma_y}$ to $\mathbf{h}_i$, one gets

$$\mathbf{\Sigma_y}\mathbf{h}_i = \sigma_i^2 \mathbf{h}_i \mathbf{h}_i^T \mathbf{h}_i = \lambda_i \mathbf{h}_i, \tag{12.7}$$

where λ_i is the corresponding eigenvalue, taking the value of $\lambda_i = \sigma_i^2 \|\mathbf{h}_i\|_2^2$. Of course, when there exists background noise, the noise's presence may create some perturbation to the eigenvector pattern. In the special case of having an uncorrelated noise (so that the noise covariance matrix is of the form $\sigma_\varepsilon^2 \mathbf{I}$), the eigenvector pattern will not be affected; just that the magnitudes of the eigenvalues change (see Exercise 12.2).

The above analysis leads to the signature-based diagnosis procedure summarized in Algorithm 12.1. In an eigenvalue analysis, most commercial software produces the set of eigenvectors in Step 5 to be unit vectors. To facilitate the comparison in Step 7, it is a good idea to normalize the column vectors in $\mathbf{H}$ while creating the signature library.

Algorithm 12.1 Linear modeling and signature-based fault diagnosis.

---

1. Establish a linear model as in Eq. 12.2 for the engineering system of interest.

2. The library of the fault signatures can be formed by taking the column vectors of the system matrix $\mathbf{H}$. This modeling process is conducted offline and based on physical and engineering principles governing the said system.

3. During the online process, collect the data of the response, $\mathbf{y}$.

4. Calculate its sample covariance matrix $\mathbf{S_y}$ and use it as the estimation of $\mathbf{\Sigma_y}$.

5. Compute the eigenvalues and eigenvectors of $\mathbf{S_y}$.

6. Locate the eigenvector corresponding to the largest eigenvalue. This eigenvector is the estimated fault signature.

7. Compare the eigenvector located in Step 6 with the column vectors in the signature library. A statistical test is usually necessary, due to the presence of background noise and the use of the sample covariance matrix $\mathbf{S_y}$. Identify the fault source based on a signature matching criterion.

12.3 SIMILARITY METRICS

In both anomaly detection and fault diagnosis, a central question is how to define the similarity (or dissimilarity) between data instances. It is evident that without a similarity metric, it is impossible to entertain the concept of anomaly, as being anomalous means different, and a data instance is an anomaly because it is so substantially different from the rest of instances in a group. The similarity metric is equally crucial in the mission of diagnosis. For the supervised learning-based approach, the similarity metric is embedded in the loss functions. For the signature-based approach, a similarity metric is used explicitly to decide the outcome in the matching and comparison step. We discuss in this section a few schools of thoughts concerning the similarity metrics.

12.3.1 Norm and Distance Metrics

In an n-dimensional vector space, $\mathcal{H}_{n \times 1}$, the length of a vector or the distance between two points is defined through the concept of norm, which is a function mapping from the vector space to the nonnegative half of the real axis, i.e.,

$$\mathcal{H}_{n \times 1} \longmapsto [0, +\infty).$$

Consider a vector $\boldsymbol{x}$. Its p-norm, $p \geq 1$, is defined as

$$\|\boldsymbol{x}\|_p := (|x_1|^p + |x_2|^p + \ldots + |x_n|^p)^{\frac{1}{p}}. \tag{12.8}$$

When $p = 2$, the above definition is the 2-norm, also known as the Euclidean distance, that we use repeatedly throughout the book. When $p = 1$, the above definition gives the 1-norm, also nicknamed the Manhattan distance. The definition of a p-norm is valid when $p = \infty$, known as the ∞-norm, defined as

$$\|\boldsymbol{x}\|_\infty := \max\{|x_1|, |x_2|, \ldots, |x_n|\}. \tag{12.9}$$

When $0 < p < 1$, the expression in Eq. 12.8 is no longer a norm, because the triangular inequality condition, required in the definition of a valid norm, is not satisfied. When $p = 0$, the expression in Eq. 12.8 is called the 0-norm, which is also not a valid norm. Nevertheless, analysts use the 0-norm as a convenient notation to denote the number of non-zero elements in a vector.

The norm, $\|\boldsymbol{x}\|_p$, is the length of vector $\boldsymbol{x}$ and can be considered as the distance between the point, $\boldsymbol{x}$, and the origin. For two points in the vector space, $\boldsymbol{x}_i$ and $\boldsymbol{x}_j$, the distance between them follows the same definition as in Eq. 12.8 or Eq. 12.9 after replacing $\boldsymbol{x}$ by $\boldsymbol{x}_i - \boldsymbol{x}_j$.

The p-norm has a nice geometric interpretation. Fig. 12.4 illustrates the boundaries defined by $\|\boldsymbol{x}\|_p = 1$ in a 2-dimensional space. The boundary of the 1-norm, $\|\boldsymbol{x}\|_1 = 1$, is the diamond shape, that of the 2-norm, $\|\boldsymbol{x}\|_2 = 1$, is the circle, and that of the ∞-norm, $\|\boldsymbol{x}\|_\infty = 1$, is the square. When $p > 2$, the boundary is a convex shape between the circle and the square. When $p < 1$, even though $\|\boldsymbol{x}\|_p$ is no longer a proper norm, the boundary of $\|\boldsymbol{x}\|_p = 1$ can be visualized on the same plot, as the concave shapes inside the diamond.

The 2-norm, corresponding to the Euclidean distance, is the shortest distance between two points in a Euclidean space. This 2-norm metric measures

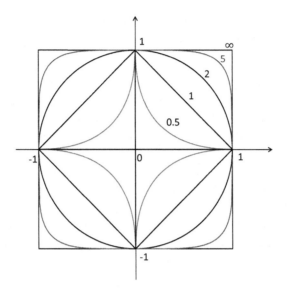

FIGURE 12.4 Boundaries defined by $\|\boldsymbol{x}\|_p = 1$.

the distance in our daily sense. It is widely used in many machine learning and data science methods as the metric defining the loss functions. It is arguably the most widely used similarity metric in anomaly detection.

12.3.2 Inner Product and Angle-Based Metrics

A similarity metric can also be defined as the angle between two vectors. This angle-based metric is particularly popular in the signature-based fault diagnosis.

To define the angle, the concept of inner product needs to be added to a vector space. In an n-dimensional vector space, $\mathcal{H}_{n\times 1}$, the inner product of two vectors, $\boldsymbol{x}$ and $\boldsymbol{y}$, is defined as

$$\langle \boldsymbol{x}, \boldsymbol{y} \rangle := \boldsymbol{x}^T \boldsymbol{y} = \sum_{i=1}^{n} x_i y_i, \tag{12.10}$$

where $\langle \cdot, \cdot \rangle$ is the notation used to denote an inner product. Given this definition, it is established that $\langle \boldsymbol{x}, \boldsymbol{x} \rangle = \|\boldsymbol{x}\|_2^2$.

In a Euclidean space, the angle, θ, formed by a pair of vectors can be defined by using the inner product, such that

$$\theta = \arccos \left(\frac{\langle \boldsymbol{x}, \boldsymbol{y} \rangle}{\|\boldsymbol{x}\|_2 \|\boldsymbol{y}\|_2} \right). \tag{12.11}$$

See Fig. 12.5, left panel, for an illustration.

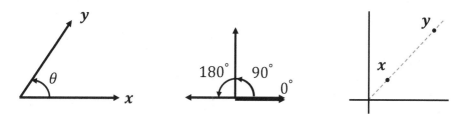

FIGURE 12.5 Angle-based similarity metric. Left panel: θ between two vectors; middle panel: three angular relationships; right panel: two data points on the same line in a vector space.

This angle, θ, can be used as a measure of similarity between two vectors. Take a look at Fig. 12.5, middle panel. If $\theta = 0°$, meaning that the two vectors are parallel and they point to the same direction, then these two vectors are considered the same, subject possibly to a difference in magnitude. If $\theta = 90°$, then the two vectors are said to be orthogonal to each other and they bear no similarity. If $\theta = 180°$, meaning that the two vectors are parallel but they point to the opposite directions, these vectors could still be considered the same, if the pointing direction does not matter in the context of an application. For this reason, some of the angle-based similarity criteria considers only the acute angles formed by two vectors.

The distance-based similarity metric and the angle-based similarity metric may serve different purposes in detection and diagnosis. The distance between two vectors depends on the lengths of them, but the angle does not. Look at the two data points in Fig. 12.5, right panel, which are on the same line but at different locations. The elements in x are proportional to those in y, so that the angle between x and y is zero. If the two data points are considered different, then, the angle-based metric cannot signal such difference; the distance-based metric must be used instead. In some applications, such as in the signature-based diagnosis, however, what matters is the pattern exhibited by the relative magnitudes among the elements in a vector, rather than the absolute magnitudes. Recall that in the signature-based diagnosis, eigenvectors are normalized to be unit vectors, so that the vector lengths are neutralized. In that circumstance, the angle-based measure is a better metric. The distance-based metric can still be used if the vectors involved are normalized before comparison.

One advantage of using the angle-based similarity metric is its robustness in a high-dimensional space, as compared to the distance-based metric. When comparing two vectors, x and y, in an n-dimensional space, the Euclidean distance, $\|x - y\|_2$, is affected more by the background noise embedded in the two vectors than the angle between them.

The distance-based and angle-based metrics can be connected. Recall the kernel function $K(\boldsymbol{x}_i, \boldsymbol{x}_j)$ used in the formulation of SVM—revisit Eqs. 2.47 and 2.48. Note that the kernel function is the exponential of the Euclidean distance between $\boldsymbol{x}_i$ and $\boldsymbol{x}_j$. Consider a reproducing kernel Hilbert space induced by the defined kernel function, $K(\cdot, \cdot)$. This RKHS is spanned by a possibly infinite set of basis functions, denoted as $\boldsymbol{\phi}(\boldsymbol{x}) = (\phi_1(\boldsymbol{x}), \phi_2(\boldsymbol{x}), \ldots, \phi_\ell(\boldsymbol{x}), \ldots)$. The theory of RKHS [86] tells us that

$$K(\boldsymbol{x}_i, \boldsymbol{x}_j) = \langle \boldsymbol{\phi}(\boldsymbol{x}_i), \boldsymbol{\phi}(\boldsymbol{x}_j) \rangle,$$

which connects the distance-based metric in the left-hand side with the angle-based metric in the right-hand side. This above result underlies the well-known *kernel trick*. The RKHS basis functions on the right-hand side provides theoretical foundation for how an unknown function is reconstructed by learning through the training data. But the basis functions themselves are difficult to express analytically in closed forms. On the other hand, the kernel functions, such as the radial basis kernel in Eq. 2.48, can be easily expressed in simple, closed forms. With the equality above, one does not have to worry about the RKHS basis function, $\boldsymbol{\phi}(\cdot)$, but can simply use the corresponding kernel function, $K(\cdot, \cdot)$, instead. This substitute is the trick referred to as the kernel trick.

12.3.3 Statistical Distance

The Mahalanobis distance [140] used in Chapter 7 is also known as the statistical distance. A statistical distance is to measure the distance between a data point from a distribution or two data points in a vector space by accounting for the variance structure associated with the vector space.

Consider an $\boldsymbol{x}$ and a $\boldsymbol{y}$ from the multivariate normal distribution, $\mathcal{N}(\boldsymbol{\mu}, \boldsymbol{\Sigma})$. The statistical distance between them is defined as

$$\text{MD}(\boldsymbol{x}, \boldsymbol{y}) := \sqrt{(\boldsymbol{x} - \boldsymbol{y})^T \boldsymbol{\Sigma}^{-1} (\boldsymbol{x} - \boldsymbol{y})}. \tag{12.12}$$

This expression is equivalent to Eq. 7.3.

The statistical distance, $\text{MD}(\boldsymbol{x}, \boldsymbol{\mu})$, measures the distance between the observation of $\boldsymbol{x}$ and the distribution, $\mathcal{N}(\boldsymbol{\mu}, \boldsymbol{\Sigma})$. Its interpretation is that the likelihood of obtaining $\boldsymbol{x}$ as an observation from $\mathcal{N}(\boldsymbol{\mu}, \boldsymbol{\Sigma})$ can be quantified by this statistical distance—the smaller the distance is, more likely that $\boldsymbol{x}$ will be observed (or the larger the likelihood of the observation).

The statistical distance between two samples is a weighted distance, whereas the Euclidean distance is an un-weighted distance. Given the same Euclidean distance between two points, their respective statistical distance could be different and is in fact re-scaled by the variance along the direction of the distance in question. Intuitively speaking, variance implies uncertainty. The vector space embodying the data are re-shaped by the level of uncertainty. Along the axis of low uncertainty, the scale is magnified (an old one mile could count as ten), whereas along the axis of high uncertainty, the scale is suppressed (an old ten miles may count only as one).

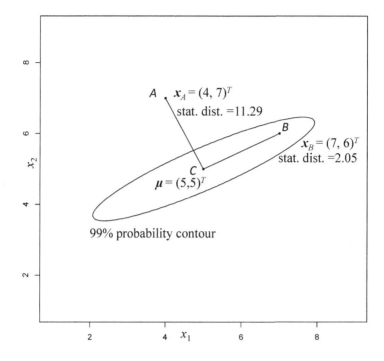

FIGURE 12.6 Statistical distance versus Euclidean distance. The elliptic contour is the 99% probability contour of a bivariate normal distribution.

As illustrated in Fig. 12.6, points A and B have the same Euclidean distance from the distribution center, C, but the respective statistical distances are very different. $\mathrm{MD}(A, C)$ is a whole lot greater than $\mathrm{MD}(B, C)$, because the vector, AC, aligns with the direction of a much smaller variance than the vector, BC. The distance between any point on the 99% probability contour and C is the same, although the respective Euclidean distance varies.

12.3.4 Geodesic Distance

When a vector space is unstructured, so that any pair of points in the space can reach each other in a straight line, that straight line is the shortest path between the pair of points, and the distance between them is measured by the corresponding Euclidean distance. But when a space is structured or curved, meaning that certain pathways are no longer possible, then, the shortest path between two points may not be a straight line anymore. One example is the shortest flight route between two cities on the surface of the earth. Because the flight route is constrained by the earth's surface and the shortest route

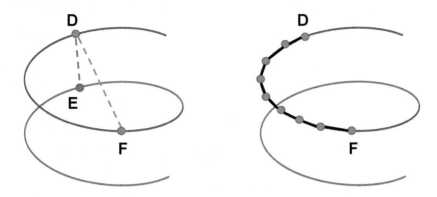

FIGURE 12.7 Left panel: geodesic distance versus Euclidean distance in a space with a swirl structure; right panel: the geodesic distance is approximated by the summation of a series of Euclidean distances associated with the short hops.

between two cities is a curved line, not a straight line. The distance in a structured space is measured by a geodesic distance.

Fig. 12.7 presents an example, inspired by the Swiss roll example presented by Tenenbaum et al. [213, Fig. 3]. Suppose that the data are constrained along the swirl structure in the space. Given this structural constraint, the distance between D and F is shorter than that between D and E, because along the curve and going from D, one reaches F first before reaching E. The distance between D and F along the curve, or that between D and E, is the geodesic distance between them. Should the space is unstructured, then the straight line linking D and E is shorter than that linking D and F. The implication is that using the pairwise Euclidean distance in all circumstances could mislead a data science algorithm to wrongly rank the similarity between data instances—in this case, using the Euclidean distance deems E more similar to D, in spite of the fact that the opposite is true. Tenenbaum et al. [213] raise the awareness of the existence of structures in data spaces and its impact on machine learning tasks.

Computing the geodesic distance can be complicated. This distance is usually approximated by the summation of the short hops using a series of intermediate points in between; see Fig. 12.7, right panel. The distance of any of the short hops is still calculated using a Euclidean distance. But the summation is no longer Euclidean. In Section 12.5, a minimum spanning tree (MST) is used to capture the structure of the underlying data space. The geodesic distance between any two points is approximated by the shortest path linking them through the MST.

12.4 DISTANCE-BASED METHODS

In this section, a few methods focusing on detecting local, pointwise anomalies are introduced.

12.4.1 Nearest Neighborhood-based Method

The concept of k-nearest neighborhood is explained in Section 5.3.1. Ramaswamy et al. [172] base their definition of anomaly on the distance of a point's k-th nearest neighbor.

Denote by $D^k(x_i)$ the distance between the k-th nearest neighbor of x_i and x_i itself. Ramaswamy et al.'s method is to compute $D^k(x_i)$ for every data point in a dataset, rank them from the largest to the smallest, and declare the first N_o data instances as anomalies. In this method, the neighborhood size, k, and the anomaly cut-off, N_o, are prescribed. The resulting method is referred to as the kNN anomaly detection.

Angiulli and Pizzuti [9] follow this kNN idea but argue that instead of using the distance between the k-th nearest neighbor and the target data point, one should use the summation of all distances from the most nearest neighbor to the k-th nearest neighbor. Angiulli and Pizzuti call this summation the weight of data point x_i. Same as in the kNN anomaly detection, this weight is used to rank all data points and classify the top N_o points as anomalies. This revised nearest neighborhood-based method is referred to as kNNW, with the "W" implying "weight."

12.4.2 Local Outlier Factor

Breunig et al. [22] introduce a local outlier factor (LOF) method that makes use of the k-th nearest neighbor distance. First, for a given neighborhood size k, Breunig et al. introduce the concept of a reachability distance. Denoted by RD^k, the reachability distance of the data point, x_*, with respect to an arbitrary point in the dataset, x_i, is

$$\mathrm{RD}^k(x_*, x_i) = \max\{D^k(x_i), D(x_*, x_i)\}, \qquad (12.13)$$

where $D(x_*, x_i)$ is the Euclidean distance between the two data points. Basically, the reachability distance is a lower bound truncated distance. When the two points are too close, their reachability distance is no smaller than the k-th nearest neighbor distance, whereas when the two data points are far away enough, their reachability distance is the actual distance between them. In a sense, the concept of reachability distance is like putting a shield on x_i. The point x_* can reach to x_i up to its k-th nearest neighbor but not nearer. Breunig et al. state that using the reachability distance reduces the statistical fluctuation of the actual distance and exerts a smoothing effect.

Next, Breunig et al. [22] want to quantify the reachability density of points in the neighborhood of x_*. Points that have a lower reachability density than its neighbors are deemed anomalies. Note that here the term "density" does not mean a probability density but rather the number of data points per unit volume.

A local reachability density (LRD) is defined as

$$\text{LRD}(\boldsymbol{x}_*) = \frac{1}{\left(\sum_{\boldsymbol{x}_i \in \mathfrak{N}_k(\boldsymbol{x}_*)} \text{RD}^k(\boldsymbol{x}_*, \boldsymbol{x}_i)\right) / |\mathfrak{N}_k(\boldsymbol{x}_*)|}, \tag{12.14}$$

where $\mathfrak{N}_k(\boldsymbol{x}_*)$ is the k-nearest neighborhood of $\boldsymbol{x}_*$ and $|\cdot|$ takes the cardinality of a set. LRD of $\boldsymbol{x}_*$ is the inverse of the average reachability distance using data points in the k-nearest neighborhood of the same point. When the average reachability distance is large, the density is low.

Finally, Breunig et al. [22] define their anomaly score as the ratio of the average local reachability density of $\boldsymbol{x}_*$'s k-nearest neighbors over the local rechability density of $\boldsymbol{x}_*$ and label it as LOF, such as

$$\text{LOF}(\boldsymbol{x}_*) = \frac{\left(\sum_{\boldsymbol{x}_i \in \mathfrak{N}_k(\boldsymbol{x}_*)} \text{LRD}(\boldsymbol{x}_i)\right) / |\mathfrak{N}_k(\boldsymbol{x}_*)|}{\text{LRD}(\boldsymbol{x}_*)}. \tag{12.15}$$

The smaller the local density of $\boldsymbol{x}_*$, the higher its LOF, and more likely it is an anomaly. The LOF scores, once computed for all data points, are used to rank the data instances. The tope N_o instances are declared anomalies.

12.4.3 Connectivity-based Outlier Factor

Tang et al. [207] argue that the reachability density proposed by Breunig et al. [22] only considers the distances but does not consider the connectivity among neighborhood points. Yet, a low density does not always imply an anomaly. Rather, one should look at the degree of isolation of the said data point, which can be measured by the lack of connectivity. In other words, a data point that is less connected to other data points in a neighborhood is more likely an anomaly. Tang et al. state that "*isolation can imply low density, but the other direction is not always true.*"

Tang et al. [207] introduce a connectivity-based outlier factor (COF) score, which is in spirit similar to the LRD ratio used in Eq. 12.15, but the respective LRD is replaced with a connectivity-based distance metric.

Tang et al. [207] first define the distance between two non-empty sets, $\mathcal{X}$ and $\mathcal{Y}$, that are also disjoint, i.e., $\mathcal{X} \bigcap \mathcal{Y} = \emptyset$, such that

$$D(\mathcal{X}, \mathcal{Y}) = \min\{D(\boldsymbol{x}, \boldsymbol{y}) : \forall \boldsymbol{x} \in \mathcal{X} \text{ and } \boldsymbol{y} \in \mathcal{Y}\}. \tag{12.16}$$

Consider a target point, $\boldsymbol{x}_*$, to be evaluated. Tang et al. [207] iteratively build a k-nearest neighborhood for $\boldsymbol{x}_*$ and establish the sequence of connection. The procedure is outlined in Algorithm 12.2. In this algorithm, $\mathcal{G}_t$ records all the neighbor points and $\mathcal{E}_t$ records the local, pairwise connection steps linking $\boldsymbol{x}_*$ from the nearest point to the farthest point in the neighborhood. This neighborhood, $\mathcal{G}_t$, is different from $\mathfrak{N}_k(\boldsymbol{x}_*)$ in principle, as $\mathfrak{N}_k(\boldsymbol{x}_*)$ is decided based purely on pairwise distances without considering the sequence of connection. This sequence of connection information is what Tang et al. argue makes all the difference between COF and LOF.

Algorithm 12.2 Build the locally connected k-nearest neighborhood for $\boldsymbol{x}_*$. Let $\mathcal{D} = \{\boldsymbol{x}_i\}_{i=1}^N$ be the original dataset and k be the prescribed neighborhood size.

1. Let $t = 1$, $\mathcal{G}_t = \{\boldsymbol{x}_*\}$, $\boldsymbol{x}_{(0)} = \boldsymbol{x}_*$, $\mathcal{D}_t = \mathcal{D}\backslash\boldsymbol{x}_{(0)}$, and $\mathcal{E}_t = \emptyset$.

2. Find $\boldsymbol{x}_{(t)} \in \mathcal{D}_t$, such that $D(\mathcal{G}_t, \boldsymbol{x}_{(t)})$ is minimized.

3. Augment $\mathcal{G}_t$ such that $\mathcal{G}_t = \mathcal{G}_t \cup \{\boldsymbol{x}_{(t)}\}$.

4. Let $\mathcal{E}_t = \mathcal{E}_t \cup \{(\boldsymbol{x}_{(t-1)}, \boldsymbol{x}_{(t)})\}$.

5. Let $\mathcal{D}_t = \mathcal{D}_t\backslash\boldsymbol{x}_{(t)}$.

6. Let $t = t + 1$, and repeat from Step 2 until there are $k + 1$ elements in $\mathcal{G}_t$ (or k elements besides $\boldsymbol{x}_*$).

Once the neighborhood and its connectivity are established, Tang et al. [207] introduce the following connectivity-based distance, or as they call it, the *chaining distance*, denoted by $\mathrm{CD}_{\mathcal{G}}(\boldsymbol{x}_*)$, such that

$$\mathrm{CD}_{\mathcal{G}}(\boldsymbol{x}_*) = \frac{1}{k}\sum_{i=1}^{k}\frac{2(k+1-i)}{k+1}D(\boldsymbol{x}_{(i-1)}, \boldsymbol{x}_{(i)}), \qquad (12.17)$$

where $(\boldsymbol{x}_{(i-1)}, \boldsymbol{x}_{(i)})$ is the i-th element in $\mathcal{E}$. Obviously, CD above is a weighted average distance, with a higher weight given to the connections closer to $\boldsymbol{x}_*$ and a lower weight given to the connections farther away from $\boldsymbol{x}_*$. Tang et al. choose the weight such that when $D(\boldsymbol{x}_{(i-1)}, \boldsymbol{x}_{(i)})$ is the same for all i's, the weight coefficients are summed to one (see Exercise 12.9). Tang et al. [207] define their COF score, under a given k, as

$$\mathrm{COF}(\boldsymbol{x}_*) = \frac{\mathrm{CD}_{\mathcal{G}}(\boldsymbol{x}_*)}{\sum_{\boldsymbol{x}_i \in \mathcal{G}(\boldsymbol{x}_*)}\mathrm{CD}_{\mathcal{G}}(\boldsymbol{x}_i)/|\mathcal{G}(\boldsymbol{x}_*)|}. \qquad (12.18)$$

The use of COF follows that of LOF. The larger a COF, the more likely the corresponding data instance is deemed an anomaly.

12.4.4 Subspace Outlying Degree

To deal with high-dimensional data problems, analysts choose to consider a subset of the original features, an action commonly known as *dimension reduction*. The potential benefit of looking into a subspace is that data points distributed indistinguishably in the full dimensional space could deviate significantly from others when examined in a proper subspace. On the other hand, the danger of using a subspace approach is that if not chosen properly, the difference between a potential anomaly and normal points may disappear altogether in the subspace. It is obvious that the tricky part of a subspace approach is how to find the right subspace.

Kriegel et al. [125] present a subspace outlying degree (SOD) method. The method works as follows. First, Kriegel et al. compute the variance of the set of the reference points in $\mathcal{D}$ as

$$\text{VAR} = \frac{1}{|\mathcal{D}|} \sum_{\boldsymbol{x}_* \in \mathcal{D}} D(\boldsymbol{x}_*, \boldsymbol{\mu})^2, \tag{12.19}$$

where $\boldsymbol{\mu}$ is the average position of the points in $\mathcal{D}$. Similarly, compute the variance along the i-th attribute as

$$\text{VAR}_i = \frac{1}{|\mathcal{D}|} \sum_{\boldsymbol{x}_* \in \mathcal{D}} D((\boldsymbol{x}_*)_i, \boldsymbol{\mu}_i)^2, \tag{12.20}$$

where $(\boldsymbol{x}_*)_i$ and $\boldsymbol{\mu}_i$ are, respectively, the i-th element in $\boldsymbol{x}_*$ and $\boldsymbol{\mu}$.

Then, create a subspace vector based on the following criterion, where n is the dimension of the original data space and α is a constant,

$$\nu_i = \begin{cases} 1, & \text{if } \text{VAR}_i < \alpha \cdot \frac{\text{VAR}}{n}, \\ 0, & \text{otherwise.} \end{cases} \tag{12.21}$$

Kriegel et al. [125] suggest setting $\alpha = 0.8$. When ν_i in Eq. 12.21 is one, the corresponding variable is selected to construct the subspace; otherwise, the corresponding variable is skipped over. Denote the resulting subspace by $\mathcal{S}$, which is represented by the vector $\boldsymbol{\nu} = (\nu_1, \nu_2, \ldots, \nu_n)$. In a three-dimensional space, for instance, $\boldsymbol{\nu} = (1, 0, 1)$ indicates that the selected subspace is spanned by the first and third axes.

To measure the deviation of a data point, $\boldsymbol{x}_*$, from a subspace, Kriegel et al. [125] use the following formula,

$$D(\boldsymbol{x}_*, \mathcal{S}) := \sqrt{\sum_{i=1}^{n} \nu_i((\boldsymbol{x}_*)_i - \boldsymbol{\mu}_i)^2}. \tag{12.22}$$

Kriegel et al. further define their SOD score as

$$\text{SOD}(\boldsymbol{x}_*) := \frac{D(\boldsymbol{x}_*, \mathcal{S})}{\|\boldsymbol{\nu}\|_1}, \tag{12.23}$$

where $\|\boldsymbol{\nu}\|_1$ is the number of dimensions of the selected subspace. A higher SOD score means that $\boldsymbol{x}_*$ deviates from the selected subspace a lot and is thus likely an anomaly.

12.5   GEODESIC DISTANCE-BASED METHOD

As explained in Section 12.3.4, the Euclidean-based similarity metric works well in an unstructured data space but could mislead a learning method when there are intrinsic structures in a data space restricting certain pathways connecting data points. In the circumstances of structured data spaces, a geodesic distance ought to be used. The methods introduced in Section 12.4 rely heavily on the use of Euclidean distance to define similarity, with the exception of COF. COF, through the use of the connection sequence, bears certain characteristics of the geodesic distance. In the benchmark case study of Section 12.6.1, COF does perform rather competitively. More recently, Ahmed et al. [4] develop an MST-based unsupervised anomaly detection method that takes full advantage of a geodesic distance-based similarity metric.

12.5.1   Graph Model of Data

The MST-based anomaly detection method employs a minimum spanning tree to approximate the geodesic distances between data points in a structured space and then uses the distance approximation as the similarity metric. The data are modeled as a network of nodes through a graph. Consider a connected undirected graph $G = (U, E)$, where U denotes the collection of vertices or nodes and E represents the collection of edges connecting these nodes as pairs. For each edge $e \in E$, there is a weight associated with it. It could be either the distance between the chosen pair of nodes or the cost to connect them.

A minimum spanning tree is a subset of the edges in E that connects all the nodes together, without any cycles and with the minimum possible total edge weight. This total edge weight, also known as the total length or total cost of the MST, is the summation of the weights of the individual edges. If one uses the Euclidean distance between a pair of nodes as the edge weight, the resulting spanning tree is called a Euclidean MST.

Consider the example in Fig. 12.8, where $U=\{1, 2, 3, 4\}$ and $E =\{e_{12}, e_{13}, e_{14}, e_{23}, e_{24}, e_{34}\}$. All edges in E are all different in length and the edge length order is specified in the left panel of Fig. 12.8. If one wants to connect all the nodes in U without forming a cycle, there could be 16 such combinations with only one having the minimum total edge length. That one is the MST for this connected graph, shown in the right panel of Fig. 12.8. Note that some of the edges in Fig. 12.8 look like having the same length. The edge e_{13} looks even longer than e_{34} and e_{23}. One way to imagine a layout satisfying the edge length order specified in Fig. 12.8 is to envision node #3 not in the same plane formed by node #1, #2, and #4, but hovering in the space and being close to node #1.

Ahmed et al. [4] consider data instances as nodes and the Euclidean distance between any pairs of data points as the edge weight and then construct an MST to connect all the nodes. Specifically, they use the algorithm in [168] to construct an MST. Although the distance between a pair of immediately

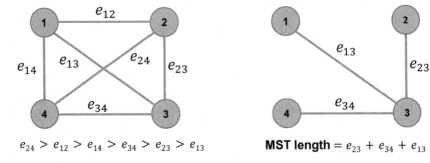

$$e_{24} > e_{12} > e_{14} > e_{34} > e_{23} > e_{13}$$

MST length $= e_{23} + e_{34} + e_{13}$

FIGURE 12.8 Formation of a minimum spanning tree. Left panel: the complete graph; right panel: the minimum spanning tree. (Reprinted with permission from Ahmed et al. [4].)

connected nodes is still Euclidean, the distance between a general pair of nodes (i.e., data points) is not. Rather, it is the summation of many small-step, localized Euclidean distances hopping from one data point to another point. The MST reflects the connectedness among data points in a structured space and the MST-based distance approximates the geodesic distance between two data points.

12.5.2 MST Score

Ahmed et al.'s method [4] focuses on detecting the local, pointwise anomalies. But the MST nature enables the method to incorporate a preprocessing step that can identify potential anomalous clusters. The idea is simple. First, build a global MST using all the data points. After the formation of the global MST, one can look for an unusual long edge and deem it as the connecting edge between an anomalous cluster and the rest of the MST. Once the long edge is disconnected, it separates the MST into two groups, and the smaller group is considered an anomalous cluster. The "unusual" aspect can be verified through a statistical test, say, longer than the 99th percentile of all edge lengths in the original MST. This preprocessing step can be iteratively applied to the remaining larger group, until no more splitting.

For detecting the local anomalies, one needs to go into the neighborhood level. Same as the local anomaly detection methods introduced in Section 12.4, two parameters are prescribed for the MST-based approach: the neighborhood size, k, and the cut-off threshold, N_o, for declaring anomalies.

Denote by $\mathcal{D}$ the set of data points to be examined, where $\mathcal{D}$ could be the whole original dataset or could be the remaining set of data after the anomalous cluster is removed. For a given data point in $\mathcal{D}$, first isolate its k nearest neighbors and treat them as this data point's neighborhood. Then, build an MST in this neighborhood. The localized, neighborhood-based MSTs

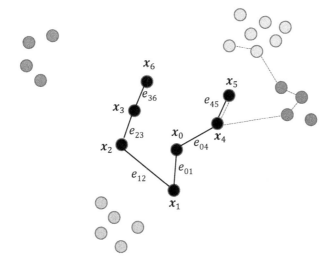

FIGURE 12.9 Local MST and LoMST score. The total edge weight of the local MST for x_0 is its LoMST score, i.e., $W_{x_0} = e_{01} + e_{12} + e_{23} + e_{36} + e_{04} + e_{45}$. (Reprinted with permission from Ahmed et al. [4].)

are referred to as *local MSTs* (LoMST). The total edge length of the LoMST associated a data point is called the LoMST score for this data point and is considered the metric measuring its connectedness with the rest of the points in the neighborhood as well as how far away it is from its neighbors. The LoMST is then used as the differentiating metric to signal the possibility that the said data point may be an anomaly.

Consider the illustrating example in Fig. 12.9. Suppose that one has chosen $k = 6$ and start with data point x_0. Then, one can locate its neighbors as x_1, x_2, x_3, x_4, x_5 and x_6. The MST construction algorithm connects x_0 to its neighbors in the way as shown in Fig. 12.9. For x_0, the total edge weight is $W_{x_0} = e_{01} + e_{12} + e_{23} + e_{36} + e_{04} + e_{45}$, which is used as the LoMST score for x_0. This procedure will be repeated for other data points. Fig. 12.9 does show another MST, which is for x_5 in the dotted edges.

The LoMST score for a selected data instance is compared with its neighbor's score. The steps of comparison are to be repeated to cover all nodes in $\mathcal{D}$. Then the comparison scores are normalized to be between zero and one. The resulting normalized scores are also referred to as the LoMST scores, as long as there is no ambiguity in the context. The normalized LoMST scores are sorted in decreasing order, so that the top N_o instances are flagged as anomalies. The method is summarized in Algorithm 12.3.

Algorithm 12.3 MST-based anomaly detection method. Input: dataset $\mathcal{D}$, rows represent observations and columns represent attributes, the neighborhood size, k, and the cut-off level for identifying anomalies, N_o. Output: the anomaly index set, $\widehat{\mathcal{O}}$.

---

1. Preprocess to remove obvious anomalous clusters, if necessary.

2. Set $\mathcal{T} = \emptyset$, $\widehat{\mathcal{O}} = \emptyset$, $i = 1$.

3. For $\boldsymbol{x}_i \in \mathcal{D}$, determine its k nearest neighbors and save them in U_i.

4. Construct a complete graph using nodes in U_i. The resulting edges are in the set, E_i.

5. Construct a local MST using the edges in E_i.

6. Calculate the total length of $\boldsymbol{x}_i$'s local MST and denote it as $W_{\boldsymbol{x}_i}$.

7. Calculate the average of the total length of the LoMSTs associated with all nodes in U_i, and denote the average as $\overline{W}_i$.

8. Calculate the LoMST score for $\boldsymbol{x}_i$ as $T_i = W_{\boldsymbol{x}_i} - \overline{W}_i$.

9. Let $\mathcal{T} = \mathcal{T} \cup \{T_i\}$ and $i = i + 1$. Re-iterate from Step 3 until all data points in $\mathcal{D}$ are visited.

10. Normalize the scores stored in $\mathcal{T}$ to be between 0 and 1.

11. Rank the normalized scores in $\mathcal{T}$ in descending order.

12. Identify the top N_o scores and store the corresponding observations as point anomalies in $\widehat{\mathcal{O}}$.

12.5.3 Determine Neighborhood Size

For the neighborhood-based methods, including LoMST, an important parameter to be specified prior to the execution of a respective algorithm is the neighborhood size k. The difficulty in choosing k in an unsupervised setting is that methods like cross validation that work for supervised learning do not apply here. Ahmed et al. [4] advocate an approach based on the following observations, illustrated in Fig. 12.10 using two benchmark datasets.

When Ahmed et al. [4] plot the average LoMST scores for a broad range of k (here 1–100), they observe that at small k values, the average LoMST score tends to fluctuate, but as they keep increasing k, the average LoMST score tends to become stable at certain point. This leads to the understanding that when a proper k is chosen and the structure of the data is revealed, the label of the instances become fixed; such stability is reflected in a less fluctuating LoMST score. If one keeps increasing k, there is the possibility that the data structure becomes mismatch with the assigned number of clusters, so that the current assignments of anomalies and normal instances become destabilized. Consequently, the average LoMST score could fluctuate again. Based on this observation, a sensible strategy in choosing k is to select a range of k where the average LoMST scores are stable. If there are more than one stable ranges, analysts are advised to select the first one.

Let us look at the examples in Fig. 12.10. For the `Cardiotocography` dataset, Ahmed et al. [4] choose a k ranging from 27–47 and for the `Glass` dataset, they choose a k ranging from 70–95. Within the identified stable range, which k to choose matters less. What Ahmed et al. suggest is to select the k value that returns the maximum standard deviation of the LoMST scores, because by maximizing the standard deviation among the LoMST scores, it increases the separation between the normal instances and anomalous instances and facilitates the detection mission.

12.6 CASE STUDY

12.6.1 Benchmark Cases

As mentioned early in this chapter, one profound difficulty in assessing the performance of anomaly detection method is due to the lack of knowledge of the ground truth. Luckily, Campos et al. [30] published a comprehensive survey on the topic of anomaly detection and collected and shared 20 benchmark datasets of wide varieties, for each of which the anomalies are known. Readers are directed to the following website to retrieve the datasets, i.e., `http://www.dbs.ifi.lmu.de/research/outlier-evaluation/`, that hosts the supplemental material and online repository of [30]. Several versions of these datasets are available. These versions mainly differ in terms of the preprocessing steps used. What is used in this section is the normalized version of the datasets in which all the missing values are removed and categorical variables are converted into numerical format.

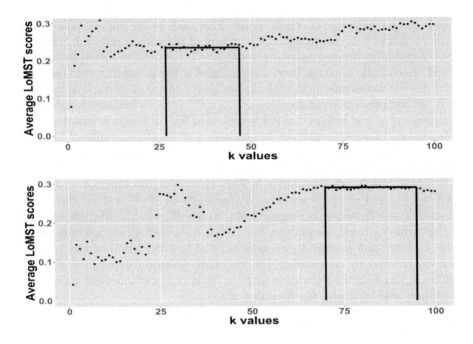

FIGURE 12.10 Selection of the neighborhood size k. Top panel: the `Cardiotocography` dataset; bottom panel: the `Glass` dataset. (Reprinted with permission from Ahmed et al. [4].)

TABLE 12.1   Benchmark datasets and the performance of LoMST under different data-to-attribute ratios.

Dataset	N/n	N	n	$|\mathcal{O}|$	Rank (Best k)	Rank (Practical k)
Arrhythmia	2	450	259	12	1	1
SpamBase	81	4,601	57	280	1	1
KDDcup99	1,479	60,632	41	200	3	2
WPBC	6	198	33	47	1	3
Ionosphere	11	351	32	126	2	3
WBC	51	367	30	10	1	1
ALOI	1,852	50,000	27	1,508	7	7
Parkinson	9	195	22	5	1	1
Annthyroid	343	7,200	21	347	11	10
Waveform	164	3,443	21	100	1	1
Cardiotocography	102	2,126	21	86	2	3
Lymphography	8	148	19	6	1	1
Pendigits	617	9,868	16	20	1	2
HeartDisease	21	270	13	7	1	1
PageBlocks	548	5,473	10	99	4	7
Stamps	38	340	9	16	1	6
Shuttle	113	1,013	9	13	1	1
WDBC	13	454	9	10	2	2
Pima	96	768	8	26	1	1
Glass	31	214	7	9	1	1

Source: Ahmed et al. [4]. With permission.

Table 12.1 summarizes the basic characteristics of the 20 benchmark datasets, for each of which N is the total number of data amount, n is the number of attributes, and $|\mathcal{O}|$ is the number of anomalies. Table 12.1 also presents the data-to-attribute ratio, N/n, which is rounded to the nearest integer. The two columns under the "Rank" headings in Table 12.1 are to be explained later.

Ahmed et al. [4] focus on the neighborhood-based approaches. These approaches include LoMST, a few others introduced in Section 12.4, and additional approaches that are not explained in this chapter but their details can be found in [30]. All these methods have a common parameter, which is the neighborhood size, k. Since the SPC method can also be used for the purpose of anomaly detection, it is included in the study as well. The specific SPC technique used is the Hotelling T^2 control chart. In total, 14 competitive methods are evaluated in this benchmark case study. The unexplained acronyms used in the subsequent tables are: local density factor (LDF), outlier detection using indegree number (ODIN), simplified local outlier factor (SLOF), local outlier probabilities (LoOP), influenced outlierness (INFLO), local distance-based outlier factor (LDOF), fast angle-based outlier detection (FABOD), and kernel density estimation outlier score (KDEOS).

TABLE 12.2 Performance comparison based on the best k value.

	LoMST	COF	LDF	kNN	ODIN	LOF	kNNW
Better	6	0	3	1	1	0	1
Equal	7	5	5	2	1	3	2
Close	5	10	6	7	8	8	7
Worse	2	5	6	10	10	9	10
Rank	2.2	3.3	3.8	5.0	7.7	5.1	4.5

	SLOF	LoOP	INFLO	LDOF	FABOD	KDEOS	SPC
Better	0	0	0	0	0	0	1
Equal	2	2	2	1	2	2	0
Close	6	6	5	7	5	1	1
Worse	12	12	13	12	13	17	18
Rank	5.9	5.8	6.2	7.9	7.0	8.9	11.7

Source: Ahmed et al. [4]. With permission.

Campos et al. [30] do not specify how to select k. They try a range of k values (from 1 to 100) to obtain all the results and then choose the best k value for each method. In the first comparison, Ahmed et al. [4] follow the same approach and label this as the "best k" comparison. The performance criterion used here is the precision at N_o, $P@N_o$, explained in Section 12.1.3. The results are presented in Table 12.2. Please note that the best k value in Table 12.2 may be different for respective methods. To better reflect the detection capability as they are compared to one another, Ahmed et al. break down the comparative performance into four major categories, namely *Better*, *Equal*, *Close* and *Worse*. Their meanings are as follows:

- *Better*, if a method is uniquely the best among all candidates.

- *Equal*, if a method ties with other methods for being the best.

- *Close*, if a method's correct detections are within 20% of the best alternative(s).

- *Worse*, if a method's correct detections are more than 20% lower than that of the best alternative(s).

If analysts rank each of the 14 methods in a scale of 1 to 14 according to its actual performance in relative to others, then an average relative rank can be calculated for each method. The average ranks are reported in the row below the four categories—the smaller, the better. The LoMST method's average rank is 2.2, with some closest competitors being COF (3.3), LDF (3.8), kNNW (4.5), kNN (5.0) and LOF (5.1).

Understandably, the "best k" is not practical, as analysts in reality do not know the anomalies while selecting k. Using the strategy devised in Section 12.5.3 to select a practical value of k for LoMST, Ahmed et al. [4] apply

TABLE 12.3   Performance comparison based on the practical k value.

	LoMST	COF	LDF	kNN	ODIN	LOF	kNNW
Better	5	2	1	1	2	0	0
Equal	5	1	4	5	1	3	4
Close	7	11	4	6	8	7	9
Worse	3	6	11	8	9	10	7
Rank	2.8	4.2	5.7	5.3	6.7	6.1	4.3

	SLOF	LoOP	INFLO	LDOF	FABOD	KDEOS	SPC
Better	0	0	0	0	2	0	1
Equal	1	1	2	1	2	0	0
Close	8	7	7	6	6	3	3
Worse	11	12	11	13	10	17	16
Rank	6.5	5.6	5.8	7.6	4.9	11.7	8.7

Source: Ahmed et al. [4]. With permission.

the same k to the other 12 alternative methods that need this value (SPC does not need to know k). The performance comparison based on the practical k is presented in Table 12.3, arranged in the same way as Table 12.2. Under the "practical k," the average rank of LoMST is 2.8, with some closest competitors being COF(4.2), kNNW (4.3), FABOD(4.9), kNN (5.3), and LDF (5.7).

Ahmed et al. [4] summarize LoMST's performance with respect to the data size and report the values under the last two columns in Table 12.1. Looking at the two extremes, the case of the highest number of observations ($N = 60,632$, KDDcup99), which is the one having the second highest N/n ratio, versus the case of the highest number of attributes ($n = 259$, Arrhythmia), which is also the one having the lowest N/n ratio, LoMST performs on top in both cases. It can also be noticed that on two of the datasets when the number of anomalies are too numerous (over a few hundreds to more than one thousand), LoMST does not do well enough. In hindsight, it makes intuitive sense, as LoMST is designed to find the local, pointwise anomalies, which, when existing, should be of a relatively small amount.

Another note is about the computational complexity of LoMST, which comes from two major sources. First, one needs to conduct the k-nearest neighbor search based on the chosen k. Then, for each observation, one needs to build a local MST using its k-nearest neighbors. For the first step, Ahmed et al. [4] use the fast approximate nearest neighbor searching approach [10, 16] with a time complexity of $O(nN \log N) + O(kn \log N)$. The first time complexity component, $O(nN \log N)$, represents the time to build the tree structure, whereas the second component, $O(kn \log N)$, represents the k-nearest neighborhood query time for a single observation. In the second step, building the local MST has the time complexity of $O(|U| \log |E|)$. Because LoMST is a localized, neighborhood-based MST, the values of $|U|$ and $|E|$ depend on k but usually remain small. The neighborhood search and the local MST step are repeated N times, whereas the tree structure building is a one-time action. As such, the total complexity of the LoMST algorithm is approximately $O(nN \log N) + O(N[kn \log N + |U| \log |E|])$.

12.6.2 Hydropower Plant Case

The hydropower data initially received are time-stamped for a duration of seven months. The data are collected from different functional areas in the plant, such as the turbines, generators, bearings, and so on. The data records are collected at 10-minute intervals. Missing data are common. There are a total of 9,508 observations (rows in a data table) and 222 attribute variables (columns in a data table). Each row has a time stamp assigned to it. Attribute variables are primarily temperatures, vibrations, pressure, harmonic values, active power, etc. Before applying the anomaly detection method, some basic preprocessing and statistical analysis are performed in order to clean the data. To maintain the similarity with the 20 benchmark datasets, the hydropower data are normalized. After the preprocessing, the total number of observations comes down to 9,219. In summary, for the hydropower plant dataset, $N = 9,219$, $n = 222$, and $|\mathcal{O}|$ is unknown. The details of data preprocessing can be found in [4].

Besides LoMST, Ahmed et al. [4] apply two other popular anomaly detection methods to the same hydropower data. The two other methods are LOF [22] and SOD [125]. Here, LOF is used as a representative of the neighborhood-based methods, while SOD is a representative of the subspace methods. Note that SOD is not included in the benchmark case study, because the methods included in Section 12.6.1 are primarily neighborhood based. Had SOD been included in the benchmark study, under the practical k setting, SOD would have an average rank of 6.7 and the number of instances of its detection in the four categories would be 0, 0, 8, and 12, respectively.

For all three methods, one needs to specify the value of the nearest neighbors k. In this case, it would be great to get some suggestions from the domain experts about the possible size of an anomaly cluster based on their knowledge of the system. Ahmed et al. [4] indeed receive advice from their industrial collaborators, consider the value of k in a range of 10–20, and find the anomaly scores for each k in the range. Then they take the average of the resulting anomaly scores as the final anomaly score for each of the data instances. For SOD, one needs to select two parameters instead of one—one parameter is k, while the other one is the number of reference points for forming the subspace. To maintain the comparability with LoMST and LOF, Ahmed et al. choose $k = 15$ for SOD, which is the middle point of the above-suggested range. Concerning the number of reference points, it should be smaller than k but not too small, because an overly small number of reference points may render instability in SOD. Ahmed et al. explore a few options and finally settle on ten. Below ten, the SOD method becomes unstable.

By applying the three methods, the top 100 anomalies identified are shown in Table 12.4. It can be observed that after the top 30 time stamps, no new anomaly-prone days emerge. Rather similar data patterns keep repeating themselves with slight differences in the time stamps. We therefore skip some rows after the top 30 stamps in Table 12.4.

The performance of the three methods are reasonably consistent, as 14 out of the top 30 probable anomalies identified by these methods are common, represented by an asterisk (*) in Table 12.4. This similarity continues even if one goes beyond the top 30 time stamps. By looking closely at these top 100 time stamps, one may find that there are some particular days and certain time chunks on these days which are more prone to anomaly. Such a pattern makes sense, as for most of engineering systems, random anomalies happen sparsely, while systematic anomalies take place in a cluster.

These three methods work differently, especially since SOD is from another family of methods. In spite of their differences, they have returned similar results for the hydropower dataset. This serves as a way to cross validate the detection outcomes while the true anomalies are unknown.

The three methods do have differences in their detection outcomes. The LOF method completely misses the 4th of July time stamps, although almost half of the 100 top anomaly-prone time stamps returned by both SOD and LoMST methods belong to that day. Ahmed et al. [4] investigate the issue and find that most of the time stamps in July correspond to low active power, whereas the time stamps from the 4th of July are marked with abnormally high active power. The rest of the attributes behave almost identically as other days of July. When the number of attributes increases, the nearest neighborhood methods usually fall short of detecting anomalies if abnormal values only happen to one or a few dimensions. This is where the subspaces method can do better (assuming that the abnormal value subspace is successfully identified). It is therefore not surprising to see that SOD detects these anomalies correctly. It is also encouraging to see that LoMST is capable of detecting these anomalies as well, even though LoMST is a neighborhood-based method. On the other hand, LOF and LoMST, being local methods, successfully identified point anomalies on the 16th of April, whereas the SOD method fails to identify them. In a nutshell, the LoMST method attains the merit of subspace-based methods without losing the benefits of local neighborhood-based methods.

Anomaly detection does not immediately reveal the root causes of the anomalies. Finding out which variables contribute to the anomalies helps the domain expert to verify the root causes and then fix them. This calls for a diagnostic follow-up. As presented in Section 12.2.1, after the anomalies are identified, the original unsupervised learning problem is translated into a supervised learning problem. While doing so, Ahmed et al. [4] discard the July 4th time stamps from the top 100 time stamps, as the reason for that happening is straightforward. They proceed with the remaining of the top 100 time stamps and assign them a response value of one (meaning an anomaly), and all other data records outside the top 100 time stamps a response value of zero (meaning a normal condition). Ahmed et al. then build a CART using the

TABLE 12.4   Summary of the top 100 anomalies in the hydropower dataset. Events followed by asterisk (*) are the common ones identified by all three methods in the top 30 time stamps.

LoMST	LOF	SOD
1/12/2016 11:20*	1/12/2016 11:30*	9/14/2015 8:00
9/14/2015 1:00*	9/14/2015 1:00*	1/12/2016 11:30*
1/2/2016 9:10*	9/14/2015 1:10*	9/13/2015 7:00*
1/11/2016 12:00*	1/12/2016 11:20*	7/4/2015 8:30
1/2/2016 9:30*	1/9/2016 6:50*	7/4/2015 8:20
7/4/2015 11:20	1/2/2016 9:10*	9/14/2015 1:50
7/4/2015 11:10	9/14/2015 8:00	7/4/2015 5:40
7/4/2015 11:30	1/2/2016 9:20*	1/11/2016 12:00*
1/9/2016 6:50*	1/9/2016 18:30	9/14/2015 1:00*
7/4/2015 10:40	9/14/2015 8:10*	10/3/2015 14:40
1/2/2016 9:20*	9/13/2015 7:00*	7/4/2015 5:50
7/4/2015 9:40	9/14/2015 2:00	10/13/2015 8:15*
9/13/2015 7:00*	1/11/2016 14:40	9/14/2015 1:10*
1/11/2016 1:30*	1/11/2016 13:50	11/2/2015 9:56
7/4/2015 9:50	1/11/2016 12:00*	7/4/2015 6:30
9/16/2015 10:50*	1/11/2016 13:00	7/4/2015 4:30
9/14/2015 14:10	9/16/2015 10:50*	1/2/2016 9:20*
9/14/2015 13:50	9/17/2015 11:30	9/14/2015 2:00
7/4/2015 5:20	10/3/2015 14:40	9/14/2015 8:10*
9/14/2015 1:10*	1/2/2016 21:40	7/4/2015 4:20
1/12/2016 11:40	4/16/2015 23:10	1/11/2016 1:30*
1/12/2016 11:30*	10/4/2015 3:10	1/2/2016 21:40
9/14/2015 13:20	10/13/2015 8:15*	7/4/2015 4:40
7/4/2015 4:50	10/14/2015 23:35	9/16/2015 10:50*
9/14/2015 8:10*	10/14/2015 23:15	1/2/2016 1:30*
4/16/2015 23:10	1/2/2016 9:30*	1/11/2016 14:40
4/16/2015 16:00	4/16/2015 16:00	1/2/2016 9:10*
10/13/2015 8:15*	11/2/2015 9:56	1/12/2016 11:20*
7/4/2015 5:30	1/11/2016 1:30*	1/9/2016 6:50*
7/4/2015 9:10	1/11/2016 11:50	9/14/2015 13:05
....	....	....
9/13/2015 19:10	10/13/2015 5:45	7/4/2015 0:00
7/4/2015 4:40	1/2/2016 21:00	7/4/2015 5:30
7/4/2015 6:20	1/9/2016 18:20	7/4/2015 6:20
7/4/2015 5:00	1/9/2016 18:40	7/4/2015 6:50
7/4/2015 13:50	9/14/2015 0:40	7/4/2015 7:00
....	....	....
9/14/2015 8:00	9/14/2015 2:10	7/4/2015 7:50
1/9/2016 18:30	9/14/2015 8:20	10/13/2015 5:45
1/11/2016 13:00	9/14/2015 8:30	9/16/2015 11:00
1/11/2016 11:50	9/14/2015 8:40	10/13/2015 6:35
7/4/2015 9:30	10/14/2015 8:15	10/13/2015 8:25
....	....	....
10/14/2015 7:25	1/9/2016 18:00	10/4/2015 4:30
10/14/2015 7:35	1/11/2016 11:40	10/4/2015 4:20
7/4/2015 10:10	10/13/2015 6:35	1/2/2016 21:50
7/4/2015 10:20	10/4/2015 23:10	1/11/2016 13:50
7/4/2015 10:30	9/13/2015 19:30	9/13/2015 21:40
....	....	....
7/4/2015 10:50	1/9/2016 18:10	1/11/2016 12:10
1/11/2016 13:40	1/11/2016 13:40	1/9/2016 18:30
1/11/2016 13:50	9/13/2015 19:40	10/4/2015 3:10
10/4/2015 3:10	10/14/2015 7:55	1/11/2016 11:30
1/9/2016 18:40	1/11/2016 11:30	10/14/2015 7:25

Source: Ahmed et al. [4]. With permission.

R package `rpart` with the package's default parameter values. The resulting CART is in fact presented in Fig. 12.3, and the interpretation of the tree model and how it helps with fault diagnosis is discussed in Section 12.2.1.

GLOSSARY

ANN: Artificial neural network

CART: Classification and regression tree

CD: Connectivity-based distance, or chaining distance

COF: Connectivity-based outlier factor

CSV: Comma-separated values Excel file format

FABOD: Fast angle-based outlier detection

INFLO: Influenced outlierness

KDEOS: Kernel density estimation outlier score

kNN: k-th nearest neighbor distance-based anomaly detection

kNNW: k nearest neighborhood distances summation

LDF: Local density factor

LDOF: Local distance-based outlier factor

LOF: Local outlier factor

LoMST: Local minimum spanning tree

LoOP: Local outlier probabilities

LRD: Local reachability density

MARS: Multivariate adaptive regression splines

MD: Mahalanobis distance or statistical distance

MST: Minimum spanning tree

ODIN: Outlier detection using indegree number

PCA: Principal component analysis

RD: Reachability distance

RKHS: Reproducing kernel Hilbert space

SLOF: Simplified local outlier factor

SOD: Subspace outlying degree

SPC: Statistical process control

SQC: Statistical quality control

SVM: Support vector machine

EXERCISES

12.1 Speaking of the types of anomaly, one type of anomaly is called the contextual anomaly, meaning that an observation may be an anomaly when its covariates take certain values, but the same observation may not be an anomaly when its covariates are under a different condition. Please come up with some examples explaining the contextual anomaly.

12.2 Given Eq. 12.4, derive the expression of Eq. 12.5. Also prove that the eigenvector of $\boldsymbol{\Sigma_y}$ remains $\mathbf{h}_i$ if the covariance matrix of ε is $\sigma_\varepsilon^2 \mathbf{I}$.

12.3 Consider the p-norm in Section 12.3.1.

    a. Recall that $\|\boldsymbol{x}\|_0$ is used to represent the number of nonzero elements in $\boldsymbol{x}$. Show that the 0-norm is not a valid norm. Which requirement for a valid norm is not satisfied by the 0-norm?

    b. Prove that

$$\|\boldsymbol{x}\|_\infty \le \|\boldsymbol{x}\|_2 \le \|\boldsymbol{x}\|_1 \le \sqrt{n}\|\boldsymbol{x}\|_2 \le n\|\boldsymbol{x}\|_\infty,$$

    where n is the dimension of $\boldsymbol{x}$.

12.4 One kernel function used in SVM or other machine learning methods is the dth-degree polynomial kernel, defined as

$$K(\boldsymbol{x}, \boldsymbol{x}') = (1 + \langle \boldsymbol{x}, \boldsymbol{x}' \rangle)^d.$$

For the polynomial kernel, it is possible to write down the mapping functions, $\phi(\boldsymbol{x})$, explicitly and in a closed form, so that the relationship,

$$K(\boldsymbol{x}, \boldsymbol{x}') = \langle \phi(\boldsymbol{x}), \phi(\boldsymbol{x}') \rangle,$$

can be proven. Please do this for $d = 2$ by writing explicitly the expression of $K(\boldsymbol{x}, \boldsymbol{x}')$ and that of $\phi(\boldsymbol{x})$, thereby showing and confirming the equality in the above equation.

12.5 To construct the plot of Fig. 12.6, consider a bivariate Gaussian distribution, $\mathcal{N}(\boldsymbol{\mu}, \boldsymbol{\Sigma})$, where

$$\boldsymbol{\mu} = \begin{pmatrix} 5 \\ 5 \end{pmatrix}, \quad \text{and}$$

$$\boldsymbol{\Sigma} = \begin{pmatrix} 1 & 0.45 \\ 0.45 & 0.25 \end{pmatrix}.$$

The position of point A is $\boldsymbol{x}_A = (4, 7)^T$ and that of point B is $\boldsymbol{x}_B = (7, 6)^T$. The mean of the Gaussian distribution is represented by point C, whose position is $\boldsymbol{\mu} = (5, 5)^T$.

a. Please compute both the Euclidean and statistical distances of AC and BC. Do your results confirm the statement in Section 12.3.3?

b. Please compute the statistical distance between any point on the 99% probability contour to point C.

c. Can you drive the general formula for the statistical distance between any point on the $100(1 - \alpha)\%$ probability contour and point C for a given $\alpha \in [0, 1]$?

12.6 Can you please come up with an example illustrating a circumstance for which the kNN anomaly detection cannot correctly identify the anomaly but kNNW could?

12.7 Consider the anomaly examples presented in Fig. 12.1. Let us remove points, A_1 and A_2, and the cluster of C_3 for the moment. Then, imagine that the points in C_2 are clustered more tightly than those in C_1. Suppose that the points in both C_1 and C_2 are uniformly scattered. The nearest neighbor distance in C_1 is five (whatever unit it may be), whereas the nearest neighbor distance in C_2 is one. The distance between A_3 and its nearest neighbor in C_2 is three. Consider $k = 1$ and $k = 2$.

a. Use this example to show that both kNN and kNNW anomaly detection methods are ineffective to flag A_3 as an anomaly.

b. Show that LOF is capable of detecting A_3 as an anomaly.

12.8 Consider an example presented by Tang et al. [207]. Figure 3 in [207] is modified and then presented below, where the distance between points #1 and #2 is five, that between #2 and #7 is three, that between #4 and #5 is six, and the distance between any other two adjacent points on the line is one.

a. Let $k = 10$, and use this example to compute the chaining distance for point #1. Please show explicitly the sets of $\mathcal{G}$ and $\mathcal{E}$ associated with point #1.

b. Select a k and show that $\mathcal{G}$ could be different from $\mathfrak{N}$ for point #1.

12.9 Prove that in Eq. 12.17, the coefficients are summed to one when $D(\boldsymbol{x}_{(i-1)}, \boldsymbol{x}_{(i)})$ is the same for all i's.

12.10 In Section 12.4.4, we state that "[T]he danger of using a subspace approach is that if not chosen properly, the difference between a potential anomaly and normal points may disappear altogether in the subspace." One popular method to select a subspace is principal component analysis (PCA). PCA is to find the subspaces that account for the largest variance in data. Please come up with an example in a two-dimensional space, such that once the one-dimensional subspace of the largest variance is selected and data projected onto that space, the intrinsic structure existing in the original data disappears. In other words, the otherwise distinguishable two classes of data in the original two-dimensional data space are no longer separable in the wrongly selected one-dimensional subspace.

12.11 In Fig. 12.8, there are sixteen options of spanning trees that can connect all nodes without forming a cycle. Please list all sixteen choices and show that the one presented in the right panel is indeed the minimum spanning tree.

Bibliography

[1] T. Ackermann. *Wind Power in Power Systems*. John Wiley & Sons, New York, 2005.

[2] M. S. Adaramola and P.-Å. Krogstad. Experimental investigation of wake effects on wind turbine performance. *Renewable Energy*, 36(8):2078–2086, 2011.

[3] P. Agarwal and L. Manuel. Extreme loads for an offshore wind turbine using statistical extrapolation from limited field data. *Wind Energy*, 11:673–684, 2008.

[4] I. Ahmed, A. Dagnino, and Y. Ding. Unsupervised anomaly detection based on minimum spanning tree approximated distance measures and its application to hydropower turbines. *IEEE Transactions on Automation Science and Engineering*, 16(2):654–667, 2019.

[5] D. Aigner, C. A. K. Lovell, and P. Schmidt. Formulation and estimation of stochastic frontier production function models. *Journal of Econometrics*, 6(1):21–37, 1977.

[6] P. Ailliot and V. Monbet. Markov-switching autoregressive models for wind time series. *Environmental Modelling & Software*, 30:92–101, 2012.

[7] H. Akaike. A new look at the statistical model identification. *IEEE Transactions on Automatic Control*, 19(6):716–723, 1974.

[8] T. W. Anderson. *The Statistical Analysis of Time Series*. John Wiley & Sons, New York, 1971.

[9] F. Angiulli and C. Pizzuti. Outlier mining in large high-dimensional datasets. *IEEE Transactions on Knowledge and Data Engineering*, 17(2):203–215, 2005.

[10] S. Arya, D. M. Mount, N. S. Netanyahu, R. Silverman, and A. Y. Wu. An optimal algorithm for approximate nearest neighbor searching fixed dimensions. *Journal of the ACM*, 45(6):891–923, 1998.

[11] R. D. Banker, A. Charnes, and W. W. Cooper. Some models for estimating technical and scale inefficiencies in data envelopment analysis. *Management Science*, 30(9):1078–1092, 1984.

[12] R. J. Barthelmie, K. S. Hansen, S. T. Frandsen, O. Rathmann, J. G. Schepers, W. Schlez, J. Phillips, K. Rados, A. Zervos, E. S. Politis, and P. K. Chaviaropoulos. Modelling and measuring flow and wind turbine wakes in large wind farms offshore. *Wind Energy*, 12(5):431–444, 2009.

[13] R. J. Barthelmie and L. E. Jensen. Evaluation of wind farm efficiency and wind turbine wakes at the Nysted offshore wind farm. *Wind Energy*, 13(6):573–586, 2010.

[14] R. J. Barthelmie, S. C. Pryor, S. T. Frandsen, K. S. Hansen, J. G. Schepers, K. Rados, W. Schlez, A. Neubert, L. E. Jensen, and S. Neckelmann. Quantifying the impact of wind turbine wakes on power output at offshore wind farms. *Journal of Atmospheric and Oceanic Technology*, 27(8):1302–1317, 2010.

[15] O. Belghazi and M. Cherkaoui. Pitch angle control for variable speed wind turbines using genetic algorithm controller. *Journal of Theoretical and Applied Information Technology*, 39:5–10, 2012.

[16] J. L. Bentley. Multidimensional binary search trees used for associative searching. *Communications of the ACM*, 18(9):509–517, 1975.

[17] R. J. Bessa, V. Miranda, A. Botterud, J. Wang, and E. M. Constantinescu. Time adaptive conditional kernel density estimation for wind power forecasting. *IEEE Transactions on Sustainable Energy*, 3(4):660–669, 2012.

[18] A. Betz. *Introduction to the Theory of Flow Machines*. Pergamon Press, 1966.

[19] F. D. Bianchi, H. De Battista, and R. J. Mantz. *Wind Turbine Control Systems*. Springer-Verlag, London, 2007.

[20] G. E. P. Box, G. M. Jenkins, and G. C. Reinsel. *Time Series Analysis: Forecasting and Control*. John Wiley & Sons, New York, 4th edition, 2008.

[21] L. Breiman. Random forests. *Machine Learning*, 45(1):5–32, 2001.

[22] M. M. Breunig, H.-P. Kriegel, R. T. Ng, and J. Sander. LOF: Identifying density-based local outliers. In *Proceedings of the 2000 ACM SIGMOD International Conference on Management of Data*, volume 29, pages 93–104. ACM, 2000.

[23] B. G. Brown, R. W. Katz, and A. H. Murphy. Time series models to simulate and forecast wind speed and wind power. *Journal of Climate and Applied Meteorology*, 23:1184–1195, 1984.

[24] E. Byon. Wind turbine operations and maintenance: A tractable approximation of dynamic decision-making. *IIE Transactions*, 45:1188–1201, 2013.

[25] E. Byon and Y. Ding. Season-dependent condition-based maintenance for a wind turbine using a partially observed Markov decision process. *IEEE Transactions on Power Systems*, 25:1823–1834, 2010.

[26] E. Byon, L. Ntaimo, and Y. Ding. Optimal maintenance strategies for wind turbine systems under stochastic weather conditions. *IEEE Transactions on Reliability*, 59:393–404, 2010.

[27] E. Byon, L. Ntaimo, C. Singh, and Y. Ding. Wind energy facility reliability and maintenance. In P. M. Pardalos, S. Rebennack, M. V. F. Pereira, N. A. Iliadis, and V. Pappu, editors, *Handbook of Wind Power Systems: Optimization, Modeling, Simulation and Economic Aspects*, pages 639–672. Springer, Berlin Heidelberg, 2013.

[28] E. Byon, E. Pérez, Y. Ding, and L. Ntaimo. Simulation of wind farm operations and maintenance using DEVS. *Simulation–Transactions of the Society for Modeling and Simulation International*, 87:1093–1117, 2011.

[29] E. Byon, A. K. Shrivastava, and Y. Ding. A classification procedure for highly imbalanced class sizes. *IIE Transactions*, 42(4):288–303, 2010.

[30] G. O. Campos, A. Zimek, J. Sander, R. J. G. B. Campello, B. Micenková, E. Schubert, I. Assent, and M. E. Houle. On the evaluation of unsupervised outlier detection: Measures, datasets, and an empirical study. *Data Mining and Knowledge Discovery*, 30(4):891–927, 2016.

[31] J. A. Carta, P. Ramírez, and S. Velázquez. Influence of the level of fit a density probability function to wind-speed data on the WECS mean power output estimation. *Energy Conversion and Management*, 49:2647–2655, 2008.

[32] J. A. Carta, P. Ramírez, and S. Velázquez. A review of wind speed probability distributions used in wind energy analysis—case studies in the Canary Islands. *Renewable and Sustainable Energy Reviews*, 13:933–955, 2009.

[33] A. Carvalho, M. C. Gonzalez, P. Costa, and A. Martins. Issues on performance of wind systems derived from exploitation data. In *Proceedings of the 35th Annual Conference of IEEE Industrial Electronics*, pages 3599–3604, Porto, Portugal, 2009.

[34] J. E. Cavanaugh. Unifying the derivations of the Akaike and corrected Akaike information criteria. *Statistics and Probability Letters*, 31:201–208, 1997.

[35] N. V. Chawla, K. W. Bowyer, L. O. Hall, and W. P. Kegelmeyer. SMOTE: Synthetic minority over-sampling technique. *Journal of Artificial Intelligence Research*, 16:321–357, 2002.

[36] H. A. Chipman, E. I. George, and R. E. McCulloch. BART: Bayesian additive regression trees. *The Annals of Applied Statistics*, 4:266–298, 2010.

[37] Y. Choe, E. Byon, and N. Chen. Importance sampling for reliability evaluation with stochastic simulation models. *Technometrics*, 57:351–361, 2015.

[38] S. G. Coles. *An Introduction to Statistical Modeling of Extreme Values.* Springer, New York, 2001.

[39] N. Conroy, J. P. Deane, and B. P. Ó. Gallachóir. Wind turbine availability: Should it be time or energy based? A case study in Ireland. *Renewable Energy*, 36(11):2967–2971, 2011.

[40] A. Crespo, J. Hernández, and S. Frandsen. Survey of modelling methods for wind turbine wakes and wind farms. *Wind Energy*, 2(1):1–24, 1999.

[41] N. A. C. Cressie. *Statistics for Spatial Data.* John Wiley & Sons, New York, 1991.

[42] N. A. C. Cressie, S. Burden, W. Davis, P. Krivitsky, P. Mokhtarian, T. Suesse, and A. Zammit-Mangion. Capturing multivariate spatial dependence: Model, estimate and then predict. *Statistical Science*, 30(2):170–175, 2015.

[43] N. A. C. Cressie and C. K. Wikle. *Statistics for Spatio-Temporal Data.* John Wiley & Sons, New York, 2011.

[44] P. Crochet. Adaptive Kalman filtering of 2-metre temperature and 10-metre wind-speed forecasts in Iceland. *Meteorological Applications*, 11(2):173–187, 2004.

[45] P. De Boer, D. P. Kroese, S. Mannor, and R. Y. Rubinstein. A tutorial on the cross-entropy method. *Annals of Operations Research*, 134:19–67, 2005.

[46] D. G. T. Denison, C. C. Holmes, B. K. Mallick, and A. F. M. Smith. *Bayesian Methods for Nonlinear Classification and Regression.* John Wiley & Sons, New York, 2002.

[47] D. G. T. Denison, B. K. Mallick, and A. F. M Smith. Bayesian MARS. *Statistics and Computing*, 8:337–346, 1998.

[48] M. Derby. DOE's perspective. In *The 2011 National Renewable Energy Laboratory Workshop on Wind Turbine Condition Monitoring*, Broomfield, CO, 2011. September 19.

[49] Y. Ding, D. Ceglarek, and J. Shi. Fault diagnosis of multi-station manufacturing processes by using state space approach. *Transactions of ASME, Journal of Manufacturing Science and Engineering*, 124:313–322, 2002.

[50] Y. Ding, J. Tang, and J. Z. Huang. Data analytics methods for wind energy applications. In *Proceedings of ASME Turbo Expo 2015: Turbine Technical Conference and Exposition, GT2015-43286*, pages 1–9, Montreal, Canada, 2015. June 15–19.

[51] A. G. Drachmann. Heron's windmill. *Centaurus*, 7:145–151, 1961.

[52] S. D. Dubey. Normal and Weibull distributions. *Naval Research Logistics Quarterly*, 14:69–79, 1967.

[53] V. Dubourg, B. Sudret, and F. Deheeger. Metamodel-based importance sampling for structural reliability analysis. *Probabilistic Engineering Mechanics*, 33:47–57, 2013.

[54] J. Durbin. The fitting of time-series models. *Review of the International Statistical Institute*, 28(3):233–244, 1960.

[55] B. Efron and R. Tibshirani. *An Introduction to the Bootstrap*. Chapman & Hall/CRC Press, New York, 1993.

[56] A. Emami and P. Noghreh. New approach on optimization in placement of wind turbines within wind farm by genetic algorithms. *Renewable Energy*, 35(7):1559–1564, 2010.

[57] Y. Ephraim and N. Merhav. Hidden Markov processes. *IEEE Transactions on Information Theory*, 48:1518–1569, 2002.

[58] B. Everitt. *The Cambridge Dictionary of Statistics*. Cambridge University Press, Cambridge, UK, 1998.

[59] A. A. Ezzat, M. Jun, and Y. Ding. Spatio-temporal asymmetry of local wind fields and its impact on short-term wind forecasting. *IEEE Transactions on Sustainable Energy*, 9(3):1437–1447, 2018.

[60] A. A. Ezzat, M. Jun, and Y. Ding. Spatio-temporal short-term wind forecast: A calibrated regime-switching method. *The Annals of Applied Statistics*, in press, 2019.

[61] J. Fan and T. H. Yim. A cross-validation method for estimating conditional densities. *Biometrika*, 91:819–834, 2004.

[62] F. Felker. The status and future of wind energy. Technical report, National Wind Technology Center, Boulder, CO, 2009. Available at http://www.ncsl.org/documents/energy/Felker0609.pdf.

[63] L. M. Fitzwater, C. A. Cornell, and P. S. Veers. Using environmental contours to predict extreme events on wind turbines. In *Proceedings of the 2003 ASME Wind Energy Symposium, WIND2003-865*, pages 244–285, 2003. Reno, Nevada.

[64] R. Fletcher. *Practical Methods of Optimization*. John Wiley & Sons, New York, 2th edition, 1987.

[65] J. Fogle, P. Agarwal, and L. Manuel. Towards an improved understanding of statistical extrapolation for wind turbine extreme loads. *Wind Energy*, 11:613–635, 2008.

[66] S. Fortunato. Community detection in graphs. *Physics Reports*, 486:75–174, 2010.

[67] K. Freudenreich and K. Argyriadis. Wind turbine load level based on extrapolation and simplified methods. *Wind Energy*, 11:589–600, 2008.

[68] J. Friedman. Multivariate adaptive regression splines. *Annals of Statistics*, 19(1):1–67, 1991.

[69] M. Fuentes. A high frequency kriging approach for non-stationary environmental processes. *Environmetrics*, 12(5):469–483, 2001.

[70] P. M. O. Gebraad, F. W. Teeuwisse, J. W. Wingerden, P. A. Fleming, S. D. Ruben, J. R. Marden, and L. Y. Pao. Wind plant power optimization through yaw control using a parametric model for wake effects—a CFD simulation study. *Wind Energy*, 19(1):95–114, 2016.

[71] G. Giebel, R. Brownsword, G. Kariniotakis, M. Denhard, and C. Draxl. The state-of-the-art in short-term prediction of wind power: A literature overview. Technical report, Risø National Laboratory, Roskilde, Denmark, 2011. Available at http://www.anemos-plus.eu/images/pubs/deliverables/aplus.deliverable_d1.2.stp_sota_v1.1.pdf.

[72] T. Gneiting. Nonseparable, stationary covariance functions for space-time data. *Journal of the American Statistical Association*, 97(458):590–600, 2002.

[73] T. Gneiting. Making and evaluating point forecasts. *Journal of the American Statistical Association*, 106:746–762, 2011.

[74] T. Gneiting, M. Genton, and P. Guttorp. Geostatistical space-time models, stationarity, separability and full symmetry. In B. Finkenstadt, L. Held, and V. Isham, editors, *Statistical Methods for Spatio-Temporal Systems*, chapter 4. Chapman & Hall/CRC, 2007.

[75] T. Gneiting, K. Larson, K. Westrick, M. G. Genton, and E. Aldrich. Calibrated probabilistic forecasting at the Stateline wind energy center: The regime-switching space-time method. *Journal of the American Statistical Association*, 101:968–979, 2006.

[76] T. Gneiting and A. E. Raftery. Strictly proper scoring rules, prediction, and estimation. *Journal of the American Statistical Association*, 102:359–378, 2007.

[77] M. Goldstein and S. Uchida. A comparative evaluation of unsupervised anomaly detection algorithms for multivariate data. *PLoS ONE*, 11(4):e0152173:1–31, 2016.

[78] P. J. Green. Reversible jump Markov chain Monte Carlo computation and Bayesian model determination. *Biometrika*, 82:711–732, 1995.

[79] C. Gu. *Smoothing Spline ANOVA*. Springer-Verlag, New York, 2013.

[80] H. Guo, S. Watson, P. Tavner, and J. Xiang. Reliability analysis for wind turbines with incomplete failure data collected from after the date of initial installation. *Reliability Engineering and System Safety*, 94:1057–1063, 2009.

[81] S. T. Hackman. *Production Economics: Integrating the Microeconomic and Engineering Perspectives*. Springer-Verlag, Heidelberg, 2008.

[82] B. Hahn, M. Durstewitz, and K. Rohrig. Reliability of wind turbines—experiences of 15 years with 1,500 WTs. In J. Peinke, P. Schaumann, and S. Barth, editors, *Wind Energy: Proceedings of the Euromech Colloquium*, pages 329–332. Springer, 2007.

[83] P. Hall, J. Racine, and Q. Li. Cross-validation and the estimation of conditional probability. *Journal of the American Statistical Association*, 99:154–163, 2004.

[84] P. Hall and L. Simar. Estimating a changepoint, boundary, or frontier in the presence of observation error. *Journal of the American Statistical Association*, 97(458):523–534, 2002.

[85] K. S. Hansen, R. J. Barthelmie, L. E. Jensen, and A. Sommer. The impact of turbulence intensity and atmospheric stability on power deficits due to wind turbine wakes at Horns Rev wind farm. *Wind Energy*, 15(1):183–196, 2012.

[86] T. Hastie, R. Tibshirani, and J. Friedman. *The Elements of Statistical Learning: Data Mining, Inference, and Prediction*. Springer, New York, 2nd edition, 2009.

[87] T. J. Hastie and R. J. Tibshirani. *Generalized Additive Models*. Chapman & Hall/CRC, 1990.

[88] M. He, L. Yang, J. Zhang, and V. Vittal. A spatio-temporal analysis approach for short-term forecast of wind farm generation. *IEEE Transactions on Power Systems*, 29(4):1611–1622, 2014.

[89] P. Heidelberger. Fast simulation of rare events in queueing and reliability models. *ACM Transactions on Modeling and Computer Simulation*, 5:43–85, 1995.

[90] E. J. Henley and H. Kumamoto. *Reliability Engineering and Risk Assessment*. Prentice-Hall, 1981.

[91] A. S. Hering and M. G. Genton. Powering up with space-time wind forecasting. *Journal of the American Statistical Association*, 105:92–104, 2010.

[92] C. Hildreth. Point estimates of ordinates of concave functions. *Journal of the American Statistical Association*, 49(267):598–619, 1954.

[93] D. Hinkley. On quick choice of power transformation. *Journal of the Royal Statistical Society: Series C (Applied Statistics)*, 26(1):67–69, 1977.

[94] T. Hofmann, B. Schlkopf, and A. J. Smola. Kernel methods in machine learning. *Annals of Statistics*, 36(3):1171–1220, 2008.

[95] H. Hwangbo, Y. Ding, O. Eisele, G. Weinzierl, U. Lang, and G. Pechlivanoglou. Quantifying the effect of vortex generator installation on wind power production: An academia-industry case study. *Renewable Energy*, 113:1589–1597, 2017.

[96] H. Hwangbo, A. L. Johnson, and Y. Ding. A production economics analysis for quantifying the efficiency of wind turbines. *Wind Energy*, 20:1501–1513, 2017.

[97] H. Hwangbo, A. L. Johnson, and Y. Ding. Power curve estimation: Functional estimation imposing the regular ultra passum law. *Working Paper*, 2018. Available at SSRN: http://ssrn.com/abstract=2621033.

[98] H. Hwangbo, A. L. Johnson, and Y. Ding. Spline model for wake effect analysis: Characteristics of single wake and its impacts on wind turbine power generation. *IISE Transactions*, 50(2):112–125, 2018.

[99] B. J. Hyndman, D. M. Bashtannyk, and G. K. Grunwald. Estimating and visualizing conditional densities. *Journal of Computational and Graphical Statistics*, 5:315–336, 1996.

[100] International Electrotechnical Commission (IEC). *IEC TS 61400-1 Ed. 2: Wind Turbines – Part 1: Design Requirements*. IEC, Geneva, Switzerland, 1999.

[101] International Electrotechnical Commission (IEC). *IEC TS 61400-1 Ed. 3, Wind Turbines – Part 1: Design Requirements*. IEC, Geneva, Switzerland, 2005.

[102] International Electrotechnical Commission (IEC). *IEC TS 61400-12-1 Ed. 1, Wind Turbines – Part 12-1: Power Performance Measurements of Electricity Producing Wind Turbines*. IEC, Geneva, Switzerland, 2005.

[103] International Electrotechnical Commission (IEC). *IEC TS 61400-26-1 Ed. 1, Wind Turbines – Part 26-1: Time-based Availability for Wind Turbine Generating Systems*. IEC, Geneva, Switzerland, 2011.

[104] International Electrotechnical Commission (IEC). *IEC TS 61400-12-2 Ed. 1, Wind Turbines – Part 12-2: Power Performance of Electricity Producing Wind Turbines Based on Nacelle Anemometry*. IEC, Geneva, Switzerland, 2013.

[105] International Electrotechnical Commission (IEC). *IEC TS 61400-26-2 Ed. 1, Wind Turbines – Part 26-2: Production-based Availability for Wind Turbines*. IEC, Geneva, Switzerland, 2014.

[106] S. R. Jammalamadaka and A. SenGupta. *Topics in Circular Statistics*. World Scientific, 2001.

[107] A. H. Jazwinski. Adaptive filtering. *Automatica*, 5:475–485, 1969.

[108] N. O. Jensen. A note on wind generator interaction. Technical report Risø-M, No. 2411, Risø National Laboratory, Roskilde, Denmark, 1983. Available at http://orbit.dtu.dk/files/55857682/ris_m_2411.pdf.

[109] J. Jeon and J. W. Taylor. Using conditional kernel density estimation for wind power density forecasting. *Journal of the American Statistical Association*, 107:66–79, 2012.

[110] P. Jirutitijaroen and C. Singh. The effect of transformer maintenance parameters on reliability and cost: A probabilistic model. *Electric Power System Research*, 72:213–234, 2004.

[111] I. T. Jolliffe. *Principal Component Analysis*. Springer, New York, 2nd edition, 2002.

[112] B. J. Jonkman. *TurbSim User's Guide: Version 1.50*. National Renewable Energy Laboratory, Golden, CO, 2009.

[113] B. J. Jonkman and M. L. Buhl Jr. *FAST User's Guide*. National Renewable Energy Laboratory, Golden, CO, 2005.

[114] M. Jun and M. Stein. An approach to producing space-time covariance functions on spheres. *Technometrics*, 49(4):468–479, 2007.

[115] M. S. Kaiser, M. J. Daniels, K. Furakawa, and P. Dixon. Analysis of particulate matter air pollution using Markov random field models of spatial dependence. *Environmetrics*, 13:615–628, 2002.

[116] R. E. Kalman. A new approach to linear filtering and prediction problems. *Transactions of ASME, Journal of Basic Engineering*, 82(1):35–45, 1960.

[117] G. Kariniotakis. *Renewable Energy Forecasting: From Models to Applications*. Woodhead Publishing, 2017.

[118] R. E. Kass and L. Wasserman. A reference Bayesian test for nested hypotheses and its relationship to the Schwarz criterion. *Journal of the American Statistical Association*, 90:928–934, 1995.

[119] K. Kazor and A. S. Hering. The role of regimes in short-term wind speed forecasting at multiple wind farms. *Stat*, 4(1):271–290, 2015.

[120] N. D. Kelley and B. J. Jonkman. Overview of the TurbSim stochastic inflow turbulence simulator. NREL/TP-500-41137, Version 1.21, National Renewable Energy Laboratory, Golden, Colorado, 2003. Available at https://nwtc.nrel.gov/system/files/TurbSimOverview.pdf.

[121] M. G. Khalfallah and A. M. Koliub. Effect of dust on the performance of wind turbines. *Desalination*, 209(1):209–220, 2007.

[122] R. Killick and I. Eckley. Changepoint: An R package for changepoint analysis. *Journal of Statistical Software*, 58(3):1–19, 2014.

[123] J. Kjellin, F. Bülow, S. Eriksson, P. Deglaire, M. Leijon, and H. Bernhoff. Power coefficient measurement on a 12 kW straight bladed vertical axis wind turbine. *Renewable Energy*, 36(11):3050–3053, 2011.

[124] J. P. C. Kleijnen. *Design and Analysis of Simulation Experiments*. Springer-Verlag, New York, 2008.

[125] H.-P. Kriegel, P. Krger, E. Schubert, and A. Zimek. Outlier detection in axis-parallel subspaces of high dimensional data. In *Advances in Knowledge Discovery and Data Mining: Proceedings of the 13th Pacific-Asia Conference on Knowledge Discovery and Data Mining*, pages 831–838. Springer, Berlin Heidelberg, 2009.

[126] P.-Å. Krogstad and J. A. Lund. An experimental and numerical study of the performance of a model turbine. *Wind Energy*, 15(3):443–457, 2012.

[127] T. Kuosmanen. Representation theorem for convex nonparametric least squares. *The Econometrics Journal*, 11(2):308–325, 2008.

[128] A. Kusiak and Z. Song. Design of wind farm layout for maximum wind energy capture. *Renewable Energy*, 35(3):685–694, 2010.

[129] M. P. Laan, N. N. Sørensen, P.-E. Réthoré, J. Mann, M. C. Kelly, N. Troldborg, J. G. Schepers, and E. Machefaux. An improved k-ϵ model applied to a wind turbine wake in atmospheric turbulence. *Wind Energy*, 18(5):889–907, 2015.

[130] T. J. Larsen and A. M. Hansen. *How 2 HAWC2, the User's Manual.* Risø National Laboratory, Roskilde, Denmark, 2007.

[131] G. Lee, E. Byon, L. Ntaimo, and Y. Ding. Bayesian spline method for assessing extreme loads on wind turbines. *The Annals of Applied Statistics*, 7:2034–2061, 2013.

[132] G. Lee, Y. Ding, M. G. Genton, and L. Xie. Power curve estimation with multivariate environmental factors for inland and offshore wind farms. *Journal of the American Statistical Association*, 110(509):56–67, 2015.

[133] G. Lee, Y. Ding, L. Xie, and M. G. Genton. Kernel Plus method for quantifying wind turbine upgrades. *Wind Energy*, 18:1207–1219, 2015.

[134] G. Li and J. Shi. Application of Bayesian model averaging in modeling long-term wind speed distributions. *Renewable Energy*, 35:1192–1202, 2010.

[135] H. Link, W. LaCava, J. van Dam, B. McNiff, S. Sheng, R. Wallen, M. McDade, S. Lambert, S. Butterfield, and F. Oyague. Gearbox reliability collaborative project report: Findings from phase 1 and phase 2 testing. Technical report, National Renewable Energy Laboratory, Golden, CO, 2011. Available at https://www.nrel.gov/docs/fy11osti/51885.pdf.

[136] P. Louka, G. Galanis, N. Siebert, G. Kariniotakis, P. Katsafados, I. Pytharoulis, and G. Kallos. Improvements in wind speed forecasts for wind power prediction purposes using Kalman filtering. *Journal of Wind Engineering and Industrial Aerodynamics*, 96(12):2348–2362, 2008.

[137] W. Lovejoy. Computationally feasible bounds for partially observed Markov decision processes. *Operations Research*, 39:162–175, 1991.

[138] P. Lynch. *The Emergence of Numerical Weather Prediction: Richardson's Dream.* Cambridge University Press, Cambridge, UK, 2006.

[139] M. Maadooliat, J. Z. Huang, and J. Hu. Integrating data transformation in principal components analysis. *Journal of Computational and Graphical Statistics*, 24(1):84–103, 2015.

[140] P. C. Mahalanobis. On the generalized distance in statistics. *Proceedings of the National Institute of Sciences (Calcutta)*, 2:49–55, 1936.

[141] L. M. Maillart. Maintenance policies for systems with condition monitoring and obvious failures. *IIE Transactions*, 38:463–475, 2006.

[142] L. Manuel, P. S. Veers, and S. R. Winterstein. Parametric models for estimating wind turbine fatigue loads for design. *Transactions of ASME, Journal of Solar Energy Engineering*, 123:346–355, 2001.

[143] M. D. Marzio, A. Panzera, and C. C. Taylor. Smooth estimation of circular cumulative distribution functions and quantiles. *Journal of Nonparametric Statistics*, 24:935–949, 2012.

[144] M. D. Marzio, A. Panzera, and C. C. Taylor. Nonparametric regression for circular responses. *Scandinavian Journal of Statistics*, 40:238–255, 2013.

[145] M. D. Marzio, A. Panzera, and C. C. Taylor. Nonparametric regression for spherical data. *Journal of the American Statistical Association*, 109:748–763, 2014.

[146] P. McKay, R. Carriveau, and D. S.-K. Ting. Wake impacts on downstream wind turbine performance and yaw alignment. *Wind Energy*, 16(2):221–234, 2013.

[147] D. C. Montgomery. *Introduction to Statistical Quality Control*. John Wiley & Sons, New York, 6th edition, 2009.

[148] J. M. Morales, A. J. Conejo, H. Madsen, P. Pinson, and M. Zugno. *Integrating Renewables in Electricity Markets—Operational Problems*. Springer, New York, 2014.

[149] P. Moriarty. Database for validation of design load extrapolation techniques. *Wind Energy*, 11:559–576, 2008.

[150] P. Moriarty, W. E. Holley, and S. Butterfield. Effect of turbulence variation on extreme loads prediction for wind turbines. *Transactions of ASME, Journal of Solar Energy Engineering*, 124:387–395, 2002.

[151] H. Mueller-Vahl, G. Pechlivanoglou, C. N. Nayeri, and C. O. Paschereit. Vortex generators for wind turbine blades: A combined wind tunnel and wind turbine parametric study. In *Proceedings of ASME Turbo Expo 2012: Turbine Technical Conference and Exposition*, volume 6, pages 899–914, Copenhagen, Denmark, 2012. June 11–15.

[152] E. Nadaraya. On estimating regression. *Theory of Probability and Its Applications*, 9:141–142, 1964.

[153] A. Natarajan and W. E. Holley. Statistical extreme load extrapolation with quadratic distortions for wind turbines. *Transactions of ASME, Journal of Solar Energy Engineering*, 130:031017:1–7, 2008.

[154] M. E. J. Newman. Modularity and community structure in networks. *Proceedings of the National Academy of Sciences*, 103(23):8577–8582, 2006.

[155] B. Niu, H. Hwangbo, L. Zeng, and Y. Ding. Evaluation of alternative efficiency metrics for offshore wind turbines and farms. *Renewable Energy*, 128:81–90, 2018.

[156] O. B. Olesen and J. Ruggiero. Maintaining the regular ultra passum law in data envelopment analysis. *European Journal of Operational Research*, 235(3):798–809, 2014.

[157] S. Øye. The effect of vortex generators on the performance of the ELKRAFT 1000 kW turbine. In *Aerodynamics of Wind Turbines: 9th IEA Symposium*, pages 9–14, Stockholm, Sweden, 1995. December 11–12.

[158] C. Pacot, D. Hasting, and N. Baker. Wind farm operation and maintenance management. In *Proceedings of the PowerGen Conference Asia*, pages 25–27, Ho Chi Minh City, Vietnam, 2003.

[159] C. Park, J. Z. Huang, and Y. Ding. A computable plug-in estimator of minimum volume sets for novelty detection. *Operations Research*, 58(5):1469–1480, 2010.

[160] J. M. Peeringa. Extrapolation of extreme responses of a multi-megawatt wind turbine. ECN-C-03-131, Energy Research Centre of the Netherlands, Petten, Netherlands, 2003. Available at http://www.ecn.nl/docs/library/report/2003/c03131.pdf.

[161] E. Pérez, Y. Ding, and L. Ntaimo. Multi-component wind turbine modeling and simulation for wind farm operations and maintenance. *Simulation–Transactions of the Society for Modeling and Simulation International*, 91:360–382, 2015.

[162] S. Pieralli, M. Ritter, and M. Odening. Efficiency of wind power production and its determinants. *Energy*, 90:429–438, 2015.

[163] P. Pinson. Wind energy: Forecasting challenges for its operational management. *Statistical Science*, 28:564–585, 2013.

[164] P. Pinson, L. Christensen, H. Madsen, P. E. Sørensen, M. H. Donovan, and L. E. Jensen. Regime-switching modelling of the fluctuations of offshore wind generation. *Journal of Wind Engineering and Industrial Aerodynamics*, 96(12):2327–2347, 2008.

[165] P. Pinson, H. A. Nielsen, H. Madsen, and T. S. Nielsen. Local linear regression with adaptive orthogonal fitting for wind power application. *Statistics and Computing*, 18:59–71, 2008.

[166] A. Pourhabib, J. Z. Huang, and Y. Ding. Short-term wind speed forecast using measurements from multiple turbines in a wind farm. *Technometrics*, 58(1):138–147, 2016.

[167] A. Pourhabib, B. K. Mallick, and Y. Ding. Absent data generating classifier for imbalanced class sizes. *Journal of Machine Learning Research*, 16:2695–2724, 2015.

[168] R. C. Prim. Shortest connection networks and some generalizations. *Bell Labs Technical Journal*, 36(6):1389–1401, 1957.

[169] J. M. Prospathopoulos, E. S. Politis, K. G. Rados, and P. K. Chaviaropoulos. Evaluation of the effects of turbulence model enhancements on wind turbine wake predictions. *Wind Energy*, 14(2):285–300, 2011.

[170] M. Puterman. *Markov Decision Processes*. John Wiley & Sons, New York, 1994.

[171] A. E. Raftery. Bayesian model selection in social research. *Sociological Methodology*, 25:111–163, 1995.

[172] S. Ramaswamy, R. Rastogi, and K. Shim. Efficient algorithms for mining outliers from large datasets. In *Proceedings of the 2000 ACM SIGMOD International Conference on Management of Data*, volume 29, pages 427–438. ACM, 2000.

[173] C. E. Rasmussen and C. K. I. Williams. *Gaussian Processes for Machine Learning*. The MIT Press, 2006.

[174] P. Regan and L. Manuel. Statistical extrapolation methods for estimating wind turbine extreme loads. *Transactions of ASME, Journal of Solar Energy Engineering*, 130:031011:1–15, 2008.

[175] S. Rehman and N. M. Al-Abbadi. Wind shear coefficients and their effect on energy production. *Energy Conversion and Management*, 46:2578–2591, 2005.

[176] G. Reikard. Using temperature and state transitions to forecast wind speed. *Wind Energy*, 11(5):431–443, 2008.

[177] J. Ribrant. *Reliability Performance and Maintenance—A Survey of Failures in Wind Power Systems*. Master's Thesis, School of Electrical Engineering and Computer Science, KTH Royal Institute of Technology, Stockholm, Sweden, 2006.

[178] M. Riedmiller and H. Braun. A direct adaptive method for faster backpropagation learning: The RPROP algorithm. In *Proceedings of the 1993 IEEE International Conference on Neural Networks*, San Francisco, CA, 1993. March 28–April 1.

[179] R. A. Rigby and D. M. Stasinopoulos. Generalized additive models for location, scale and shape. *Applied Statistics, Series C*, 54:507–554, 2005.

[180] Risø-DTU. Wind Turbine Load Data: http://www.winddata.com.

[181] D. Robb. Improved wind station profits through excellence in O&M. *Wind Stats Report*, 24:5–6, 2011.

[182] D. Robb. Wind technology: What is working and what is not? *Wind Stats Report*, 24:4–5, 2011.

[183] K. O. Ronold and G. C. Larsen. Reliability-based design of wind-turbine rotor blades against failure in ultimate loading. *Engineering Structures*, 22:565–574, 2000.

[184] M. Rosenblatt. Conditional probability density and regression estimates. In P. Krishnaiah, editor, *Multivariate Analysis II*, pages 25–31. Academic Process, New York, 1969.

[185] D. B. Rubin. Matching to remove bias in observational studies. *Biometrics*, 29(1):159–183, 1973.

[186] D. B. Rubin. Using propensity scores to help design observational studies: Application to the tobacco litigation. *Health Services and Outcomes Research Methodology*, 2:169–188, 2001.

[187] D. Ruppert, S. J. Sheather, and M. P. Wand. An effective bandwidth selector for local least squares regression. *Journal of the American Statistical Association*, 90:1257–1270, 1995.

[188] Y. Saatçi, R. Turner, and C. E. Rasmussen. Gaussian process change point models. In *Proceedings of the 27th International Conference on Machine Learning (ICML-10)*, pages 927–934, Haifa, Israel, 2010. June 21–24.

[189] J. Sacks, S. B. Schiller, and W. J. Welch. Designs for computer experiments. *Technometrics*, 31:41–47, 1989.

[190] J. Sacks, W. J. Welch, T. J. Mitchell, and H. P. Wynn. Design and analysis of computer experiments. *Statistical Science*, 4:409–423, 1989.

[191] I. Sanchez. Short-term prediction of wind energy production. *International Journal of Forecasting*, 22:43–56, 2006.

[192] B. Sanderse, S. P. Pijl, and B. Koren. Review of computational fluid dynamics for wind turbine wake aerodynamics. *Wind Energy*, 14(7):799–819, 2011.

[193] T. J. Santner, B. J. Williams, and W. I. Notz. *The Design and Analysis of Computer Experiments*. Springer-Verlag, New York, 2003.

[194] M. Sathyajith. *Wind Energy: Fundamentals, Resource Analysis and Economics*. Springer, Berlin Heidelberg, 2006.

[195] M. Schlater. Some covariance models based on normal scale mixtures. *Bernoulli*, 16(3):780–797, 2010.

[196] B. Schlkopf, J. C. Platt, J. Shawe-Taylor, A. J. Smola, and R. C. Williamson. Estimating the support of a high-dimensional distribution. *Neural Computation*, 13(7):1443–1471, 2001.

[197] G. E. Schwarz. Estimating the dimension of a model. *Annals of Statistics*, 6(2):461–464, 1978.

[198] Y. E. Shin, Y. Ding, and J. Z. Huang. Covariate matching methods for testing and quantifying wind turbine upgrades. *The Annals of Applied Statistics*, 12(2):1271–1292, 2018.

[199] G. L. Smith. Sequential estimation of observation error variances in a trajectory estimation problem. *AIAA Journal*, 5(11):1964–1970, 1967.

[200] R. L. Smith. Extreme value theory. In W. Ledermann, E. Lloyd, S. Vajda, and C. Alexander, editors, *Handbook of Applicable Mathematics*, volume 7, pages 437–472. John Wiley & Sons, 1990.

[201] Z. Song, Y. Jiang, and Z. Zhang. Short-term wind speed forecasting with Markov-switching model. *Applied Energy*, 130:103–112, 2014.

[202] J. D. Sørensen and S. R. K. Nielsen. Extreme wind turbine response during operation. *Journal of Physics: Conference Series*, 75:012074:1–7, 2007.

[203] I. Staffell and R. Green. How does wind farm performance decline with age? *Renewable Energy*, 66:775–786, 2014.

[204] M. Stein. Space-time covariance functions. *Journal of the American Statistical Association*, 100(469):310–321, 2005.

[205] B. Stephen, S. J. Galloway, D. McMillan, D. C. Hill, and D. G. Infield. A copula model of wind turbine performance. *IEEE Transactions on Power Systems*, 26:965–966, 2011.

[206] E. A. Stuart. Matching methods for causal inference: A review and a look forward. *Statistical Science*, 25(1):1–21, 2010.

[207] J. Tang, Z. Chen, A. W. Fu, and D. W. Cheung. Enhancing effectiveness of outlier detections for low density patterns. In *Advances in Knowledge Discovery and Data Mining: Proceedings of the Sixth Pacific-Asia Conference on Knowledge Discovery and Data Mining (PAKDD)*, pages 535–548. Springer, Berlin Heidelberg, 2002.

[208] J. Tatsu, P. Pinson, P. Trombe, and H. Madsen. Probabilistic forecasts of wind power generation accounting for geographically dispersed information. *IEEE Transactions on Smart Grid*, 5(1):480–489, 2014.

[209] P. Tavner, S. Faulstich, B. Hahn, and G. J. W. van Bussel. Reliability and availability of wind turbine electrical and electronic components. *EPE Journal*, 20(4):45–50, 2010.

[210] P. J. Tavner, C. Edwards, A. Brinkman, and F. Spinato. Influence of wind speed on wind turbine reliability. *Wind Engineering*, 30:55–72, 2006.

[211] P. J. Tavner, J. Xiang, and F. Spinato. Reliability analysis for wind turbines. *Wind Energy*, 10:1–8, 2007.

[212] C. C. Taylor. Automatic bandwidth selection for circular density estimation. *Computational Statistics and Data Analysis*, 52:3493–3500, 2008.

[213] J. B. Tenenbaum, V. De Silva, and J. C. Langford. A global geometric framework for nonlinear dimensionality reduction. *Science*, 290:2319–2323, 2000.

[214] J. L. Torres, A. García, M. De Blas, and A. De Francisco. Forecast of hourly average wind speed with ARMA models in Navarre Spain. *Solar Energy*, 79(1):65–77, 2005.

[215] N. Troldborg, G. C. Larsen, H. A. Madsen, K. S. Hansen, J. N. Sørensen, and R. Mikkelsen. Numerical simulations of wake interaction between two wind turbines at various inflow conditions. *Wind Energy*, 14(7):859–876, 2011.

[216] O. Uluyol, G. Parthasarathy, W. Foslien, and K. Kim. Power curve analytic for wind turbine performance monitoring and prognostics. In *Annual Conference of the Prognostics and Health Management Society*, volume 2, page 049, Montreal, Canada, 2011. August 19.

[217] U.S. Department of Energy. Wind Vision: A New Era for Wind Power in the United States: http://www.energy.gov/sites/prod/files/WindVision_Report_final.pdf.

[218] U.S. Department of Energy. 20% Wind energy by 2030—Increasing wind energy's contribution to U.S. electricity supply. DOE/GO-102008-2567, Office of Energy Efficiency and Renewable Energy, Washington, D.C., 2008. Available at https://www.nrel.gov/docs/fy08osti/41869.pdf.

[219] U.S. Energy Information Agency. Annual Energy Review 2011: https://www.eia.gov/totalenergy/data/annual/pdf/aer.pdf.

[220] U.S. Energy Information Agency. Electric Power Annual 2016, Chapter 1 National Summary Data: https://www.eia.gov/electricity/annual/pdf/epa.pdf.

[221] V. N. Vapnik. *The Nature of Statistical Learning Theory*. Springer-Verlag, New York, 1995.

[222] H. R. Varian. The nonparametric approach to demand analysis. *Econometrica*, 50:945–973, 1982.

[223] P. S. Veers and S. Butterfield. Extreme load estimation for wind turbines: Issues and opportunities for improved practice. In *Proceedings of the 2001 ASME Wind Energy Symposium, AIAA-2001-0044*, Reno, Nevada, 2001.

[224] C. M. Velte, M. O. L. Hansen, K. E. Meyer, and P. Fuglsang. Evaluation of the performance of vortex generators on the DU 91-W2-250 profile using stereoscopic PIV. In *Proceedings of the 12th World Multi-Conference on Systemics, Cybernetics and Informatics (WMSCI 2008)*, volume 2, pages 263–267, Orlando, Florida, 2008. June 29-July 2.

[225] Y. Wan, M. Milligan, and B. Parsons. Output power correlation between adjacent wind power plants. *Transactions of the ASME, Journal of Solar Energy Engineering*, 125:551–555, 2003.

[226] G. Watson. Smooth regression analysis. *Sankhyā: The Indian Journal of Statistics, Series A*, 26:359–372, 1964.

[227] J. Wen, Y. Zheng, and D. Feng. A review on reliability assessment for wind power. *Renewable and Sustainable Energy Reviews*, 13:2485–2494, 2009.

[228] S. N. Wood. Thin plate regression splines. *Journal of the Royal Statistical Society: Series B (Statistical Methodology)*, 65(1):95–114, 2003.

[229] Y. Xia, K. H. Ahmed, and B. W. Williams. Wind turbine power coefficient analysis of a new maximum power point tracking technique. *IEEE Transactions on Industrial Electronics*, 60(3):1122–1132, 2013.

[230] N. Yampikulsakul, E. Byon, S. Huang, and S. Sheng. Condition monitoring of wind turbine system with non-parametric regression-based analysis. *IEEE Transactions on Energy Conversion*, 29(2):288–299, 2014.

[231] J. Yan, K. Li, E. Bai, J. Deng, and A. Foley. Hybrid probabilistic wind power forecasting using temporally local Gaussian process. *IEEE Transactions on Sustainable Energy*, 7(1):87–95, 2016.

[232] M. You, E. Byon, J. Jin, and G. Lee. When wind travels through turbines: A new statistical approach for characterizing heterogeneous wake

effects in multi-turbine wind farms. *IISE Transactions*, 49(1):84–95, 2017.

[233] B. Zeigler, T. Kim, and H. Praehofer. *Theory of Modeling and Simulation*. Academic Press, Orlando, FL, 2000.

[234] F. Zwiers and H. Von Storch. Regime-dependent autoregressive time series modeling of the Southern Oscillation. *Journal of Climate*, 3(12):1347–1363, 1990.

Index

Printed in the United States
by Baker & Taylor Publisher Services